SOME APPLICATIONS OF TOPOLOGICAL K-THEORY

NORTH-HOLLAND
MATHEMATICS STUDIES **45**

Notas de Matemática (74)
Editor: Leopoldo Nachbin
Universidade Federal do Rio de Janeiro
and University of Rochester

Some Applications of Topological K-Theory

N. MAHAMMED
University of Sciences and Technical Studies
Lille, France

R. PICCININI
Memorial University of Newfoundland
Newfoundland, Canada

U. SUTER
University of Neuchâtel
Neuchâtel, Switzerland

1980

NORTH-HOLLAND PUBLISHING COMPANY – AMSTERDAM • NEW YORK • OXFORD

ISBN: 0 444 86113 0

Publishers:
NORTH-HOLLAND PUBLISHING COMPANY
AMSTERDAM • NEW YORK • OXFORD

Sole distributors for the U.S.A. and Canada:
ELSEVIER NORTH-HOLLAND, INC.
52 VANDERBILT AVENUE, NEW YORK, N.Y. 10017

Library of Congress Cataloging in Publication Data

Mahammed, N 1944-
Some applications of topological K-theory.

(Notas de matemática ; 74) (North-Holland mathematics studies ; 45)
Bibliography: p.
Includes index.
1. K-theory. I. Piccinini, Renzo A., 1933- joint author. II. Suter, U., 1935- joint author. III. Title. IV. Series.
QA1.N86 no. 74 [QA612.33] 510s [514'.23]
ISBN 0-444-86113-0 80-23219

PRINTED IN THE NETHERLANDS

PREFACE

In the intervening years since its conception and early development by Grothendieck, Atiyah and Hirzebruch there has been a tremendous expansion in the knowledge and applications of Topological K - Theory. Among its impressive applications we cite Adam's solution of the famous (real) Vector Field problem on spheres and the considerable simplification of the solution of the Hopf Invariant One problem. Perhaps the real power of Topological K - Theory lies in the depth and diversity of its applications.

In this book we attempt to present systematically some applications which are more or less accessible to a graduate student or the non-specialist in Algebraic Topology who has some feeling for the concepts and techniques of this branch of Mathematics.

With the above philosophy in mind we have presented the material in Chapters one to seven in such a way that it is almost self-contained. We have also included a descriptive Chapter (Chapter 8) about the Atiyah-Singer Index Theorem, a theorem heavily based in K - Theory and undoubtedly, one of the most important mathematical results of the last two decades. Again, in the interest of self-sufficiency, we have included an introductory Chapter (Chapter 0) on the results of Topological K - Theory which are pertinent to the development of our book. This Chapter is descriptive in nature; it is not intended to be an exhaustive study but rather that it may, together with the indicated literature, form the basis of, and motivation for, later detailed study of this subject.

The important results within each Chapter (with the exception of Chapter 0) are identified by a pair of positive integers, the first of which refers to the Section in which the result is contained; if a reference is given by a triple, the first number indicates the Chapter. Chapter 0 is not divided up into Sections and all references to it are given by a pair whose first entry is 0 .

Blaise Pascal once observed that no book is actually written by a single author, since every book contains ideas and results of others. The large bibliography should convince the reader that this is particularly true in the book she or he is presently reading. However, Pascal's observation is true in our case also for other reasons: firstly, because we are three authors and secondly, because our book would not have been written were it not for the help, encouragement and suggestions of many friends among whom we wish to note B.Eckmann, R.Fritsch, H.Glover, D.Gottlieb, P.Heath, P.Hilton, D.Lehmann, G.Mislin, D.Rideout, F.Sigrist and A.Zabrodsky. To all of them our heartily thanks.

We wish further to express our gratitude to the three Universities with which we are connected, to the Natural Science and Engineering Research Council of Canada and to the Mathematics Research Institute of the Swiss Federal Institute of Technology in Zurich for the encouragement and material support.

Many thanks are due to Frau H.Jordan of the University of Konstanz for the excellent typing work.

Lastly, we wish to thank L.Nachbin for accepting our book into his prestigious collection.

N.Mahammed
R.Piccinini
U.Suter

TABLE OF CONTENTS

CHAPTER 0

A REVIEW OF K - THEORY

In what follows, $\mathbb{F}$ will represent the field $\mathbb{R}$ of real numbers, the field $\mathbb{C}$ of complex numbers or the skew-field $\mathbb{H}$ of quaternions. We use *CW* to denote the category of finite CW - complexes and continuous functions (maps).

(0.1) - For a given $X \in \text{obj } CW$, let $K\mathbb{F}(X)$ be the group obtained by symmetrization of the semi-group $\text{Vect}_{\mathbb{F}}(X)$ of all isomorphic classes of $\mathbb{F}$ - vector bundles over X, with addition induced by the Whitney sum. It follows that all the elements of $K\mathbb{F}(X)$ are of the type $[\xi] - [\eta]$, where ξ and η are $\mathbb{F}$ - vector bundles over X (*). Notice that since there is, for every vector bundle ξ over X, a vector bundle ξ' such that $\xi \oplus \xi'$ is trivial, the elements of $K\mathbb{F}(X)$ can be written $[\xi] - n$, where n is an integer. On the other hand, it also follows from the appropriate universal property that $K\mathbb{F}(X)$ is unique up to isomorphism. Pull-back of vector bundles shows that $K\mathbb{F}$ is a contravariant functor from *CW* to the category of abelian groups.

(0.2) - If G is a compact Lie group and $X \in \text{obj } CW$ is a G - space, considering G - vector bundles over X we obtain, analogously to $K\mathbb{F}$, a functor $K\mathbb{F}_G$. Hence, all constructions and properties relative to $K\mathbb{F}$ which respect G - actions can be transported to $K\mathbb{F}_G$.

(0.3) - If X is based, say with base point x_o, define $\tilde{K}\mathbb{F}(X)$ as the kernel of the homomorphism

$$i^! : K\mathbb{F}(X) \longrightarrow K\mathbb{F}(\{x_o\}) ,$$

(*) On several occasions we shall identify a vector bundle ξ with the element $[\xi]$ in $K\mathbb{F}$ - theory.

induced by the inclusion map $i : \{x_o\} \longrightarrow X$;
sing map $X \longrightarrow \{x_o\}$ shows that there is a spl

$$K\mathbb{F}(X) \cong \tilde{K}\mathbb{F}(X) \otimes K\mathbb{F}(\{x_o\})$$

which is natural; in this splitting, if X is con
identify $x = [\xi] - n \in K\mathbb{F}(X)$ with

$$([\xi] - \dim \xi, \dim \xi - n) = ([\xi] - \dim \xi,$$

Furthermore, if for every (unbased) object X we
as the union of X and some point $*$ - which will
as the base point of X^+ - then,

$$K\mathbb{F}(X) \cong \tilde{K}\mathbb{F}(X^+) .$$

Given a pair of finite CW - complexes (X,Y), def
as $\tilde{K}\mathbb{F}(X/Y)$; the isomorphism $K\mathbb{F}(X) \cong \tilde{K}\mathbb{F}(X^+)$ is
by writing $X/\emptyset = X^+$.

(0.4) - The cofunctors $K\mathbb{F}$ will be denoted by
KSp for $\mathbb{F} = \mathbb{R}, \mathbb{C}$ and $\mathbb{H}$, respectively; likewise,
corresponding reduced functors. The functors $K\mathbb{F}$
are corepresentable; in fact,

$$KO = [\ ,\mathbb{Z} \times BO] ,$$
$$KU = [\ ,\mathbb{Z} \times BU] ,$$
$$KSp = [\ ,\mathbb{Z} \times BSp] ;$$

the reduced functors are given by the based homoto
maps. This is a consequence of the *Classification*
which, in particular, shows that for every connect
plex such that $\dim X \leq d(n+1) - 2$ ($d = \dim \mathbb{F}$ a
vector space)

$$\tilde{K}\mathbb{F}(X) \cong [X, G_n(\mathbb{F}^{2n})] ,$$

where $G_n(\mathbb{F}^{2n})$ is the appropriate Grassmann mani

(0.5) For $n \geq 0$, define $\tilde{K}\mathbb{F}^{-n}(X) = \tilde{K}\mathbb{F}(\Sigma^n X)$ fo
where $\Sigma^n X$ is the n - fold suspension of X ; fo
$K\mathbb{F}^{-n}(X) = \tilde{K}\mathbb{F}(\Sigma^n X^+)$ and for pairs, $K\mathbb{F}^{-n}(X,Y) =$
Then, according to Atiyah-Hirzebruch [28], we hav
sequences

$$\ldots \longrightarrow K\mathbb{F}^{-n}(X,Y) \longrightarrow K\mathbb{F}^{-n}(X) \longrightarrow K\mathbb{F}^{-n}$$

$$\ldots \longrightarrow K\mathbb{F}^{o}(X,Y) \longrightarrow K\mathbb{F}^{o}(X) \longrightarrow K\mathbb{F}$$

for any pair of CW - complexes (X,Y), and

$$\ldots \longrightarrow K\mathbb{F}^{-n}(X,Y) \longrightarrow \tilde{K}\mathbb{F}^{-n}(X) \longrightarrow \tilde{K}\mathbb{F}^{-n}(Y) \longrightarrow$$

$$\ldots \longrightarrow K\mathbb{F}^{0}(X,Y) \longrightarrow \tilde{K}\mathbb{F}^{0}(X) \longrightarrow \tilde{K}\mathbb{F}^{0}(Y)$$

if X,Y are based CW - complexes, $Y \subset X$ and have the same base point.

(0.6) Actually the previous exact sequences can be extended infinitely to the right; the argument goes as follows. The Bott Periodicity Theorem ([46],[134]) states that

$$\mathbb{Z} \times BU \simeq \Omega^2 BU ,$$
$$\mathbb{Z} \times BO \simeq \Omega^4 BSp ,$$
$$\mathbb{Z} \times BSp \simeq \Omega^4 BO ;$$

the first homotopy equivalence gives rise to a Ω - spectrum $KU_{2n} = \mathbb{Z} \times BU$, $KU_{2n+1} = \Omega BU$, $n \in \mathbb{Z}$; the other two homotopy equivalences define the Ω - spectra $KO = \{KO_n\}$ and $KSp = \{KSp_n\}$, with $KO_n = \Omega^k BO$, $KSp_n = \Omega^k BSp$, $n \equiv 8 - k \pmod 8$, $0 \le k \le 7$, and with the convention that $\Omega^0 BO = \mathbb{Z} \times BO$, $\Omega^0 BSp = \mathbb{Z} \times BSp$. These spectra define "generalized cohomology theories" [252] and thus, the claim made at the beginning of the paragraph.

Notice that the coefficient groups of $\widetilde{KO}$ and $\widetilde{KSp}$ are the same but with a shift of 4 in dimension. The previous observations give readily the $\tilde{K}\mathbb{F}$ - theory of the spheres, which we can summarize in the following table:

n mod 8	0	1	2	3	4	5	6	7
$\widetilde{KO}(S^n)$	$\mathbb{Z}$	$\mathbb{Z}_2$	$\mathbb{Z}_2$	0	$\mathbb{Z}$	0	0	0
$\widetilde{KU}(S^n)$	$\mathbb{Z}$	0	$\mathbb{Z}$	0	$\mathbb{Z}$	0	$\mathbb{Z}$	0
$\widetilde{KSp}(S^n)$	$\mathbb{Z}$	0	0	0	$\mathbb{Z}$	$\mathbb{Z}_2$	$\mathbb{Z}_2$	0

$(\widetilde{KSp}(S^n) \cong \widetilde{KO}(S^{n+4}))$.

The theories KO and KSp differ also in another significant way: if $\mathbb{F} = \mathbb{R}$ or $\mathbb{C}$, the tensor product of vector bundles over X induces a ring structure on KO and KU; since the tensor product of two quaternionic vector spaces is a real vector space, we do not obtain a multiplicative structure in KSp. We also observe that the theories KO and KSp are related to KU by the Bott Homomorphisms [49]

$$KU(X) \xrightarrow{r} KO(X) \xrightarrow{c} KU(X)$$

$$KU(X) \xrightarrow{q} KSp(X) \xrightarrow{h} KU(X) \ ;$$

these satisfy the relations

$$rc = 2 = qh \ , \quad cr = hq = 1 + t \ ,$$

where 2 is multiplication by 2 in KO or KSp and t is the homomorphism induced in KU by conjugation of complex vector bundles. In this context we notice that if $X = S^{4n}$ then c (resp. r) is an automorphism of $\mathbb{Z}$ whenever n is even (resp. odd) and multiplication by 2 whenever n is odd (resp. even).

(0.7) - If ξ is a real vector n - bundle over $X \in \text{Obj}\ CW$, the *Stiefel-Whitney classes* of ξ are elements $w_i(\xi) \in H^i(X;\mathbb{Z}_2)$, $1 \leq i \leq n$; if ξ is a complex vector n - bundle over X , the *Chern classes* of ξ are elements $c_i(\xi) \in H^{2i}(X;\mathbb{Z})$. These classes can be defined axiomatically [134]; their existence and uniqueness can be shown from the Leray-Hirsh Theorem and the *Splitting Principle*: given a (real or complex) vector bundle $E \longrightarrow X$, there is a space F(E) and a map $i : F(E) \longrightarrow X$ such that the induced bundle over F(E) splits off as a Whitney sum of line bundles and such that i induces a monomorphism in cohomology with the appropriate coefficients (i.e., $\mathbb{Z}$ in the complex case and $\mathbb{Z}_2$ in the real one) [49], [134].

(0.8) A real vector n - bundle is said to be *orientable* if its structural group reduces from O(n) to SO(n); notice that a complex bundle can be viewed as a real bundle and as such it is orientable since $U(n) \subseteq SO(2n)$. Now let ξ be a real vector bundle $E \longrightarrow X$; for a riemannian metric on E let $D(E) \longrightarrow X$ and $S(E) \longrightarrow X$ be respectively the associated disc and sphere bundles. Set $T(\xi) = D(E)/S(E)$, the *Thom space* of ξ . We use integral cohomology if ξ is orientable; otherwise, we use cohomology with coefficients in $\mathbb{Z}_2$. Then, the *Thom Isomorphism Theorem* [134] asserts that

$$H^i(X) \cong \tilde{H}^{i+n}(T(\xi)) \ .$$

This theorem has a counter-part in K - theory: if ξ is a complex vector bundle over X, we have that $KU^i(X) \cong \widetilde{KU}^i(T(\xi))$ [17]; however, for a real vector bundle ξ, a Thom isomorphism in KO - theory exists only if ξ has a Spin - reduction.

(0.9) - In this book we shall use some interesting conseqences of the Thom Isomorphism Theorem in equivariant K - theory, which we explain presently. Suppose that X is a point and that G acts freely on $S(E)$; then the exact sequence of the pair $(D(E),S(E))$ in KO_G - theory gives rise to the following exact sequence:

$$\ldots \longrightarrow KO_G^{-4n}(*) \xrightarrow{\Phi} KO_G^0(*) \xrightarrow{\theta} KO^0(S(E)/G) \longrightarrow \ldots$$

Since

$$KO_G^{4n}(*) = \begin{cases} RO(G), & \text{if } n = \text{even}, \\ RSp(G), & \text{if } n = \text{odd}, \end{cases}$$

where $RO(G)$ (resp. $RSp(G)$) is the ring (resp. group) of the orthogonal (resp. quaternionic) representations of G, we have the exact sequences (see [196]) :

$$RO(G) \xrightarrow{\Phi} RO(G) \xrightarrow{\theta} KO^0(S(E)/G), \text{ if } n = \text{even}$$
$$RSp(G) \xrightarrow{\Phi} RO(G) \xrightarrow{\theta} KO^0(S(E)/G), \text{ if } n = \text{odd}.$$

For the convenience of the reader we recall that θ is the natural homomorphism which assigns to a representation $\rho \in RO(G)$ of degree d the equivalence class of the real vector bundle ξ_ρ

$$S(E) \times_G \mathbb{R}^d \longrightarrow S(E)/G$$

where G acts over each factor of $S(E) \times \mathbb{R}^d$ via ρ. Let us observe that if the group G has a central element acting on $S(E)$ as an antipodal map, then θ is an epimorphism [148]. Finally, in the complex case, with only the hypothesis that G acts freely on $S(E)$, the same kind of considerations give rise to the *Atiyah Exact Sequence* [17; page 103]

$$0 \longrightarrow KU^1(S(E)/G) \longrightarrow RU(G) \xrightarrow{\Phi} RU(G) \xrightarrow{\theta} KU^0(S(E)/G) \longrightarrow 0$$

where Φ is the multiplication in $RU(G)$ by $\sum_{i=0}^{\dim \rho} (-1)^i \Lambda^i \rho$.

(0.10) - Since KO^* and KU^* are generalized cohomology theories they possess an *Atiyah-Hirzebruch Spectral Sequence* (E_r, d_r) such that for every finite CW - complex X ,

$$E_2^{p,q} = H^p(X;\tilde{K}\,\mathbb{F}^q(S^o))$$

and

$$E_\infty^{p,q} \cong \frac{\ker\left[K\,\mathbb{F}^{p+q}(X) \to K\,\mathbb{F}^{p+q}(X_{p-1})\right]}{\ker\left[K\,\mathbb{F}^{p+q}(X) \to K\,\mathbb{F}^{p+q}(X_p)\right]} = \mathcal{G}^{p,q}K\,\mathbb{F}(X)$$

where $\mathbb{F} = \mathbb{R}$ or $\mathbb{C}$ and X_p is the p - skeleton of X [28], [106]. For $\mathbb{F} = \mathbb{C}$ the spectral sequence collapses if $H^{odd}(X;\mathbb{Z}) = 0$ and hence, $H^{even}(X;\mathbb{Z})$ is isomorphic to the graded ring $\mathcal{G}KU(X)$ associated to KU(X). (Later in the chapter we shall present another situation in which the spectral sequence collapses; further examples will be given in the book, notably in chapter 5).

(0.11) - The Chern classes of complex vector bundles over a finite CW - complex X are used to define a graded ring homomorphism

$$ch(X) : KU^*(X) = KU^o(X) \oplus KU^1(X) \longrightarrow H^*(X;\mathbb{Q}) .$$

Given a complex n - bundle ξ over X , let $s_1, s_2, \ldots, s_k, \ldots$ be the functions of $c_1(\xi), \ldots, c_n(\xi)$ defined inductively by the relations

$$s_k - c_1(\xi)s_{k-1} + c_2(\xi)s_{k-2} - \ldots + (-1)^k k c_k(\xi) = 0 ;$$

then define

$$ch(\xi) = n + \sum_{k \geq 1} \frac{1}{k!} s_k$$

This is the *Chern Character* of ξ . The Chern character can be defined on KU as a ring homomorphism (as one can see, for example, using the Splitting Principle); in addition, ch is a natural transformation of functors.

If we introduce the condition that $H_*(X;\mathbb{Z})$ is torsion-free, the Atiyah-Hirzebruch spectral sequence for $KU^*(X)$ collapses and furthermore, the following *Integrality Theorem* holds: "the Chern character

$$ch(X) : KU^*(X) \longrightarrow H^*(X;\mathbb{Q})$$

induces an isomorphism of $\mathcal{F}^p KU^*(X)/\mathcal{F}^{p+1}KU^*(X)$ onto $H^p(X;\mathbb{Z})$, viewed as a subgroup of $H^p(X;\mathbb{Q})$, where $\mathcal{F}^p KU^*(X) = \ker[KU^*(X) \to KU^*(X_{p-1})]$", [106]. This means that for every element $z \in KU^*(X)$ the first non-zero component of $ch(z)$ of strictly positive degree is an integral class; on the other hand, given any $a \in H^i(X;\mathbb{Z})$, $i > 0$, $a \neq 0$, there is a $z \in KU^*(X)$ such that $ch(z) = a +$ higher terms. Finally, the Integrality Theorem shows that under the stated hypothesis on X, $ch(X)$ is a ring monomorphism.

(0.12) - If ξ is a real or complex vector bundle $E \to X$, define $\Lambda^i\xi$ (i = non-negative integer) to be the vector bundle whose fibre at $x \in X$ is the ith exterior power $\Lambda^i E_x$. The exterior power operations on bundles are used to define *operations* in $K\mathbb{F}$, that is to say, natural transformations of the functor $K\mathbb{F}$ on itself (see [17,Chapter III]). To this end, let $1 + K\mathbb{F}(X)[[t]]$ be the multiplicative group of formal power series in t with coefficients in $K\mathbb{F}(X)$ and starting with 1. For every vector bundle ξ as before, let $\overline{\lambda}_t(\xi) \in 1 + K\mathbb{F}(X)[[t]]$ be defined by $\sum_{i \geq o} [\Lambda^i\xi]t^i$, where $[\Lambda^i\xi]$ is the element of $K\mathbb{F}(X)$ defined by the vector bundle $\Lambda^i\xi$; the universal property of $K\mathbb{F}(X)$ shows that there is a unique homomorphism

$$\lambda_t : K\mathbb{F}(X) \longrightarrow 1 + K\mathbb{F}(X)[[t]]$$

defined, for every $x \in K\mathbb{F}(X)$, by

$$\lambda_t(x) = 1 + \sum_{i \geq 1} \lambda^i(x)t^i$$

and such that $\lambda_t(x+y) = \lambda_t(x) \cdot \lambda_t(y)$. Observe that if ϵ^n is the trivial $\mathbb{F}$ - vector bundle of dimension n over X, then

$$\lambda^i([\epsilon^n]) = \binom{n}{i}.$$

The *Grothendieck Operations* γ^i are defined by the conditions

$$\gamma_t(z) = \sum_{i \geq o} \gamma^i(z)t^i$$

$$\gamma_t(z) = \lambda_{t/(1-t)}(z)$$

and the *Adams Operations* ψ^i by the conditions

$$\psi_t(z) = \sum_{i\geq 1} \psi^i(z)t^i$$

$$\psi_{-t}(z) = -t\left(\frac{d}{dt}\lambda_t(z)\right)\left(\lambda_t(z)\right)^{-1} .$$

(Note: We shall write $\psi^i_{\mathbb{F}}$, $\mathbb{F} = \mathbb{R}$, $\mathbb{C}$ or $\mathbb{H}$, whenever we want to explicit the $K\mathbb{F}$ - theory involved).

The Grothendieck operations have the following properties:

1) for every $z \in K\mathbb{F}(X)$, $\gamma^o(z) = 1$, $\gamma^1(z) = z$;

2) for every $z,z' \in K\mathbb{F}(X)$, $\gamma^k(z+z') = \sum_{i+j=k} \gamma^i(z)\gamma^j(z')$.

We should observe that the γ^k and λ^k operations are related by the following formulas: for every $z \in K\mathbb{F}(X)$,

a) $\gamma^k(z) = \lambda^k(z) + \sum_{i<k} a_{i,k}\, \lambda^i(z)$

b) $\lambda^k(z) = \gamma^k(z) + \sum_{i<k} b_{i,k}\, \gamma^i(z)$,

where $a_{i,k}$ and $b_{i,k}$ are integers [134; 12.3.2] .

The Adams operations,

$$\psi^k : K\mathbb{F}(X) \longrightarrow K\mathbb{F}(X)$$

are *ring homomorphisms* and furthermore,

1) $\psi^k(z) - \lambda^1(z)\psi^{k-1}(z) + \lambda^2(z)\psi^{k-2}(z) - \ldots + (-1)^k k\lambda^k(z) = 0$;

2) $\psi^k[\xi] = [\xi^k]$, if ξ is a line bundle;

3) $\psi^k \circ \psi^m = \psi^{km} = \psi^m \circ \psi^k$;

4) $\psi^p(z) \equiv z^p \pmod p$, p prime number, $z \in \tilde{K}\mathbb{F}(X)$;

5) $\psi^k_{\mathbb{C}}(z) = k^q z$, $z \in \tilde{K}U(S^{2q})$.

We also define ψ^k for $k \leq 0$, setting

$$\psi^o(z) = \text{rank } z , \quad z \in K\mathbb{F}(X) ,$$

$\psi^k[\xi] = \psi^{-k}[\xi^*]$, where ξ^* is the dual of ξ [17].

Finally, the additivity properties of $\psi^k_{\mathbb{C}}$ and ch , and the Splitting Principle, show that if $ch(z) = a_o + a_1 + \ldots + a_j + \ldots$, with $a_j \in H^{2j}(X;\mathbb{Q})$, then

$$ch\left(\psi^k_{\mathbb{C}}(z)\right) = a_o + ka_1 + \ldots + k^j a_j + \ldots .$$

(0.13) The Adams operations do not commute with the Thom Isomorphism; indeed, there are certain corrective factors $\theta_k^{\mathbb{F}}$ due to Bott. The characteristic class $\theta_k^{\mathbb{C}}$ is defined on $\mathrm{Vect}_{\mathbb{C}}(X)$, taking values in $KU(X)$; its real counterpart $\theta_k^{\mathbb{R}}$ is defined for Spin (8n) - bundles. For a complex line bundle λ the class $\theta_k^{\mathbb{C}}$ is given by

$$\theta_k^{\mathbb{C}}(\lambda) = 1 + [\lambda^*] + \ldots + [\lambda^*]^{k-1} \qquad (k \geq 1)$$

and extended multiplicatively by the Splitting Principle to an arbitrary vector bundle, so that

$$\theta_k^{\mathbb{C}}(\xi \oplus \eta) = \theta_k^{\mathbb{C}}(\xi) \cdot \theta_k^{\mathbb{C}}(\eta), \; \theta_{pq}^{\mathbb{C}}(\xi) = \theta_p^{\mathbb{C}}(\xi) \cdot \psi_{\mathbb{C}}^{p}\left(\theta_q^{\mathbb{C}}(\xi)\right) .$$

These relations also hold in the real case, whenever $\theta_k^{\mathbb{R}}$ is defined. If ξ is a complex vector bundle or a real 8n - bundle which admits a reduction to Spin(8n) and

$$i_! : K\mathbb{F}^*(X) \longrightarrow \widetilde{K}\mathbb{F}^*(T(\xi))$$

is the Thom Isomorphism, for every $z \in K\mathbb{F}(X)$,

$$\psi^k i_!(z) = i_! \theta_k(\xi) \psi^k(z) \qquad [49] .$$

Finally, we mention that if ξ is a complex 4n - vector bundle such that $\Lambda^{4n}\xi = 1$ (i.e. the structure group of ξ can be reduced to SU(4n) and hence $r\xi$ admits a reduction to Spin(8n)), then

$$\theta_k^{\mathbb{C}}(\xi) = c\, \theta_k^{\mathbb{R}}(r\xi) .$$

It is possible to extend the definition of $\theta_k^{\mathbb{C}}$ to KU(X), provided we interpret $\theta_k^{\mathbb{C}}(\xi)$ as an element of $KU(X) \otimes \mathbb{Q}_k$, where $\mathbb{Q}_k$ is the additive group of fractions p/k^q , with $p,q \in \mathbb{Z}$. Similarly, $\theta_k^{\mathbb{R}}$ can be defined on the subgroup of KO(X) formed by the elements of even virtual dimension and having trivial first Stiefel-Whitney class [4, Part II, § 5] .

(0.14) - Let T(X) be the subgroup of KO(X) generated by elements of the form $[\xi] - [\eta]$ where ξ and η are orthogonal bundles whose associated sphere bundles are fibre homotopy equivalent; define the group J(X) to be KO(X)/T(X). As it is the case for KO(X), J(X) splits off as a direct sum

$$J(X) \cong \tilde{J}(X) \oplus \mathbb{Z} ;$$

according to Atiyah [12], the group $\tilde{J}(X)$ is *finite* for a connected finite CW-complex X. In order to compute the groups $J(X)$, Adams introduced two other quotient groups of $KO(X)$:

$$J'(X) = KO(X)/V(X)$$

and

$$J''(X) = KO(X)/W(X) .$$

Here $V(X)$ is the subgroup of the elements $x \in KO(X)$ of even virtual dimension, $w_1(x) = 0$ and for which there exists $y \in \widetilde{KO}(X)$ such that, for all $k \geq 1$

$$\theta_k(x) = \frac{\psi^k(1+y)}{1+y}$$

in $KO(X) \otimes \mathbb{Q}_k$. The subgroup $W(X)$ is defined as follows. Let e be a function of $\mathbb{Z} \times KO(X)$ into the set of non-negative integers and let $KO(X)_e$ be the subgroup of $KO(X)$ defined by the elements

$$k^{e(k,y)}(\psi^k-1)\,y ;$$

then, $W(X) = \bigcap_e KO(X)_e$, where the intersection runs over all functions e. It turns out that for each finite CW-complex X, $J'(X) = J''(X)$ [4, Part III, Theorem 1.1]; furthermore, $J'(X)$ is a lower bound for $J(X)$ in the sense that $T(X) \subseteq V(X)$ and so, the quotient map $KO(X) \longrightarrow J'(X)$ factors through an epimorphism $J(X) \longrightarrow J'(X)$ [4, Part II, Theorem 6.1]. On the other hand, $J''(X)$ is an upper bound for $J(X)$ in the sense that $W(X) \subseteq T(X)$ and the quotient map $KO(X) \longrightarrow J(X)$ factors through an epimorphism $J''(X) \longrightarrow J(X)$; the reader should notice that these results put together imply that $J(X) = J'(X) = J''(X)$. The fact that $J''(X)$ is an upper bound for $J(X)$ is a consequence of the *Adams Conjecture* which asserts that for any $k \in \mathbb{Z}$, X a finite CW-complex and $y \in KO(X)$, there is a non-negative integer $e = e(k,y)$ such that $k^e(\psi^k-1)y$ maps to zero in $J(X)$. This conjecture was proved by Quillen in [199]; more recently, a proof based on the notion of transfer was obtained by Becker and Gottlieb [41].

If $A \xrightarrow{i} X \xrightarrow{q} X/A$ is a cofibration of finite connected CW - complexes, the corresponding J - sequence is not exact at X in general. However, the following special result holds (see [4, Part II, § 3]) :

"if $i^!: \widetilde{KO}(X) \longrightarrow \widetilde{KO}(A)$ is an *epimorphism*, then

$$\widetilde{J}(X/A) \xrightarrow{J(q)} \widetilde{J}(X) \xrightarrow{J(i)} \widetilde{J}(A) \longrightarrow 0$$

is exact".

CHAPTER 1

THE HOPF INVARIANT

1. Introduction

Let f be a map of the sphere S^3 onto the sphere S^2 and let us suppose it to be simplicial relative to some triangulations of S^2 and S^3. If q and r are any two points in the interior of a 2-simplex of S^2, $f^{-1}(q)$ and $f^{-1}(r)$ are 1-cycles of the complex S^3 (see § 2). Furthermore, as shown by Hopf in a paper of 1931 [121], the linking number of these two cycles depends only on the homotopy class of f. We call this linking number the *Hopf invariant* of f and denote it by $\gamma(f)$. Hopf also proved that the function

$$\gamma : \pi_3(S^2) \longrightarrow \mathbb{Z}$$

defined by $\gamma[f] = \gamma(f)$ for every class $[f] \in \pi_3(S^2)$ is a group epimorphism. These ideas can be generalized to maps from S^{2n-1} onto S^n, $n \geq 2$; as proved by Hopf in a successive paper [122], one obtains a group homomorphism

$$\gamma : \pi_{2n-1}(S^n) \longrightarrow \mathbb{Z}$$

which is trivial if n is odd (because of the anti-commutativity of the linking numbers); in addition to that, Hopf proved that γ is epic if $n = 4$ or 8 and that there are always maps of S^{2n-1} onto S^n of Hopf invariant 2 for n even (this, in particular, shows that $\pi_{2n-1}(S^n)$ is never trivial for n even). The obvious problem, left unsolved by Hopf, was to find out for which values on n there were maps $f : S^{2n-1} \longrightarrow S^n$ with $\gamma(f) = 1$. A lot of effort was devoted to the solution of this question, known as the "Hopf invariant one problem"; the keen interest of mathematicians on it was also due to the fact that the Hopf invariant one problem has deep connections to other important problems like the determination of the real division algebras and the parallelizability of the spheres. The solution came about only in 1960 by work of Adams, who proved that the only values of n for which there are maps $f : S^{2n-1} \longrightarrow S^n$ with $\gamma(f) = 1$ are $n = 2$, 4 or 8. We wish to quote here two important partial results obtained before Adams' conclusive theorem. The first, proved by Adém in 1951 [9], says that there is no map $f : S^{2n-1} \longrightarrow S^n$ of Hopf invariant equal to 1 if n is not a power of 2; its proof was established via a delicate study of the Steenrod operations. The second result, due to Toda [241], showed that there is no map $f : S^{31} \longrightarrow S^{16}$ with $\gamma(f) = 1$ (thus, not all powers of 2 could give maps of Hopf invariant 1).

The Chapter is organized as follows. In sections 2 and 3 we present an account of Hopf's beautiful work in [121] and [122]. Since the homology approach devised by Hopf is not suitable to the development and ultimate solution of the Hopf invariant one problem, we devote §4 to the study of a cohomology version of the Hopf invariant. The idea of using cohomology for the definition of such invariant was first described by Steenrod [230]; our development follows that of [71]. Finally, we use §5 to give a K-theoretical proof of Adams' Theorem and to discuss some of its consequences.

2. The Hopf Invariant of Maps from S^3 onto S^2.

We recall that a *polyhedron* is a topological space M together with an abstract simplicial complex $\mathfrak{M}$ and a homeomorphism between M and the geometric realization of $\mathfrak{M}$; we shall admit that all simplices of $\mathfrak{M}$ had been coherently oriented by a local ordering of the vertices. In order to simplify the notation we shall use the same letter for both the topological space and the subjacent simplicial complex of a polyhedron; furthermore, throughout this chapter, whenever we refer to a simplex, we normally understand that we are dealing with the closed (or geometric) simplex.

Let M and N be given polyhedra, with M of dimension > 2 and N of dimension 2; we also are given a simplicial map $f : M \to N$. Let q be an interior point of a fixed

2 - simplex τ_2 of N ; then, for every 2 - simplex σ_2 of M which is properly mapped onto τ_2 by f , there is a unique $p \in \operatorname{int} \sigma_2$ such that $f(p) = q$. We write $\varphi_{\sigma_2}(q) = +p$ (or $-p$) according to that $f(\sigma_2)$ has the same (or opposite) orientation as τ_2. If $c_2 = \Sigma\, a_i\, \sigma_2^i$ is a 2 - chain of M with integral coefficients, define $\varphi_{c_2}(q) = \Sigma\, a_i\, \varphi_{\sigma_2^i}(q)$, with the proviso that $\varphi_{\sigma_2^i}(q) = 0$ whenever σ_2^i is not mapped properly onto τ_2 by f.

Notice that $\varphi_{c_2}(q)$ is a 0 - chain of a convenient subdivision of M [225; 3.3] .

Let now σ_3 be a tetrahedron of M which is mapped onto τ_2 by f . Observe that σ_3 has exactly two faces, say σ_2^1 and σ_2^2 , which are mapped properly onto τ_2 ; moreover, the set of points of σ_3 which are taken into q by f is a line segment with end-points $p_1 \in \operatorname{int} \sigma_2^1$ and $p_2 \in \operatorname{int} \sigma_2^2$. If $f(\sigma_2^1)$ and τ_2 have opposite orientations, then $f(\sigma_2^2)$ has the same orientation as τ_2 ; in this case we orient the segment (p_1, p_2) from p_1 to p_2 (otherwise, we orient it from p_2 to p_1). It is immediate to verify that if $\partial\varphi_{\sigma_3}(q)$ (respectively, $\partial\sigma_3$) is the boundary of $\varphi_{\sigma_3}(q)$ (respectively, of σ_3) then $\partial\varphi_{\sigma_3}(q) = \varphi_{\partial\sigma_3}(q)$. The previous ideas generalize to 3 - chains of M : if $c_3 = \Sigma\, a_i\, \sigma_3^i$ is an integral 3-chain of M , we define $\varphi_{c_3}(q)$ by setting it equal to $\Sigma\, a_i\, \varphi_{\sigma_3^i}(q)$; moreover,

(2.1) $$\partial\varphi_{c_3}(q) = \varphi_{\partial c_3}(q) .$$

One should also notice that

$$\varphi_{c_3^1 + c_3^2}(q) = \varphi_{c_3^1}(q) + \varphi_{c_3^2}(q) ,$$

$$\varphi_{-c_3}(q) = -\varphi_{c_3}(q)$$

and that these equalities allow us to write that

(2.2) $$\varphi_0(q) = 0 .$$

We now recall, for the sake of completeness, that a *closed, oriented, topological* n - *manifold* is a finite, connected, n - dimensional (oriented) polyhedron M such that, given any point $p \in M$,

$$H_i(M, M-p; \mathbb{Z}) = \begin{cases} \mathbb{Z}, & \text{if } i = n, \\ 0, & \text{if } i \neq n. \end{cases}$$

We also recall that each simplex of M is a face of an n-simplex of M, each (n-1) - simplex is face of exactly two n - simplices and finally, given any two n - simplices σ and τ of M, there is a finite sequence of n - simplices of M, say $\sigma^1 = \sigma, \sigma^2, \ldots, \sigma^q = \tau$ such that σ^i and σ^{i+1} have an (n-1) - face in common, $i = 1, \ldots, q-1$ [212; § 68] .

For the remainder of this section we shall assume that M is a closed, oriented, topological 3 - manifold; indeed, most of the time M will be the unit 3 - sphere of $\mathbb{R}^4$. If we regard M as a 3 - chain with coefficients ± 1 (according to orientation), (2.1) and (2.2) show that $\varphi_M(q)$ is a 1 - cycle of a convenient subdivision of M.

Let σ_2 be a triangle of M which has one point in common with $\varphi_M(q)$; this intersection point is endowed with a sign, according to the following rule: σ_2 lies in exactly two tetrahedra σ_3^1 and σ_3^2 of M, so if we take σ_2 with the orientation induced by σ_3^1, $\varphi_M(q) \cap \sigma_2$ will be given a positive (negative) sign according to that $\varphi_M(q)$ leaves (respectively, enters) σ_3^1.
The number

$$I(\sigma_2, \varphi_M(q)) = \pm 1$$

obtained is the *intersection number* of $\varphi_M(q)$ with σ_2. If $\varphi_M(q) \cap \sigma_2 = \emptyset$, we define the intersection number of $\varphi_M(q)$ with σ_2 to be zero. More generally, if $c_2 = \Sigma\, a_i\, \sigma_2^i$ is a 2 - chain of M such that $\varphi_M(q)$ has at most one point in common with each σ_2^i, we define the intersection number of $\varphi_M(q)$ with c_2 by

$$I(c_2, \varphi_M(q)) = \Sigma\, a_i\, I\left(\sigma_2^i, \varphi_M(q)\right).$$

This number has an interpretation which is directly related to f and τ_2 : if σ_2 is mapped properly onto τ_2 by f, define the *degree of* f *on* σ_2 (denoted by def $f|\sigma_2$) as the number $+1$ or -1 according to that $f(\sigma_2)$ has the same orientation as τ_2 or opposite orientation. If σ_2 is not mapped properly onto τ_2 by f, $\deg f|\sigma_2 = 0$. Then, $\deg f|c_2$ is defined by $\Sigma a_i \deg f|\sigma_2^i$ and the previous orientation convention show that

$$I(c_2, \varphi_M(q)) = \deg f|c_2 . \tag{2.3}$$

We next specialize M to be S^3 and N to be S^2 ; also, to simplify the notation, we shall write $\varphi(q)$ for $\varphi_M(q)$. Because $H_1(S^3;\mathbb{Z}) = 0$, the 1 - cycle $\varphi(q)$ of a subdivision of S^3 is homologous to 0 , that is to say, there is a 2-chain c_2 such that $\partial c_2 = \varphi(q)$. Of course, c_2 is not uniquely determined; we shall however, make a choice of c_2 which will prove itself to be very useful. If σ_3 is a tetrahedron of S^3 whose image by f covers τ_2 properly, then σ_3 contains a 1 - simplex (a,b) of $\varphi(q)$. Let e be one of the two vertices that the triangle containing a has in common with the triangle containing b ; let us replace (a,b) by the pair of segments (a,e) and (e,b). If we do this with every tetrahedron which contains a segment of $\varphi(q)$, we end up by replacing $\varphi(q)$ with a 1 - cycle w_1 which runs over triangles of S^3 ; note that w_1 together with $\varphi(q)$ bounds a 2-chain determined by triangles of type (a,e,b). Next, we replace in each triangle (e',a,e), (e,b,e''),... the pairs of edges $((e',a), (a,e))$, $((e,b), (b,e''))$,... by the 1 - simplices (e',e), (e,e''),... respectively. The edges (e',e), (e,e''),... define a 1 - cycle v_1 of S^3 , which together with $\varphi(q)$ bounds a 2 - chain c_2^1 determined by the triangles (e',a,e), (a,e,b),... . On the other hand, v_1 being homologous to 0 and being formed by 1 - simplices of S^3 , it is the boundary of a 2 - chain c_2^2 constructed with triangles of S^3. Thus, take c_2 to be $c_2^1 + c_2^2$.

We discuss now why the previous choice of c_2 is useful. Since the triangles of c_2^2 belong to the original triangulation of S^3 , they will be taken by f simplicially into triangles, edges or vertices of the triangulation of S^2. As for c_2^1, its triangles of the type (a,e,b) are taken by f into segments (q,r), where $r = f(e)$ is a vertex of τ_2; the triangles of type (e',a,e) are taken by f into triangles (r',q,r), where $r' = f(e')$ is also a vertex of τ_2 (notice that (r',q,r) may be degenerated). If we subdivide τ_2 so that q is connected to the vertices and we subdivide M to include $a,b,\ldots$ as vertices, then with respect to such triangulations of S^3 and S^2, f is simplicial on c_2.

Recall that the boundary operator commutes with a simplicial map; this implies that the (geometric) boundary of $f(c_2)$ reduces to just the vertex q and hence, $f(c_2)$ is a 2 - cycle of S^2. We conclude that $f(c_2)$ is homologous to an integral multiple of the fundamental cycle z_2 of S^2; in other words, if we represent the homology class of a cycle z by $\{z\}$, $\{f(c_2)\} = \gamma(f)\{z_2\}$, with $\gamma(f) \in \mathbb{Z}$. One should also notice that $\{f(c_2)\}$ is equal to $\deg f|c_2 \cdot \{z_2\}$ and so, $\gamma(f) = \deg f|c_2$. The integer $\gamma(f)$ is known as the *Hopf invariant* of the simplicial map f. We give next two properties of $\gamma(f)$.

I - *The number* $\gamma(f)$ *does not depend on the choice of* c_2.

In fact, if $\bar{c}_2$ is another 2 - chain such that $\partial\bar{c}_2 = \varphi(q)$, $c_2 - \bar{c}_2$ is a 2 - cycle of S^3 and so, is homologous to 0; then, $f(c_2 - \bar{c}_2)$ is also homologous to 0 and hence, $\{f(\bar{c}_2)\} = \{f(c_2)\} = \gamma(f)\{z_2\}$.

II - *The value* $\gamma(f)$ *is independent of the point* q.

Suppose first that r is an interior point of τ_2', a triangle of S^2 different from τ_2. Let c_2 and d_2 be 2 - chains of convenient subdivisions of S^3 such that $\partial c_2 = \varphi(q)$, $\partial d_2 = \varphi(r)$; notice that $\varphi(q) \cap \varphi(r) = \emptyset$. Let γ_q (respectively γ_r) be the value of $\gamma(f)$ relative to q (respectively r). Hence $\gamma_q = \deg f|c_2 = I(c_2,\varphi(r))$ and $\varphi_r = I(d_2,\varphi(q))$. Now let $c_2 \cap \varphi(r)$ (respectively, $d_2 \cap \varphi(q)$) be the 0 - chain intersection of c_2 and $\varphi(r)$ (respectively, of d_2 and $\varphi(q)$); if $\varepsilon : C_o(S^3) \to \mathbb{Z}$ is the augmentation homomorphism, $\varepsilon(c_2 \cap \varphi(r)) = \varepsilon(d_2 \cap \varphi(q))$, because $c_2 \cap \varphi(r) - d_2 \cap \varphi(q) \sim 0$. But $\varepsilon(c_2 \cap \varphi(r)) = I(c_2,\varphi(r))$, $\varepsilon(d_2 \cap \varphi(q)) =$

$= I(d_2,\varphi(q))$ and so, $\gamma_q = \gamma_r$. It is clear that if s is any point in the interior of τ_2', different from r, then $\gamma_r = \gamma_s$.

Remark. In the literature the number $I(c_2,\varphi(r)) = I(d_2,\varphi(q))$ is also called the *linking number* of $\varphi(q)$ and $\varphi(r)$ and is indicated by $L(\varphi(q),\varphi(r))$.

The simplicial approximation theorem suggests that we could possibly define the Hopf invariant of any *continuous* function $f : S^3 \to S^2$; for a given simplicial approximation f' to f, set $\gamma(f) = \gamma(f')$. It is clear that if we are to obtain a meaningful definition, the number $\gamma(f)$ must be independent of the simplicial approximation chosen. This is the substance of the following result.

Lemma 2.4. *For any continuous function* f *of* S^3 *into* S^2, *the integer* $\gamma(f)$ *is well defined and is independent of* f *within its homotopy class.*

Proof. The proof of this result is rather long and will be given in two steps.

STEP 1. Suppose that f is itself simplicial (for some basic triangulation of S^3 and S^2) and let $\bar{f}$ be another simplicial map which approximates f. We shall show that $\gamma(f) = \gamma(\bar{f})$.

Let q be an internal point of a 2 - simplex τ_2 of S^2 and let r be interior to $\tau_2' \neq \tau_2$. Construct the 1 - cycles $\varphi(q),\varphi(r)$ (respectively, $\bar{\varphi}(q),\bar{\varphi}(r)$) relative to f (respectively, $\bar{f}$); we must show the equality of the linking numbers $L(\varphi(q),\varphi(r))$ and $L(\bar{\varphi}(q),\bar{\varphi}(r))$. To this end we construct the 2 - chains c_2 and d_2 so that:

(i) c_2 is disjoint from $\varphi(r)$, $\bar{\varphi}(r)$;

(ii) d_2 is disjoint from $\varphi(q)$, $\bar{\varphi}(q)$;

(iii) $\partial c_2 = \varphi(q) - \bar{\varphi}(q)$;

(iv) $\partial d_2 = \varphi(r) - \bar{\varphi}(r)$.

These conditions on c_2 and d_2 show immediately the desired equality of the linking numbers; notice that due to symmetry, it is enough to prove the existence of c_2 with properties (i) and (iii). We observe first that since $\bar{f}$ is a simplicial approximation to f, if σ_2 is a 2-simplex of S^3 such that $f(\sigma_2)$ is non-degenerate, then $\overline{f}(\sigma_2)$ is also non-degenerate and indeed, $f(\sigma_2) = \overline{f}(\sigma_2)$. Let c_3 be the 3-chain formed by all tetrahedra of S^3 mapped onto τ_2 by f and $\overline{f}$; this 3-chain has no tetrahedra in common with the 3-chain c_3' formed by all tetrahedra of S^3 which are mapped onto τ_2' by f and $\overline{f}$. Since $\varphi(r)$ and $\overline{\varphi}(r)$ lie in c_3', the problem will be solved if we show that $\varphi(q)$ and $\overline{\varphi}(q)$ together bound a 2-chain c_2 in c_3.

Let σ_2 be a face of a tetrahedron σ_3' of c_3 such that $f(\sigma_2)$ is not degenerate. Since $I(\sigma_2,\varphi(q)) = I(\sigma_2,\overline{\varphi}(q))$, the 1-cycle $\varphi(q) - \overline{\varphi}(q)$ has intersection number 0 with σ_2, and indeed, with any face of σ_3'. We are going to show that every 1-cycle z_1 of a convenient subdivision of S^3 which has intersection number 0 with every triangle in c_3 is homologous to 0 in c_3; this shows, in particular, that there exists a 2-chain c_2 in c_3 such that $\partial c_2 = \varphi(q) - \varphi(\overline{q})$. Suppose that z_1 intersects σ_2 at a point a, which can becounted positively by a suitable choice of orientation for σ_2; since $I(\sigma_2,z_1) = 0$, z_1 must intersect σ_2 at another point b, which must be counted negatively. Now besides being a face of σ_3', the triangle σ_2 is a face of just another tetrahedron σ_3''. Let $a_1,b_1 \in \sigma_3'$ and $a_2,b_2 \in \sigma_3''$ be points of z_1 so that a_1,a_2 are close to a and b_1,b_2 are close to b; we assume that $a_1\, a\, a_2$ is the positive direction of z_1 around a and thus, $b_2\, b\, b_1$ will be the positive direction of z_1 around b (see figure).

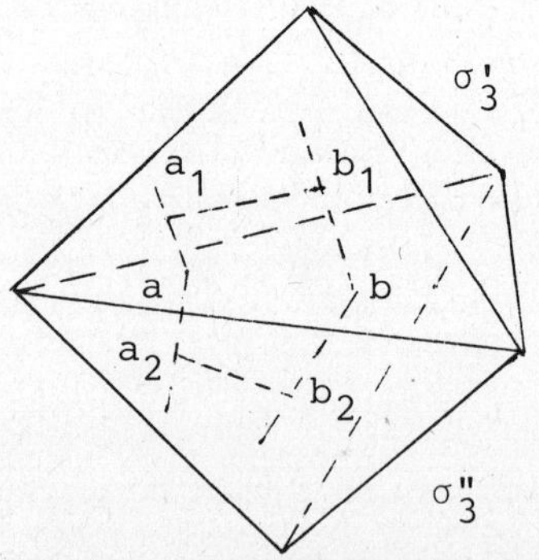

Connect a_1 to b_1 by a segment in σ_3' and connect a_2 to b_2 by a segment in σ_3'' and denote the closed, oriented polygonal line $a_1\, a\, a_2\, b_2\, b\, b_1\, a_1$ by P_1. Notice that P_1 is homologous to 0 in $\sigma_3' + \sigma_3''$ and hence, in c_3 ; consequently, $z_1^1 = z_1 - P_1$ and z_1 are homologous in c_3. The 1 - cycle z_1^1 does not contain the points a and b and moreover, its intersection with σ_2 has two points less than the intersection of z_1 and σ_2 ; finally, the intersection number of z_1^1 with any triangle of c_3 ist still 0. In this way we obtain a 1 - cycle z_1^n which is homologous to z_1 in c_3 and which is disjoint from all triangles of c_3 ; one should notice that z_1^n consists of a certain number of mutually disjoint cycles, each one contained in the interior of a tetrahedron of c_3. Hence, z_1^n is homologous to 0 in c_3. This completes the proof of Step I.

STEP 2. We show next that if f_1 and f_2 are two simplicial maps from S^3 to S^2 which belong to the same homotopy class, then $\gamma(f_1) = \gamma(f_2)$.

Let $F : S^3 \times I \to S^2$ be the homotopy $F : f_1 \simeq f_2$. Assume f_1 and f_2 to be simplicial with respect to the same triangulation of S^3; this gives rise to a triangulation of $S^3 \times I$. Let F' be a simplicial approximation to F. Then F' is a simplicial map from a convenient barycentric subdivision of $S^3 \times I$ into a triangulation of S^2. This new triangulation of $S^3 \times I$ induces a triangulation on $S^3 \times \{0\} = S_o^3$ and on $S^3 \times \{1\} = S_1^3$ and these triangulations are subdivisions of the original triangulation of S^3. Let $f_1' = F'|S_o^3$ and $f_2' = F'|S_1^3$; because $F' \simeq F$, f_1' is a simplicial approximation to f_1 and f_2' is a simplicial approximation to f_2. By the argument developed in Step I, $\gamma(f_1) = \gamma(f_1')$ and $\gamma(f_2) = \gamma(f_2')$; we have to show that $\gamma(f_1') = \gamma(f_2')$.

Let q be a point in the interior of a triangle σ_2 of S^2 and let $\varphi_{S^3\times I}(q)$, $\varphi_{S_o^3}(q)$ and $\varphi_{S_1^3}(q)$ be the chains of $S^3 \times I$ and S^3 whose carriers are the inverse images of q by F', f_1' and f_2', respectively. Since $\partial(S^3 \times I) = S_1^3 - S_o^3$,

$$\partial\varphi_{S^3\times I}(q) = \varphi_{S_1^3}(q) - \varphi_{S_o^3}(q) .$$

From this we conclude that if c_2^o and c_2^1 are 2-chains of S_o^3 and S_1^3 respectively, such that $\partial c_2^o = \varphi_{S_o^3}(q)$ and $\partial c_2^1 = \varphi_{S_1^3}(q)$ then, $c_2^1 - \varphi_{S^3 \times I}(q) - c_2^o = z_2$ is a 2-cycle of a triangulation of $S^3 \times I$. Notice that z_2 like any cycle of $S^3 \times I$, is homologous to a cycle of S_1^3, namely the projection z_2^1 of z_2 onto S_1^3. Hence, $F'(z_2)$ is homologous to $F'(z_2^1) = f_2'(z_2)$ and thus, is homologous to 0. This implies that $f_1'(c_2^o) \sim f_2'(c_2^1)$, thus $\deg f_1'|c_2^o = \deg f_2'|c_2^1$ and therefore, $\gamma(f_1') = \gamma(f_2')$, which completes the proof of the lemma.

Theorem 2.5. *The function*

$$\gamma : \pi_3(S^2) \to \mathbb{Z}$$

which takes any $[f] \in \pi_3(S^2)$ *into* $\gamma(f)$ *is a group homomorphism.*

Proof. Let $[f], [g] \in \pi_3(S^2, q_o)$ be taken arbitrarily; we may assume that these elements are represented by maps $f, g : S^3 \to S^2$ so that: (i) f and g are the geometric realizations of simplicial functions (which we represent by the same letters) relative to some triangulation of S^2 and S^3; (ii) $f(E_-^3) = g(E_+^3) = q_o$, where E_+^3, E_-^3 are respectively the northern and southern hemispheres of S^3.

Recall that $[f] + [g]$ can be represented by $h : S^3 \to S^2$ defined by $h|E_+^3 = f$, $h|E_-^3 = g$; of course, we may view h as a simplicial function. Take q in the interior of some triangle of S^2; let $\varphi^h(q)$, $\varphi^f(q)$ and $\varphi^g(q)$ be the 1-cycles of a subdivision of S^3 whose carriers are the inverse images of q by h, f and g, respectively. Let $c_2^+ \subset E_+^3$ ($c_2^- \subset E_-^3$) be the 2-chains of S^3 such that $\partial c_2^+ = \varphi^f(q)$ (respectively, $\partial c_2^- = \varphi^g(q)$). Then, $c_2^+ + c_2^-$ is a 2-chain of S^3 such that $\partial(c_2^+ + c_2^-) = \varphi^h(q)$. . All this shows that if z_2 is the fundamental cycle of S^2, $\gamma(h)\{z_2\} = \{h(c_2^+ + c_2^-)\} = \{f(c_2^+)\} + \{g(c_2^-)\} = (\gamma(f) + \gamma(g))\{z_2\}$.

We conclued this section by showing that the homomorphism γ is indeed an epimorphism; we shall achieve this goal by first exhibiting a map $f : S^3 \to S^2$ with $\gamma(f) = \pm 1$ and then, by discussing the Hopf invariant of the composition of any map from S^3 to S^2 with a map of degree c.

In what follows, the points of $\mathbb{R}^3$ will be given coordinates (x_1,x_2,x_3) and those of $\mathbb{R}^4$, the coordinates (y_1,y_2,y_3,y_4). Define

$$g : \mathbb{R}^4 \to \mathbb{R}^3$$

by the formulas

$$x_1 = 2(y_1y_3 + y_2y_4)$$

$$x_2 = 2(y_2y_3 - y_1y_4)$$

$$x_3 = y_1^2 + y_2^2 - y_3^2 - y_4^2 \quad ;$$

a simple computation shows that

$$x_1^2 + x_2^2 + x_3^2 = \left(y_1^2 + y_2^2 + y_3^2 + y_4^2\right)^2 \quad ,$$

which means that the unit sphere S^3 of $\mathbb{R}^4$ is mapped by g onto the unit sphere S^2 of $\mathbb{R}^3$. Using the stereographic projection (from the north pole) of S^2 onto the plane (x_1,x_2), we identify the points $(x_1,x_2,x_3) \in S^2$ $(x_3 \neq 1)$ with the complex number

$$q = \frac{x_1 + ix_2}{1 - x_3}$$

and identify $(0,0,1)$ with $q = \infty$. Then, if $f = g|S^3$,

$$f(y_1,y_2,y_3,y_4) = \frac{y_1 + iy_2}{y_3 + iy_4} \quad .$$

Observe that if $q = \infty$, the carrier of $\varphi(q)$ is the set of all $(y_1,y_2,y_3,y_4) \in S^3$ such that $y_3 = y_4 = 0$, which is a great circle of S^3; for any other point $q = \dfrac{x_1 + ix_2}{1 - x_3}$, the carrier of $\varphi(q)$ is the set of all $(y_1,y_2,y_3,y_4) \in S^3$ such that $\dfrac{y_1 + iy_2}{y_3 + iy_4} = \dfrac{x_1 + ix_2}{1 - x_3}$ and so

$$(1-x_3)y_1 - x_1y_3 + x_2y_4 = 0$$
$$(1-x_3)y_1 - x_1y_4 - x_2y_3 = 0 .$$

These equations show that, again, the carrier of $\varphi(q)$ is a great circle. The following result proves that $\gamma(f) = 1$.

Lemma 2.6. *Let* $f : S^3 \to S^2$ *be a map such that the carrier of* $\varphi(q)$ *is a great circle of* S^3 *for any* $q \in S^2$ *; then* $\gamma(f) = \pm 1$ *, the sign depending on the orientation of* S^3.

Proof. We develop only the geometric argument leaving the simplicial reasoning to the reader.

A 3 - dimensional subspace of $\mathbb{R}^4$ and a 2 - dimensional subspace of $\mathbb{R}^4$ intersect on a straight line if the plane is not contained in the 3 - space; hence, a great sphere and a great circle of S^3 intersect in two diammetrically opposite points, if the circle is not contained in the sphere. Hence, if H is a hemisphere of a great sphere of S^3, any great circle which is disjoint from ∂H intersects H in exactly one point. This indicates that any two great circles of S^3 which are disjoint to each other have linking number equal to ± 1 ; thus, for any two distinct points of S^2 , say q and r , $L(\varphi(q),\varphi(r)) = \pm 1$ and so, the desired result.

Theorem 2.7. *Let* $g : S_1^3 \to S^3$ *and* $h : S^2 \to S_1^2$ *be given maps of degree* c *. Then, for any map* $f : S^3 \to S^2$, $\gamma(fg) = c \cdot \gamma(f)$ *and* $\gamma(hf) = c^2 \cdot \gamma(f)$.

Proof. As usual, we may assume f, g and h to be simplicial. For any q interior to a triangle of S^2, let $\varphi^{fg}(q)$ be the 1 - cycle of S^3 obtained from the inverse image of q by fg. Let σ_3 be a tetrahedron of S^3 which contains a 1 - simplex (a,b) of $\varphi(q)$.

Suppose next that n tetrahedra of S_1^3 are mapped by g onto σ_3 in the positive fashion and that $n - c$ tetrahedra of S_1^3 are mapped negatively by g onto σ_3 : then, n respectively , n - c) segments of $\varphi^{fg}(q)$ will be mapped positively (respectively, negatively) onto (a,b) by g. This shows that $g(\varphi^{fg}(q)) = c \cdot \varphi(q)$. Let now c_2 and c_2^1 be two 2 - chains, the first of S^3 and the second of S_1^3 , such that

$\partial c_2 = \varphi(q)$ and $\partial c_2^1 = \varphi^{fg}(q)$. It follows that $\partial g(c_2^1) = c \cdot \partial c_2$ and so, $fg(c_2^1) - cf(c_2) \sim 0$ in S^2 because $g \cdot (c_2^1) - c \cdot c_2$ is a 2 - cycle of S^3. This implies that $\deg fg|c_2^1 = c \cdot \deg f|c_2$ and hence $\gamma(fg) = c \cdot \gamma(f)$.

To show the second part of the theorem, we begin by taking a point q interior to a triangle σ_2^1 of S_1^2; let $\{q_1,\ldots,q_n ; r_1,\ldots,r_{n-c}\}$ be the inverse image of q by h. We observe that each point of this set is interior to a triangle of S^3; we give a positive orientation to the triangles of S^2 which contain a point q_i in their interior and a negative orientation to these triangles which contain points r_j; it then follwos that

$$\varphi^{hf}(q) = \sum_{i=1}^{n} \varphi(q_i) - \sum_{j=1}^{n-c} \varphi(r_j) .$$

Let c_2^i, $i = 1,\ldots,n$ (respectively, d_2^j, $j = 1,\ldots,n - c$) be the 2 - chains of S^3 such that $\partial c_2^i = \varphi(q_i)$ (respectively, $\partial d_2^j = \varphi(r_j)$). By the previous characterization of $\varphi^{hf}(q)$, $\{hf(\Sigma c_2^i - \Sigma d_2^j)\} = \gamma(hf)\{z_2^1\}$, where z_2^1 is the fundamental cycle of S_1^2. On the other hand, for all $i = 1,\ldots,n$ and $j = 1,\ldots,n - c$, $\{f(c_2^i)\} = \gamma(f)\{z_2\}$, $\{f(d_2^j)\} = \gamma(f)\{z_2\}$ and so, $\{f(\Sigma c_2^i - \Sigma d_2^j)\} = c \cdot \gamma(f)\{z_2\}$. Our result is obtained from the last equality and the fact that $\{h(z_2)\} = c(\{z_2^1\}$.

Remark. The existence of a map $f : S^3 \to S^2$ of Hopf invariant 1 and (2.7) show that the homomorphism γ of (2.5) is an epimorphism and thus, $\pi_3(S^2) \neq 0$. An alternative way to prove that $\pi_3(S^2) \neq 0$ is to show that the map $f : S^3 \to S^2$ which we proved to have Hopf invariant 1 is in fact a fibration with fibre S^1; then, the exact homotopy sequence of this fibration shows that $\pi_3(S^2) \cong \mathbb{Z}$. Of course, these ideas were not avaible to Hopf when he wrote his paper.

3. The Hopf Invariant of Maps $f : S^{2n-1} \to S^n$.

In the previous section we described the construction of the Hopf invariant of a map from S^3 to S^2, along the lines traced by Hopf in [121]. In a subsequent paper [122], Hopf generalized this construction to maps from S^{2n-1} into

S^n, for $n > 2$. Such a generalization is readily obtained: given a simplicial function $f : S^{2n-1} \to S^n$ (relative to triangulations of S^n and S^{2n-1}) and a point q interior to some n-simplex of S^n, let $\varphi(q)$ be the $(n-1)$ - chain of a convenient triangulation of S^{2n-1} whose carrier is the inverse image of q by f; notice that since S^{2n-1} is a closed manifold, $\varphi(q)$ is indeed a cycle. This cycle was called by Hopf the *original cycle* of q, relative to f. Now the Hopf invariant $\gamma(f)$ is defined as the linking number $L(\varphi(q),\varphi(r))$, where r is an arbitrary point in the interior of an n - simplex of S^n, possibly different from that containing q. This number is independent of the points q and r, and as before, the Simplicial Approximation Theorem can be used to define the Hopf invariant of an arbitrary map f of S^{2n-1} onto S^n by $\gamma(f) = L(\varphi^{\bar{f}}(q),\varphi^{\bar{f}}(r))$, where $\varphi^{\bar{f}}(q)$ and $\varphi^{\bar{f}}(r)$ are the original cycles of q and r respectively, relative to a simplicial approximation $\bar{f}$ of f; again, $\gamma(f)$ is independant of f within its homotopy class. This approach gives immediately an interesting result. In fact, it is known that if z_{k-1} and z_{n-k} are non-intersecting cycles of dimensions $k - 1$ and $n - k$ respectively, belonging to a subdivision of an orientable n - dimensional topological manifold, then the linking numbers $L(z_{k-1},z_{n-k})$ and $L(z_{n-k},z_{k-1})$ can be defined and they are related by the formula: $L(z_{k-1},z_{n-k}) = (-1)^{(k-1)(n-k)+1} \cdot L(z_{n-k},z_{k-1})$ [211; § 77]. In our case, $L(\varphi(q),\varphi(r)) = (-1)^{(n-1)^2+1} L(\varphi(r),\varphi(q))$ thus, if n *is odd*, $L(\varphi(q),\varphi(r)) = 0$ and so, the Hopf invariant of any map $f : S^{2n-1} \to S^n$ with n odd is always trivial. Hence, from now on, we shall restrict n to be an *even* integer.

Recall that for $n = 2$ we have proved the existence of maps with Hopf invariant 1 (and indeed, for that value of n, there are maps $S^3 \to S^2$ with any preassigned integer as Hopf invariant); for higher values of n, Hopf has shown the existence of maps $S^{2n-1} \to S^n$ with Hopf invariant 1, if $n = 4$ and 8 [122]. We shall reproduce here these results but first, we prove the following

Theorem 3.1. *If* n *is even, there is a map* $f : S^{2n-1} \to S^n$ *with* $\gamma(f) = 2$.

Proof. Let E^n_+ be the northern hemisphere of S^n and let $g : (E^n_+, S^{n-1}) \to (S^n, q_o)$ $(q_o = (1,0,\ldots,0))$ be any map whose restriction to $E^n_+ - S^{n-1}$ is a homeomorphism (an example of such a map is the following: first take $h : E^n_+ \to S^n$ given by $h(x_1,\ldots,x_{n+1}) = (\mu x_1,\ldots,\mu x_n, 2x_{n+1} - 1)$ where

$$\mu = 0 \quad \text{if} \quad x_{n+1} = 1 \quad \text{and} \quad \mu = \sqrt{\frac{4(1-x_{n+1})x_{n+1}}{1 - x^2_{n+1}}} \quad \text{if} \quad x_{n+1} \neq 1;$$

next, take the rotation ρ of S^n given by $\rho(x_1,\ldots,x_{n+1}) = (-x_{n+1}, x_2,\ldots,x_n,x_1)$. The map ρh has the desired property [104]). Suppose now that g is simplicial and view S^{2n-1} as the boundary of $D^n \times D^n$ (where D^n is the unit n - ball of $\mathbb{R}^n$); notice that the (simplicial) boundary of $D^n \times D^n$ is also given by $\partial D^n \times D^n + D^n \times \partial D^n$, because n is even. Hence define

$$f : \partial D^n \times D^n + D^n \times \partial D^n \to S^n$$

by taking $(y,(x_1,\ldots,x_n)) \in \partial D^n \times D^n$ into $g(x_1,\ldots,x_n, \sqrt{1-(x_1^2 + \ldots + x_n^2)})$ and $((y_1,\ldots,y_n),x) \in D^n \times \partial D^n$ into $g(y_1,\ldots,y_n, \sqrt{1-(y_1^2 + \ldots + y_n^2)})$.

Let q be an interior point of some n - simplex of S^n; the carrier of the $(n-1)$ - cycle $\varphi(q)$ determined by f and q is $\partial D^n \times p \cup p \times \partial D^n$, where p is the unique point in the interior of D^n such that $f(p) = q$. Take next $p' \in \partial D^n$ and form the 0 - cycle $(p) - (p')$; let c_1 be a 1 - chain of D^n such that $\partial c_1 = (p) - (p')$. Then $\partial(D^n \times (p') - \partial D^n \times c_1 + (p') \times D^n + c_1 \times \partial D^n) = \varphi(q)$. Since $f(\partial D^n \times c_1)$, $f(c_1 \times \partial D^n)$ are homologous to 0 and $f(D^n \times (p')), f((p') \times D^n)$ are equal to the fundamental cycle of S^n, it follows that

$$\deg f|D^n \times (p') - \partial D^n \times c_1 + (p') \times D^n + c_1 \times \partial D^n = 2 ,$$

i.e., $\gamma(f) = 2$.

An immediate consequence of the previous theorem is that for n even, $\pi_{2n-1}(S^n) \neq 0$.

Now we go on to the existence of maps of Hopf invariant 1. We begin by observing that a map $f : S^{n-1} \times S^{n-1} \to S^{n-1}$ defines a map $H(f) : S^{2n-1} \to S^n$. In fact, let E_+^n and E_-^n be the northern and southern hemispheres of S^n, respectively. Since $E_+^n \cap E_-^n$ is the equator S^{n-1}, we can view E_+^n and E_-^n as cones over S^{n-1}; we shall write $E_+^n = C^+S^{n-1}$ and $E_-^n = C^-S^{n-1}$, for the sake of precision. On the other hand, the unit ball D^n is homeomorphic to CS^{n-1}, so we view S^{2n-1} as being $S^{n-1} \times CS^{n-1} \cup CS^{n-1} \times S^{n-1}$. The map $H(f)$ is then trivially defined: it is the map taking any $(x,(x',t)) \in S^{n-1} \times CS^{n-1}$ into $(f(x,x'),t) \in C^+S^{n-1}$ and any $((x,t),x') \in CS^{n-1} \times S^{n-1}$ into $(f(x,x'),t) \in C^-S^{n-1}$.

The construction of $H(f)$ we just described is known as the *Hopf construction*; we shall consider it again in Chapter 2, in connection with the study of H - spaces.

We also recall that a map $f : S^{n-1} \times S^{n-1} \to S^{n-1}$ is of *type* (c_1,c_2) if $\deg f|S^{n-1} \times q_2 = c_1$ and $\deg f|q_1 \times S^{n-1} = c_2$, where (q_1,q_2) is a base point for $S^{n-1} \times S^{n-1}$. We are now ready to prove the crucial result of this section.

Theorem 3.2. *If* $f : S^{n-1} \times S^{n-1} \to S^{n-1}$ *is of type* (c_1,c_2), $\gamma(H(f)) = \pm c_1c_2$.

Proof. We begin by decomposing S^{2n-1} into the union of the sets

$$V_1 = \left\{(x_1,\ldots,x_{2n}) \in S^{2n-1} | x_1^2 + \ldots + x_n^2 \leq x_{n+1}^2 + \ldots + x_{2n}^2\right\}$$

and

$$V_2 = \left\{(x_1,\ldots,x_{2n}) \in S^{2n-1} | x_1^2 + \ldots + x_n^2 \geq x_{n+1}^2 + \ldots + x_{2n}^2\right\},$$

and by observing that $V_1 \cong S^{n-1} \times D^n$, $V_2 \cong D^n \times S^{n-1}$ and $V_1 \cap V_2 \cong S^{n-1} \times S^{n-1}$.

For each real number c such that $c^2 \leq 1/2$ we consider the $(n-1)$ - spheres

$$S_{1,c}^{n-1} = \left\{(x_1,\ldots,x_{2n}) \in S^{2n-1} | x_1 = c, x_2 = \ldots = x_n = 0,\right.$$
$$\left. x_{n+1}^2 + \ldots + x_{2n}^2 = 1-c^2\right\} \subseteq V_1$$

and

$$S_{2,c}^{n-1} = \left\{ (x_1,\ldots,x_{2n}) \in S^{2n-1} \mid x_1^2 + \ldots + x_n^2 = 1 - c^2, \right.$$
$$\left. x_{n+1} = c, x_{n+2} = \ldots = x_{2n} = 0 \right\} \subseteq V_2 .$$

Let $z_{i-1}^{i,c}$ be the fundamental cycle of $S_{i,c}^{n-1}$, $i = 1,2$; we should notice that the homology classes $\{z_{n-1}^{1,1/\sqrt{2}}\}$ and $\{z_{n-1}^{2,1/\sqrt{2}}\}$ form a basis for the homology of $V_1 \cap V_2$ and furthermore, when viewed as $(n-1)$ - cycles of S^{2n-1}, $z_{n-1}^{1,1/\sqrt{2}}$ and $z_{n-1}^{2,1/\sqrt{2}}$ are homologous to 0 . Actually, these cycles are boundaries: for example, we can see geometrically that $z_{n-1}^{1,1/\sqrt{2}}$ bounds the n - chain c_n^2 of V_2 given by the set $\left\{ (x_1,\ldots,x_n) \in S^{2n-1} \mid x_1 \geq 1/\sqrt{2},\ x_2 = \ldots = x_n = 0, \right.$ $\left. x_{n+1}^2 + \ldots + x_{2n}^2 \leq 1/\sqrt{2} \right\}$. Now the intersection of c_n^2 with any sphere $S_{2,c}^{n-1}$ is the point $(x_1 = \sqrt{1-c^2}, x_2 = \ldots = x_n = 0$; $x_{n+1} = c,\ x_{n+2} = \ldots = x_{2n} = 0)$ and so, the intersection number $I(c_n^2, z_{n-1}^{2,c}) = \pm 1$. Since $z_{n-1}^{1,1/\sqrt{2}} \sim z_{n-1}^{1,c}$ in V_1, it follows then that

(3.3) $$L(z_{n-1}^{1,c},\ z_{n-1}^{2,c}) = \pm 1 .$$

Following our costumary procedure, we assume that $H(f)$ is simplicial; we also take q_+ and q_- to be interior points to n - simplices of E_+^n and E_-^n , respectively. Let $\varphi(q_+)$ and $\varphi(q_-)$ be the original cycles of q_+ and q_- , relative to $H(f)$. The Hopf construction shows that $\varphi(q_+)$ is an $(n-1)$ - cycle of a subdivision of V_1 and that $\varphi(q_-)$ is an $(n-1)$ - cycle of a subdivision of V_2. Now $\varphi(q_+)$ and $\varphi(q_-)$ are homologous to integral multiples of $z_{n-1}^{1,c}$ and $z_{n-1}^{2,c}$ in V_1 and V_2, respectively; say that $\varphi(q_+) \sim b_1 z_{n-1}^{1,c}$ and $\varphi(q_-) \sim b_2 z_{n-1}^{2,c}$. Since $z_{n-1}^{1,1/\sqrt{2}}$ (whose carrier $S_{1,1/\sqrt{2}}^{n-1}$ is identified to $S^{n-1} \times q_2$) is mapped by $H(f)$ into S^{n-1}

with degree c_1, deg $H(f)|c_n^2 = c_1$, which implies that $I(c_n^2, \varphi(q_-)) = \pm c_1$ and hence, $L(z_{n-1}^{1,1/\sqrt{2}}, \varphi(q_-)) = \pm c_1$. On the other hand, $L(z_{n-1}^{1,1/\sqrt{2}}, z_{n-1}^{2,c}) = \pm 1$ and $\varphi(q_-) \sim b_2 z_{n-1}^{2,c}$ imply that $L(z_{n-1}^{1,1/\sqrt{2}}, \varphi(q_-)) = \pm b_2$. Therefore, $b_2 = \pm c$, that is to say $\varphi(q_-) \sim \pm c_1 z_{n-1}^{2,c}$ in V_2 ; similarly, $\varphi(q_+) \sim \pm c_2 z_{n-1}^{1,c}$ in V_1 . Using (3.3), we conclude that $L(\varphi(q_+), \varphi(q_-)) = \pm c_1 c_2$.

Remark. The previous theorem shows that a map $f : S^{n-1} \times S^{n-1} \to S^{n-1}$ of type (c_1, c_2) determines a map $g : S^{2n-1} \to S^n$ of Hopf invariant $c_1 c_2$. In fact, if $\gamma(H(f)) = -c_1 c_2$, take $g = H(f) . f'$, where $f' : S^{2n-1} \to S^{2n-1}$ is a map of degree -1 .

Corollary 3.4. *If* $n = 2,4$ *or* 8 *there are maps from* S^{2n-1} *onto* S^n *of Hopf invariant* 1 .

Proof. According to the theorem and the previous remark, it is enough to show the existence of maps $f : S^{n-1} \times S^{n-1} \to S^{n-1}$ of type (1,1) for the values of n given in the statement. Now, this is trivial: f is just the map induced on the corresponding product of spheres by the multiplication of the complex, quaternionic and Cayley numbers.

In closing this section we want to observe that (3.4) provides a new proof for the existence of maps $S^3 \to S^2$ of Hopf invariant 1 (see § 2) ; furthermore, Hopf used (3.2) to show that for n even, there are maps $S^{2n-1} \to S^n$ of Hopf invariant 2 . The proof we gave here to (3.2) is based on [104; Theorem 1.6., Ch. IV] .

4. Cohomological Interpretation of the Hopf Invariant.

In order to describe the purely combinatorial definition of the Hopf invariant in terms of ordinary cohomology theory, we shall analyse more closely the structure of the original cycles $\varphi(q)$; to this end, we shall also review some basic ideas.

Given a (closed, oriented) topological n - manifold M , denote its ordered i - simplices by σ_i^j, $i = 0,\ldots,n$;

$j = 1,\ldots,a_i$. Select a point p^i_j in the interior of σ^j_i , for each i and j as before; this point will be called, by abuse of language, the *barycenter* of σ^j_i . Now we construct a new complex M' by defining its abstract simplices to be ordered sequences $\sigma'_k = \left(p^{i_o}_{j_o},\ldots,p^{i_k}_{j_k}\right)$ where $n \geq i_o > i_1 > \ldots > i_k \geq 0$ and $\sigma^{j_r}_{i_r}$ is a face of $\sigma^{j_{r-1}}_{i_{r-1}}$, $r = 1,\ldots,k$. Notice that M' is a subdivision of M in the sense of [225; 3.3], that M' is ordered and that M' is also a triangulation of M. The complexes M and M' are related by a simplicial map $\alpha : M' \to M$ which gives rise to a chain equivalence : α is defined by associating to each vertex p^i_j the first vertex of the simplex σ^j_i ; following [110;3.5.6] one shows that $\alpha_{\#} : C_{\#}(M') \to C_{\#}(M)$ is a chain equivalence with inverse chain map $\beta : C_{\#}(M) \to C_{\#}(M')$.

Lemma 4.1. *Let* M *and* N *be topological manifolds and let* $f : M \to N$ *be a simplicial map. Suppose that* N *is subdivided (barycentrically) by* N' *then, there is a subdivision* M' *of* M *such that* f *is simplicial with respect to* M' *and* N'.

Proof. By induction on the skeletons of M. Suppose that the ith skeleton of M is subdivided so that the restriction of f to it is simplicial. Let σ be an $(i+1)$ - simplex of M; because f is simplicial, $f(\sigma) = \tau$ is a simplex of N. Let q be the barycenter of τ ; we are going to prove that there is a point p in the interior of σ such that $f(p) = q$. In fact, if $\dim \tau = i+1$, this is so because of the simpliciality of f . If $\dim \tau \leq i$, there are at least two faces of σ , say σ^1 and σ^2 , such that $f(\sigma^1) = f(\sigma^2) = \tau$. By induction, there are points $p_1 \in \operatorname{int} \sigma^1$ and $p_2 \in \operatorname{int} \sigma^2$, such that $f(p_1) = f(p_2) = q$. By linearity, the segment p_1p_2 is entirely taken onto q; then choose p interior to the segment p_1p_2.

Given a topological n - manifold M as before, we define for each i - simplex σ^j_i the *closed dual* (n-i)-*cell* b^j_{n-i} to be the union of all simplices of M' which have p^i_j as last vertex. The cell b^j_{n-i} is an (n-i)-dimensional subcomplex of M. The set

$$\bar{M} = \left\{ b^j_{n-i} \,|\, i = 0,\dots,n \; ; \; j = 1,\dots,a_i \right\}$$

is called the *dual cell subdivision* of M. According to [71;3] and [110; 3.8.2], $\bar{M}$ is a *block dissection* of M; moreover, $H_*(\bar{M}) \cong H_*(M)$ [110; 3.8.8].

<u>Lemma 1.2.</u> *Let* M *and* N *be topological manifolds subdivided so that a given simplicial map* f *of* M *onto* N *is also simplicial with respect to these subdivisions. If* $m = \dim M \geq \dim N = n$ *and* q *is the barycenter of an* n*-simplex* τ *of* N, *then the carrier of the original cycle* $\varphi(q)$ *is an* $(m-n)$*-dimensional subcomplex of* $\bar{M}$.

<u>Proof.</u> Let σ be an n-simplex of M such that $f(\sigma) = \tau$. If ρ is any simplex of M, of which σ is a face, $f(\rho) = \tau$ and the barycenter of ρ is mapped into q by f. It follows that the entire dual cell b_{m-n} of σ is mapped into q by f.

Conversely, let p be an arbitrary point of the carrier of $\varphi(q)$. Suppose that $p \in \sigma' = \left(p^{i_0}_{j_0},\dots,p^{i_k}_{j_k}\right)$. If $f\left(p^{i_r}_{j_r}\right) \neq q$ for every $r = 0,\dots,k$, then $q \notin f(\sigma')$ and p is not in the carrier of $\varphi(p)$, contradiction. Hence, let $\sigma'^1 (\neq \emptyset)$ be the simplex of M' generated by all vertices of σ' which are taken into q by f and let σ'^2 be the face of σ' generated by the remaining vertices. There are points $p_1 \in \sigma'^1$, $p_2 \in \sigma'^2$ and integers a,b such that

$$0 \leq a,\ b \leq 1, \quad a + b = 1$$

and

$$p = ap_1 + bp_2 \,.$$

Since $f(p) = f(p_1) = q$ and $f(p_2) \neq q$, it follows that $b = 0$, $a = 1$ and $p = p_1 \in \sigma'^1$. Since f is simplicial, the vertices of σ'^1 are barycenters of simplices of dimension $\geq n$. Now, if ρ is a simplex of M such that $\dim \rho \geq n$ and f(barycenter of $\rho) = q$, then $f(\rho) = \tau$ and hence, ρ has at least one face, say σ, such that $f(\sigma) = \tau$. This shows that σ^1 is contained in the dual $(m-n)$-cell of σ.

We restrict next M and N to be respectively

S^{2n-1} and S^n. We want to study the Hopf invariant of a simplicial map $f : S^{2n-1} \to S^n$; we shall assume that both spheres are subdivided so that f is still simplicial with respect to these subdivisions (see (4.1)). Let q be interior to an n-simplex τ of S^n; since $\gamma(f)$ is independent of q, there is no loss of generality in assuming that q is indeed the barycenter of τ. Let τ' be the unique n-simplex of the subdivision $(S^n)'$ such that $\alpha(\tau') = \tau$ and let u^n be the n-cocycle of $(S^n)'$ defined by $\langle u^n, \tau'\rangle = 1$ and $\langle u^n, \rho'\rangle = 0$, for every n-simplex ρ' of $(S^n)'$ different from τ' (here $\langle\ ,\ \rangle$ is the Kronecker index). We shall also consider the fundamental cycles z_n and z_{2n-1} of S^n and S^{2n-1}, where z_n is taken so that appears in it with coefficient $+1$. Finally, to conclude this list of generalities we observe that the cohomology class $\{u^n\}$ is a generator of $H^n(S^n)$ (the Kronecker index induces an isomorphism $H^n(S^n) \to \mathrm{Hom}(H_n(S^n), \mathbb{Z})$).

<u>Theorem 4.3.</u> *Let* $f^{\#} : C^n((S^n)') \to C^{\#}((S^{2n-1})')$ *be the cochain map induced by* f. *The carrier of the* $(n-1)$-*cycle* $u = f^{\#}(u^n) \cap \beta z_{2n-1}$ *coincides with the carrier of* $\varphi(q)$.

<u>Proof.</u> Suppose that $(p^{i_o}, \ldots, p^{i_{n-1}})$ is an $(n-1)$-simplex in the carrier of u. From the definition of the cap product, we can assume the existence of a $(2n-1)$-simplex $(p^{2n-1}, \ldots, p^o)$ of $(S^{2n-1})'$ such that

$$\langle f^{\#}(u^n), (p^n, \ldots, p^o)\rangle \neq 0$$

and

$$\left(p^{i_o}, \ldots, p^{i_{n-1}}\right) = (p^{2n-1}, \ldots, p^n).$$

On the other hand, $\langle f^{\#}(u^n), (p^n, \ldots, p^o)\rangle = \langle u^n, f(p^n, \ldots, p^o)\rangle \neq 0$ implies that $f(p^n, \ldots, p^o)$ coincides, up to orientation, with τ'; this fact together with the definition of τ' shows that $f(p^n) = q$. Let σ be the n-simplex of S^{2n-1} of which p^n is the barycenter; then $f(\sigma) = \tau$ and indeed, the barycenter of any simplex of S^{2n-1} which contains σ as a face is mapped onto q by f. Hence, $f(p^{i_o}, \ldots, p^{i_{n-1}}) = f(p^n, \ldots, p^o) = q$ and therefore, $(p^{i_o}, \ldots, p^{i_{n-1}})$ is an $(n-1)$-simplex of the carrier of $\varphi(q)$.

Let now p be an arbitrary point of the carrier of $\varphi(q)$. By (4.2) p belongs to the dual $(n-1)$-cell of an n-simplex σ of S^{2n-1}. Let p^n be the barycenter of σ; then $f(p^n) = q$ and $f(\sigma) = \tau'$. Hence, there is a $(2n-1)$-simplex of $(S^{2n-1})'$, say $(p^{2n-1},\dots,p^n,\dots,p^o)$ such that $f(p^n,\dots,p^o) = \tau'$ and $p \in (p^{2n-1},\dots,p^n)$. Since $(p^{2n-1},\dots,p^o)$ is a $(2n-1)$-simplex which appears in the cycle $\beta(z_{2n-1})$ and $\langle f^{\#}(u^n), (p^n,\dots,p^o)\rangle = 1$, it follows that $(p^{2n-1},\dots,p^n)$ is an $(n-1)$-simplex of the carrier of u and therefore, p belongs to the carrier of u.

Because $H_{n-1}(S^{2n-1}) = 0$, there is an n-chain v of $(S^{2n-1})'$ such that $\partial v = u$. Now, applying (4.3) we conclude that $f_{\#}(v)$ is an n-cycle and thus, there is an integer d such that $f_{\#}(v) = d\cdot\beta(z_n)$.

On the other hand, there is $v^{n-1} \in C^{n-1}((S^{2n-1})')$ such that $f^{\#}(u^n) = \delta v^{n-1}$; notice that $v^{n-1} \cup f^{\#}(u^n)$ is a $(2n-1)$-cocycle.

Theorem 4.4. *Let* ξ *be the fundamental class of* $H^{2n-1}((S^{2n-1})') \simeq \mathbb{Z}$. *The chomology class* $\{v^{n-1} \cup f^{\#}(u^n)\}$ *is equal to* $\gamma(f)\cdot\xi$.

Proof. We begin by observing that

$$\langle v^{n-1} \cup f^{\#}(u^n),\beta(z_{2n-1})\rangle = \langle v^{n-1},f^{\#}(u^n) \cap \beta(z_{2n-1})\rangle =$$

$$\begin{aligned}\langle v^{n-1},\partial v\rangle = \langle \delta v^{n-1},v\rangle &= \langle f^{\#}(u^n),v\rangle \\ &= \langle u^n,d\cdot\beta(z_n)\rangle \\ &= d\end{aligned}$$

and hence, $\{v^{n-1} \cup f^{\#}(u^n)\} = d\cdot\xi$. We are going to prove that $\gamma(f) = d$. To this end, we take the barycenter q^* on an n-simplex τ^* of S^n, $\tau^* \neq \tau$, and we construct u^{n*} and u^* following the same argument used in the construction of u^n and u, given q. Because of (4.3), $\gamma(f) = L(u,u^*)$; the theorem is then proved if we show that $L(u,u^*) = d$.

By definition, $L(u,u^*) = I(v,u^*)$; on the other hand, since $I(\beta(z_n),q^*) = 1$ and $d\cdot\beta(z_n) = f_{\#}(v)$, it follows that $d = I(f_{\#}(v),q^*)$. We are going to show that $I(v,u^*)=I(f_{\#}(v),q^*)$.

Observe that $I(v,u^*)$ is defined: for this, we recall that the carrier of u^* is an $(n-1)$-subcomplex of S^{2n-1}, that the intersection number is bilinear and finally, that $I(x,y) \neq 0$ only whenever x and y are dual [211; § 68]. Suppose that $x = (p^n,\dots,p^o)$ and $y = (p^{2n-1},\dots,p^n)$, with $(p^{2n-1},\dots, p^n,\dots,p^o)$ a $(2n-1)$-simplex of $(S^{2n-1})'$. Now $(p^{2n-1},\dots,p^n)$ appears in u^* only if $(f(p^n,\dots,p^o) = \varepsilon\tau^{*\prime}$ $(\varepsilon = \pm 1)$, in which case it appears with coefficient $\langle f^{\#}(u^{n*}),(p^n,\dots,p^o)\rangle = \langle u^{n*},\varepsilon\tau^{*\prime}\rangle = \varepsilon$. Hence, $I(x,u^*) = \varepsilon$; on the other hand, $I(f(x),q^*) = I(\varepsilon\tau^{*\prime},q^*) = \varepsilon$.

For a *continuous function*, $f : S^{2n-1} \to S^n$ we have defined its Hopf invariant via the simplical approcimation theorem (see § 3). In view of theorem (4.4) we attempt to define $\gamma(f)$ for the continuous case directly in terms of singular cohomology. Let η and ξ be the fundamental classes of $H^n(S^n)$ and $H^{2n-1}(S^{2n-1})$, respectively. Suppose that η is represented by an n-cocycle u; the cocycle $f^{\#}(u)$ is a coboundary, i.e., there is an $(n-1)$-singular cochain v of S^{2n-1} such that $f^{\#}(u) = \delta(v)$. The obstruction to proceed with a parallel argument to that developped for the simplical case is now given by the fact that $v \cup f^{\#}(u)$ is not necessarily a cocycle; however, note that $u \cup u = \delta(\omega)$, for a certain $(2n-1)$-singular cochain, and that $v \cup f^{\#}(u) - f^{\#}(\omega)$ is a $(2n-1)$-cocycle. Let $\gamma_s(f)$ be the integer defined by

$$\{v \cup f^{\#}(u) - f^{\#}(\omega)\} = \gamma_s(f) \cdot \xi .$$

Clearly, if f is simplicial, $\gamma_s(f) = \gamma(f)$. We will now give another characterization of $\gamma_s(f)$, showing its homotopy invariance. From the simplical approximation theorem it then follows that $\gamma_s(f) = \gamma(f)$ for all continuous functions f.

Let M_f be the mapping cylinder of f; the exact singular cohomology sequence of the pair (M_f,S^{2n-1}) shows that $\tilde{H}^*(M_f,S^{2n-1})$ is free abelian of rank 2 with generators $\bar{\eta}$ in dimension n and $\bar{\xi}$ in dimension $2n$. Then,

<u>Theorem 4.5.</u> (Steenrod). $\bar{\eta} \cup \bar{\eta} = -\gamma_s(f) \cdot \bar{\xi}$.

<u>Proof:</u> We claim that the connecting homomorphism $\partial : H^{2n-1}(S^{2n-1}) \to H^{2n}(M_f,S^{2n-1})$ maps the cohomology class

$\{v \cup f^{\#}(u) - f^{\#}(\omega)\}$ onto $-\bar{\eta} \cup \bar{\eta}$; this will imply (4.5).

Consider the canonical maps $j : S^{2n-1} \to M_f$, $p : M_f \to S^n$ for which $f = p \circ j$, and choose a singular $(n-1)$ - cochain w of M_f such that $j^{\#}(w) = v$. The homomorphism $j^{\#}$ maps the cochain $c = w \cup p^{\#}(u) - p^{\#}(\omega)$ onto $v \cup f^{\#}(u) - f^{\#}(\omega)$ and we compute

$$\delta c = \delta w \cup p^{\#}(u) - p^{\#}(\delta\omega) = (\delta w - p^{\#}(u)) \cup p^{\#}(u).$$

Now, since $\delta w - p^{\#}(u)$ is a cocycle which goes to zero under $j^{\#}$, thus representing an element of $H^n(M_f, S^{2n-1})$, and since $p^{\#}(u)$ represents a generator of $H^n(M_f)$, the claim follows.

Remark: The homotopy type of (M_f, S^{2n-1}) depends only on the homotopy class of f ; hence $\gamma_s(f)$ is a homotopy invariant.

5. K - Theoretical Solution of the Hopf Invariant one Problem and Applications.

We have seen in (3.4) that for $n = 2, 4$ or 8 there are maps $f : S^{2n-1} \to S^n$ such that $\gamma(f) = 1$. The obvious question to ask is then: are there any other values of n for which there exist functions $f : S^{2n-1} \to S^n$ with Hopf invariant 1 ? Much research has been done on the question but a conclusive result was published only in 1960 by Adams :

Theorem 5.1. *The only values of* n *for which there are maps,* $f : S^{2n-1} \to S^n$ *with* $\gamma(f) = 1$ *are* 2,4 *or* 8 .

Adams' proof was established with the aid of secondary cohomology operations in singular cohomology; we present here a much simpler and shorter proof, obtained using K - theory and its primary operations. It is due to Eckmann [86]; a very similar proof was given independently by Adams-Atiyah [6].

Proof of (5.1). To start with, we observe that in view of theorem 4.5 we have to search for all integers $n = 2m$ for which there are (based) maps $f : S^{2n-1} \to S^n$ such that, if C_f is the mapping cone of f , then $\bar{\eta} \cup \bar{\eta} = \bar{\xi}$, where $\bar{\eta}$ is a generator of $H^n(C_f, \mathbb{Z})$ and $\bar{\xi}$ is a generator of $H^{2n}(C_f, \mathbb{Z})$. In other words, we are asking for what values of n there exists a finite CW - complex X such that $H^*(X;\mathbb{Z}) \simeq \mathbb{Z}[x]/\langle x^3\rangle$, with $\dim x = n$. As a consequence of the Integrality Theorem (see (0.11)), there exists $z \in KU^0(X)$ such that $ch(z) = x + qx^2$, $q \in \mathbb{Q}$; in addition, as seen at the end of (0.12), $ch(\psi^k z) = k^m x + qk^{2m}x^2$ and thus,

$$ch(\psi^k(z) - k^m z) = q(k^{2m} - k^m)x^2.$$

Again, from the Integrality Theorem it follows that

$$\mu_k = qk^m(k^m-1)$$

is an integer. Also, because ch is a monomorphism (see (0.11)),

$$\psi^k(z) - k^m z = \mu_k z^2 .$$

For $k = 2$, the previous equality and the definiton of ψ^2 show that

$$z^2(\mu_2 - 1) = -2(\lambda^2(z) + 2^{m-1}z) .$$

Now 2 does not divide z^2 otherwise, it would divide x^2 in the truncated polynomial ring $\mathbb{Z}[x]/\langle x^3\rangle$ and so, μ_2 is odd. This means that $q = \dfrac{\mu_2}{2^m(2^m-1)} = \dfrac{u}{2^m \cdot v}$ with u and v odd integers. Hence, the integer $\mu_3 = q \cdot 3^m(3^m-1)$ can be written as

$$\mu_3 = \frac{u \cdot 3^m}{2^m \cdot v}(3^m-1) ,$$

and so, necessarily 2^m divides $3^m - 1$. But this division is possible only if $m = 1,2$ or 4 (for a proof see Chapter 2, § 5).

The first application of the preceeding theorem is the determination of the dimension of the spheres which possess a *multiplication*; we recall that a *multiplication* on a sphere S^r (with base point e) is a continuous function

$$\mu : S^r \times S^r \to S^r$$

such that $\mu(e,x) = x = \mu(x,e)$, for all $x \in S^r$.

Corollary 5.2. *The only spheres with a multiplication are those of dimensions* 1,3 *and* 7 .

Proof. Clearly S^1, S^3 and S^7 have a multiplication induced by the multiplication of the complex numbers $\mathbb{C}$, the quaternions $\mathbb{H}$ and the Cayley numbers $\mathbb{O}$, respectively. Suppose now that:

$$\mu : S^{n-1} \times S^{n-1} \to S^{n-1}$$

is a multiplication, with $n > 1$. The Hopf construction gives a map $H(\mu) : S^{2n-1} \to S^n$ and since μ is of type $(1,1)$ $\gamma(H(\mu)) = 1$ (see (3.2)).

Once the dimension of the spheres with a multiplication has been established, one would like to find out when two multiplications on the same sphere are homotopic and how many homotopy classes there are in each case. This problem can be solved completely using some results due to James.

We begin by recalling the notion of *Separation element* of two maps $u,v : K \to X$ which coincide on a subspace $L \subset K$ such that $K - L = e^r$, an open r-cell. Let E^r_+ and E^r_- be the northern and southern hemispheres of the sphere S^r; the equator S^{r-1} bounds a cell V^r and suppose that e^r is the image of the interior of V^r by a map f, with $f(S^{r-1}) \subset L$. Finally, let $p_+ : V^r \to E^r_+$ and $p_- : V^r \to E^r_-$ be the orthogonal projections. The element $d(u,v) = [g] \in \pi_r(X)$ defined by the map

$$g : S^r \to X$$

such that $gp_+ = uf$ and $gp_- = vf$ is called the *Separation element* of u and v . The following properties hold [136]:

(5.3) Let $u,v : K \to X$ be as before; then $u \simeq v$, relative to L if, and only if, $d(u,v) = 0$.

(5.4) Given that $K - L = e^r$ and $u,v,w : K \to X$ are such that $u/L = v/L = w/L$ then, $d(u,v) = d(u,w) + d(w,v)$.

(5.5) Given $\delta \in \pi_r(X)$ and $u : K \to X$ with $K - L = e^r$, there is a map $v : K \to X$ with $v/L = u/L$ and $d(u,v) = \delta$.

In particular, any two multiplications f_1, f_2 of S^{n-1} define a separation element $d(f_1,f_2) \in \pi_{2n-2}(S^{n-1})$ because f_1 and f_2 coincide on $S^{n-1} \vee S^{n-1}$ and the complement of $S^{n-1} \vee S^{n-1}$ in $S^{n-1} \times S^{n-1}$ is an open $(2n-2)$-cell.

Now, if $f : S^{n-1} \times S^{n-1} \to S^{n-1}$ is a fixed multiplication, we associate to the class $[f']$ of any multiplication f' on S^{n-1} the element $d(f,f')$; notice that this correspondence is well-defined because of (5.3) and (5.4). On the other hand, if $\delta \in \pi_{2n-2}(S^{n-1})$ is given, by (5.5) there is a multiplication f' on S^{n-1} such that $d(f,f') = \delta$. The previous observations and theorem 5.2 show that:

Proposition 5.6. *For* $n - 1 = 1,3$ *or* 7, *there is a one-to-one correspondence between the homotopy classes (relative to* $S^{n-1} \vee S^{n-1}$*) of multiplications on* S^{n-1} *and the elements of* $\pi_{2n-2}(S^{n-1})$.

Observe that since $\pi_2(S^1) \cong 0$, $\pi_6(S^3) \cong Z_{12}$ and $\pi_{14}(S^7) \cong Z_{120}$ there are, respectively, one homotopy class of multiplications on S^1 (all are homotopic to the multiplication induced from $\mathbb{C}$), 12 classes of multiplications on S^3 and 120 classes of multiplications on S^7.

Theorem 5.2 has several interesting consequences which will be discussed next.

Corollary 5.7. *Let* $\mu : \mathbb{R}^n \times \mathbb{R}^n \to \mathbb{R}^n$ *be a continuous function satisfying the 'norm product rule'* $\|\mu(x,y)\| = \|x\| \cdot \|y\|$ *and having a two-sided unit: then* $n = 1,2,4$ *or* 8.

Proof. The multiplications on $\mathbb{R}, \mathbb{C}, \mathbb{H}$ and $\mathbb{O}$ show that such a map exists. Suppose now that we are given a function $\mu : \mathbb{R}^n \times \mathbb{R}^n \to \mathbb{R}^n$ with the properties stated. Then, $\mu | S^{n-1} \times S^{n-1}$ is a multiplication on S^{n-1}; the result now follows from (5.2).

A sphere $S^{n-1} \subseteq \mathbb{R}^n$ is said to be *parallelizable* if the bundle of $(n-1)$-frames on S^{n-1} has a cross-section or, in other words, if there are $n-1$ tangent vector fields on S^{n-1} which are linearly independent at each point.

Corollary 5.8. *The only parallelizable spheres are* S^1, S^3 *and* S^7.

Proof. We begin by observing that if there exists a continuous function

$$\mu : \mathbb{R}^n \times \mathbb{R}^n \to \mathbb{R}^n$$

which is bilinear, norm preserving and with a two-sided unit, then S^{n-1} is parallelizable. In fact, we may assume that e_1 is the unit and that $\{e_1,\ldots,e_n\}$ is an orthonormal basis of $\mathbb{R}^n$. Then, for every $x \in S^{n-1}$, $\{\mu(x,e_1) = x, \mu(x,e_2),\ldots,\mu(x,e_n)\}$

is an orthonormal system of independent vectors which varies continuously with x. In particular, the usual multiplications of $\mathbb{C}$, $\mathbb{H}$ and $\mathbb{O}$ show that S^1, S^3 and S^7 are parallelizable.

Suppose now that $S^{n-1} \subset \mathbb{R}^n$ is parallelizable. For each $x \in S^{n-1}$ let $f_i(x)$, $i = 1,\dots,n-1$, be the independent vectors at x, which we suppose to be orthonormal. Hence, the $(n \times n)$ - matrix

$$M(x) = (x, f_1(x), \dots, f_{n-1}(x))$$

having the columns $x, f_1(x), \dots, f_{n-1}(x)$ is *orthonormal* and depends continuously on x. Let

$$e = (1,0,\dots,0) \in S^{n-1}$$

and define

$$\mu : S^{n-1} \times S^{n-1} \longrightarrow S^{n-1}$$

by

$$\mu(x,y) = M(x) \cdot M^{-1}(e) \cdot y$$

(here y is viewed as an $(n \times 1)$ - matrix). Clearly μ is continuous and one checks readily that e is a two-sided unit. The assertion follows now from (5.2).

Remark. The statement (5.8) above was first proved by Kervaire; his proof is based on previous partial knowledge on the parallelizability of the sphere and the results of Bott on the periodicity of the stable homotopy groups of $SO(n)$ and $U(n)$ [149].

We say that $\mathbb{R}^n$ is a *division algebra* over $\mathbb{R}$ if there is a *bilinear* map

$$\mu : \mathbb{R}^n \times \mathbb{R}^n \longrightarrow \mathbb{R}^n$$

which has no *zero divisors.*

Corollary 5.9. *The only finite dimensional division algebras over* $\mathbb{R}$ *are those of dimensions* 1,2,4 *or* 8.

Proof. The usual multiplications on $\mathbb{R}$, $\mathbb{C}$, $\mathbb{H}$ and $\mathbb{O}$ show that $\mathbb{R}, \mathbb{R}^2, \mathbb{R}^4$ and $\mathbb{R}^8$ are division algebras over $\mathbb{R}$. Now assume that there exist a $\mu : \mathbb{R}^n \times \mathbb{R}^n \to \mathbb{R}^n$ which is bilinear (hence continuous) and without zero divisors. To simplify the notation we shall write $\mu(x,y) = x \cdot y$. Let $\{e_1,\dots,e_n\}$ be a basis of $\mathbb{R}^n$. The conditions on μ imply that for every $x \neq 0$ the sets $\{x \cdot e_1,\dots,x \cdot e_n\}$ and $\{e_1 \cdot x,\dots,e_n \cdot x\}$ are bases of $\mathbb{R}^n$. Indeed, if $\Sigma\lambda_i(x \cdot e_i) = 0$ we get $x \cdot \Sigma\lambda_i e_i = 0$ and so, $\lambda_i = 0$. In particular, since $\{e_1 \cdot e_1, e_2 \cdot e_1,\dots,e_n \cdot e_1\}$ is a basis, we conclude that for all $y \in S^{n-1}$ there is a *unique* element $x \in \mathbb{R}^n$ such that $y = x \cdot e_1$; moreover, x depends continuously on y. Projecting the vectors $x \cdot e_2,\dots,x \cdot e_n$ onto the tangent plane of S^{n-1} at the point $y = x \cdot e_1$, we obtain $n-1$ linearly independent vector fields on S^{n-1}. Hence, (5.9) follows from (5.8).

An *almost-complex* structure on S^{n-1} is a continuous function with domain S^{n-1} which associates to each $x \in S^{n-1}$ an endomorphism J_x of the tangent space to S^{n-1} at x, such that $J_x^2 = -1$.

Corollary 5.10. *The only spheres with an almost-complex structure are* S^2 *and* S^6.

Proof. S^2 and S^6 have almost complex structures (see [231; 41.16 and 41.21]).

Suppose that S^m $(m \geq 1)$ has an almost complex structure. Let $x,y \in S^m$ with x and y perpendicular; then the vectors y and $J_x(y)$ in the tangent plane to S^m at x are linearly independent (this follows from $J_x^2 = -1$). We define $\nu'(x,y)$ to be the unit vector orthogonal to y in the plane spanned by y and $J_x(y)$, i.e.

$$\nu'(x,y) = \frac{J_x(y) - (y|J_x(y)) \cdot y}{||J_x(y) - (y|J_x(y)) \cdot y||} \quad ; \quad x,y \in S^m,\ (x|y) = 0.$$

Clearly, ν' is continuous in both variables and $(\nu'(x,y)|x) = 0 = (\nu'(x,y)|y)$. We extend ν' to a function

$$\nu : \mathbb{R}^{m+1} \times \mathbb{R}^{m+1} \longrightarrow \mathbb{R}^{m+1}$$

as follows. For two linearly independent vectors $x,y \in \mathbb{R}^{m+1}$ let x',y' be the two orthonormal vectors obtained by a fixed

orthogonalization process, i.e. $x' = \frac{x}{\|x\|}$ and
$y' = \frac{(x|x) \cdot y - (x|y) \cdot x}{\|x\| \cdot (\|x\|^2 \|y\|^2 - (x|y)^2)^{1/2}}$. Define

$$\nu(x,y) = \begin{cases} 0 \text{ , if } x \text{ and } y \text{ are linearly dependent} \\ (\|x\|^2 \cdot \|y\|^2 - (x|y)^2)^{1/2} \cdot \nu'(x',y') \text{ ,} \\ \text{otherwise.} \end{cases}$$

The map ν is continuous and has the following further properties:

$$(5.11) \quad \begin{aligned} &(\nu(x,y)|x) = 0 = (\nu(x,y)|y) \\ &(\nu(x,y)|\nu(x,y)) = \|x\|^2 \cdot \|y\|^2 - (x|y)^2 \text{ ,} \end{aligned}$$

for every $x,y \in \mathbb{R}^{m+1}$.(*)
We complete the proof of the Corollary, showing that ν gives rise to a continuous multiplication on $\mathbb{R}^{m+2}$ which satisfies the norm product rule and has a two-sided unit, and then applying (5.7).

Let $\{e_1,\ldots,e_{m+1}\}$ be an orthonormal basis of $\mathbb{R}^{m+1}$ viewed as a subspace of $\mathbb{R}^{m+2}$ and complete it to an orthonormal basis $\{e_o,e_1,\ldots,e_{m+1}\}$ of $\mathbb{R}^{m+2}$. If X,Y are elements of $\mathbb{R}^{m+2}$ write them as

$$\begin{aligned} X &= \xi \cdot e_o + x \\ Y &= \eta \cdot e_o + y \qquad (x,y \in \mathbb{R}^{m+1} \; ; \; \xi,\eta \in \mathbb{R}) \ . \end{aligned}$$

Define $\mu : \mathbb{R}^{m+2} \times \mathbb{R}^{m+2} \to \mathbb{R}^{m+2}$ by

$$\mu(X,Y) = (\xi\eta - (x|y)) \cdot e_o + (\xi y + \eta x + \nu(x,y))$$

The map μ is continuous, and using the properties (5.11) one verifies readily that

$$\begin{aligned} &\mu(e_o,X) = X = \mu(X,e_o) \\ &\mu(X,Y) = \|X\| \cdot \|Y\| \text{ ,} \end{aligned}$$

for all $X,Y \in \mathbb{R}^{m+2}$.

(*) Eckmann calls such a function ν a *continuous vector product of two vectors* on $\mathbb{R}^{m+1}$ [86], [87] .

CHAPTER 2

TORSION FREE H - SPACES OF RANK TWO

1. Introduction

A *finite dimensional Hopf space* or, a *finite H-space* for short, is a finite, based CW-complex (X,e) endowed with a continuous *multiplication*, that is, a map

$$m : X \times X \to X$$

such that the maps $m(e,-)$ and $m(-,e)$ from X into X are homotopic to the identity of X (in other words, e acts as a two-sided homotopy unit). This notion obviously generalizes the concept of compact Lie group; notice that we do not require the CW-complex X to have a manifold structure, nor do we ask the multiplication to satisfy all group axioms, with the exception of the existence of a unit element (and this, only up to homotopy).

The sphere S^7 and the real projective space $\mathbb{R}P^7$, with multiplication induced by the Cayley product of $\mathbb{R}^8$, are examples of connected finite H-spaces which are not groups. For some time it was thought that any 1-connected finite H-space was of the homotopy type of a product $G \times S^7 \times \ldots \times S^7$, where G is a compact Lie group. Then, in 1969 Hilton and Roitberg [108] produced an example of an H-space not homotopy equivalent to a Lie group nor a product with S^7; their example was given by the total space of a principal S^3-bundle over S^7. The discovery of this Hilton-Roitberg H-space greatly stimulated the study of finite dimensional Hopf spaces; other interesting H-spaces were soon found, bringing along the problem of their classification (see [68], [109], [182], [228], [268] and [269]). In this chapter we shall study the classification problem for a special family of H-spaces, the so called "torsion free rank two H-spaces".

In [124] Hopf proved that the rational cohomology ring of a finite, connected H-space X is an exterior algebra with odd dimensional generators,

$$H^*(X;\mathbb{Q}) = \Lambda(x_1,\ldots,x_r) ,$$

$\dim x_i = 2n_i - 1$. Hopf also pointed out that if X is a compact, connected Lie group, the number r of generators is equal to the Lie group rank of X . Thus, for any finite H-space X as before, the number r is called the *rank* of X. The *type* of X is the sequence $(2n_1 - 1,\ldots,2n_r - 1)$.

A connected H-space of rank one and having no torsion in integral homology is a sphere. By a theorem of Adams (Chapter 1, Theorem 5.1), such a space has type (1), (3) or (7) (thus, in Chapter 1 we solved the classification problem for the torsion free rank one H-spaces). The case of the torsion free, rank two H-spaces was discussed by Adams [3], Douglas-Sigrist [81], Hösli [128] and Hubbuck [129]; they proved that the torsion free, connected, finite H-spaces of rank 2 have the following types: (1,1), (1,3), (1,7), (3,3), (3,5), (3,7) or (7,7) (see Theorem 2.10). The types we just listed are realized by $S^1 \times S^1$, $S^1 \times S^3$, $S^1 \times S^7$, $S^3 \times S^3$, SU(3) , $S^3 \times S^7$ and $S^7 \times S^7$, respectively. It is not hard to prove that, except for the case (3,7), the previous is a complete list of homotopy types for the torsion free, rank 2 H-spaces (see § 4). For the type (3,7) the situation is more complicated; for example, the Lie group Sp(2) and the product $S^3 \times S^7$ are two spaces of this type and they are not homotopically equivalent. Actually, a torsion free, connected, finite H-space of type (3,7) is, up to homotopy, a 3-cell complex of the form $S^3 \cup e^7 \cup e^{10}$; moreover as shown by Hilton and Roitberg, such an H-space has the homotopy type of the total space of a principal S^3-bundle over S^7. We now observe that these bundles are classified by the elements of $\pi_6(S^3) \cong \mathbb{Z}_{12}$ and that the latter group has a canonical generator ω characterized by $Sp(2) = S^3 \cup_\omega e^7 \cup e^{10}$ (see (3.3)). Let E_n be the total space of the principal S^3-bundle over S^7 which corresponds to $n\omega$; then, $E_n = S^3 \cup_{n\omega} e^7 \cup e^{10}$. Actually, the spaces E_n and E_{-n} are homeomorphic and thus, for the type (3,7) there are only seven possible homotopy types, namely E_o, E_1, E_2, E_3, E_4, E_5 and E_6. Of these, $E_o = S^3 \times S^7$, $E_1 = Sp(2)$ and E_5 (the Hilton-Roitberg H-space) are H-spaces; using Zabrodsky's method of "mixing homotopy types" [267], Stasheff [228] and Curtis-Mislin [68] proved that also E_3 and E_4 are H-spaces. So, it only remains to decide whether or not E_2 and E_6 can carry an H-space structure. It was Zabrodsky who first proved that both E_2 and E_6 are not H-spaces [268], by showing that if the CW-complex $S^3 \cup_{n\omega} e^7 \cup e^{10}$ is an H-space then $n \not\equiv 2 \pmod 4$ (see Theorem 3.4). In order to prove this result, Zabrodsky had to recur to tertiary cohomology operations; we shall present a K-theoretical proof of Zabrodsky's Theorem [218].

The results quoted before, Zabrodsky's method of "mixing homotopy types" and some classical Algebraic Topology lead to the proof of a Classification Theorem (Theorem 4.1) due to Hilton-Roitberg [109] and independently to Zabrodsky [268]. (At this point we mention that the classification problem is also solved for rank two H-spaces having torsion in integral cohomology. According to Browder - [56] and [57] - H-spaces of that sort have at most 2-torsion. Hubbuck (mimeographed notes) then proved, that an H-space of rank 2 and with 2-torsion has the same cohomology as the Lie group G_2; in particular, it is of type (3,11) and primitively generated. Finally, Mimura - Nishida - Toda [182] proved there are exactly four distinct homotopy types of primitively generated H-spaces of type (3,11)).

The chapter is organized as follows. In §2 we define the "Hopf construction" and the "projective plane" PX of an H-space X ; we then compute the KU-theory of QX - the (4n+1) - skeleton of PX - and give the type classification of torsion free rank 2 H-spaces, modulo several number-theoretical results (introduced by the KU - theory of QX), which shall be discussed in Section 5. The rationale behind this move is that although such results are directly related to K-theory, their manipulation is long and hard enough to make one loose track of the global picture, at least on a first reading.

In § 3 we investigate the H-spaces of type (3,7) and in § 4, we prove the Classification Theorem for the homotopy type of torsion free H-spaces of rank 2.

Throughout this chapter we assume (without further mentioning) that our H-spaces are *connected, finite* CW-*complexes* together with a multiplication with strict unit (this assumption is no real restriction: since the pair $(X \times X, X \vee X)$ has the homotopy extension property, any multiplication on X is homotopic to one with strict unit).

We close this introduction noting that K-theory has been used to solve other H-space problems. Refining and expanding the methods described in this chapter Hubbuck has greatly restricted the possible types of finite H-spaces having no 2-torsion in integral homology (see [129] and [130]). Furthermore, he proved (again with K-theory) that a homotopy commutative finite H-space is of the homotopy type of a product of circles [131]. (For another application of K-theory to H-spaces see also [132]). Finally, Wilkerson has applied K-theory in his investigation of spheres which are loop spaces mod p (see [257] and [258]).

2. Hopf Construction, projective Plane and Type of Torsion free Rank two H - Spaces

In chapter 1 (§§ 3 and 5) we answered the question of whether or not a sphere S^{n-1} admits an H-multiplication using the Hopf construction $H(f) : S^{n-1} * S^{n-1} \to \Sigma S^{n-1}$ and investigating the KU-theory of the mapping cone of H(f). To study our H-space problem we shall proceed in a similar way. Assuming that X carries an H-multiplication m, we define a map $H(m) : X * X \to SX$ and then study the KU-theory of its mapping cone, the so-called projective plane PX (actually, it will be enough to determine the KU-theory of a convenient subspace of PX). The cohomology of PX has a richer structure than that of X, reflecting the combination of algebra and topology involved in the definition of H-space. In this chapter we indicate the integral n^{th} cohomology group of X simply by $H^n(X)$.

We begin by putting together some facts about the

(non-reduced) *join* $X * Y$ of two connected CW - complexes X and Y. For any space Z, let $CZ = (I \times Z)/(1 \times Z)$ be the *cone* over Z; then, the join $X * Y$ is the subspace of $CY \times CY$ defined by

$$X * Y = X \times CY \cup CX \times Y .$$

We are also interested in the (non-reduced) *suspension* SZ of Z; this is the union of the cones C^+Z and C^-Z glued together along the base $0 \times Z$, i.e.,

$$SZ = C^+Z \cup C^-Z \quad \text{with} \quad C^+Z \cap C^-Z = Z .$$

There is a canonical map

$$j : X * Y \longrightarrow S(X \times Y)$$

defined by taking any $([t,z],y) \in CX \times Y$ into $[t,x,y] \in C^+(X \times Y)$, and any $(x,[s,y]) \in X \times CY$ into $[s,x,y] \in C^-(X \times Y)$. Now consider the diagram

$$X * Y \xrightarrow{\;j\;} S(X \times Y) \begin{array}{l} \xrightarrow{Sp_1} SX \\ \xrightarrow{Sq} S(X \wedge Y) \\ \xrightarrow{Sp_2} SY \end{array}$$

where p_1, p_2 and q are the obvious projection and quotient maps. As one can check in [103], the composition $Sq \circ j$ is a homotopy equivalence, while $Sp_1 \circ j$ and $Sp_2 \circ j$ are null-homotopic. Note that there is a canonical isomorphism

$$\varphi : H^n(X) \oplus H^n(X \wedge Y) \oplus H^n(Y) \cong H^n(X \times Y) , \; n \geq 1 ,$$

given by $\varphi(a,b,c) = p_1^*(a) + q^*(b) + p_2^*(c)$; similarly for the suspended spaces and maps. Hence, the kernel of j^* is equal to $\sigma p_1^*(\tilde{H}^*(X)) \oplus \sigma p_2^*(\tilde{H}^*(Y))$ and $j^*\sigma$ maps $q^*\tilde{H}^n(X \wedge Y)$ isomorphically onto $H^{n+1}(X * Y)$ (here $\sigma : \tilde{H}^k(Z) \xrightarrow[\cong]{\delta} H^{k+1}(C^-Z,Z) \xleftarrow[\cong]{} H^{k+1}(SZ,C^+Z) \xrightarrow[\cong]{} H^{k+1}(SZ)$ is the suspension isomorphism).

Next, consider the canonical maps

$$X \times Y \xrightarrow{i_1} X \times CY \xrightarrow{r_1} X ,$$

$$X \times Y \xrightarrow{i_2} CX \times Y \xrightarrow{r_2} Y$$

and the commutative diagram

$$\begin{array}{ccc}
H^n(C(X*Y), X \times CY) \otimes H^m(C(X*Y), CX \times Y) & \xrightarrow{U} & H^{n+m}(C(X*Y), X*Y) \\
 & & \uparrow \cong \delta \\
\uparrow \delta \otimes \delta & & H^{n+m-1}(X*Y) \\
 & & \uparrow j^*\sigma \\
H^{n-1}(X \times CY) \otimes H^{m-1}(CX \times Y) & \xrightarrow{U \circ i_1^* \otimes i_2^*} & H^{n+m-2}(X \times Y) \\
\uparrow r_1^* \otimes r_2^* & & \uparrow = \\
H^{n-1}(X) \otimes H^{m-1}(Y) & \xrightarrow{\times} & H^{n+m-2}(X \times Y)
\end{array} \qquad (2.1)$$

(here δ stands for the coboundary operator of the exact cohomology sequence of the appropriate pair of spaces); we wish to observe that the commutativity of (2.1) is readily established working with the exact triad $(C(X*Y), X \times CY, CX \times Y)$, and using the basic properties of the cup product (see [79,VII,§8]) plus the fact that the composition $H^k(X \times Y) \xrightarrow{\delta} H^{k+1}(X \times CY, X \times Y) \xleftarrow{\cong} H^{k+1}(X*Y, CX \times Y) \longrightarrow H^{k+1}(X*Y)$ is equal to $j^*\sigma$.

Definition 2.2. *Let* $f : X \times Y \longrightarrow Z$ *be a map between connected* CW - *complexes. The composition* $H(f) = Sf \circ j$

$$X * Y \xrightarrow{j} S(X \times Y) \xrightarrow{Sf} SZ$$

is called the Hopf construction of f.

In particular, the Hopf construction of the multiplication $m : X \times X \to X$ of an H-space X gives a map $H(m) : X * X \to SX$ (cf. Chapter 1, § 3). The mapping cone of $H(m)$,

$$C_{H(m)} = SX \cup_{H(m)} C(X * X) = PX$$

is the *projective plane* of the H-space X. As we said before, we wish to compute the KU - theory of a certain subspace of PX and thus, as a preliminary move, we investigate the cohomology ring of PX. Consider the commutative diagram

$$\begin{array}{ccccccccc}
X * X & \xrightarrow{H(m)} & SX & \xrightarrow{i} & PX & \xrightarrow{p} & S(X * X) & \xrightarrow{SH(m)} & S^2X \\
& {\scriptstyle j}\searrow & \nearrow{\scriptstyle Sm} & & & & {\scriptstyle Sj}\searrow & & \nearrow{\scriptstyle S^2m} \\
& & S(X \times X) & & & & & S^2(X \times X) &
\end{array}$$

whose horizontal line is part of Puppe's sequence of $H(m)$. Given arbitrarily $a,b \in \tilde{H}^*(X)$, the Künneth Theorem identifies $a \otimes b$ with the element $p_1^*(a) \cup p_2^*(b) \in \tilde{H}^*(X \times X)$; we define, $a * b = j^*(a \otimes b) \in \tilde{H}^*(X * X)$ (σ is the suspension isomorphism). With this notation, the reader should observe that if $\tilde{H}^*(X)$ is torsion free and has a basis $\{x_1,\ldots,x_n\}$, then $\{x_r * x_s | r,s = 1,\ldots,n\}$ is a basis of the free abelian group $\tilde{H}^*(X * X)$.

The following theorem provides the basic information on the cup products in $H^*(PX)$.

<u>Theorem 2.3.</u> *Let* u,v *be arbitrary elements of* $\tilde{H}^*(PX)$ *and let* $a,b \in \tilde{H}^*(X)$ *be such that* $i^*(u) = \sigma(a)$, $i^*(v) = \sigma(b)$. *Then,* $u \cup v = p^*\sigma(a * b)$, *i.e., one has the following situation*

$$\begin{array}{ccccc}
\tilde{H}^*(S(X * X)) & \xrightarrow{p^*} & \tilde{H}^*(PX) & \xrightarrow{i^*} & \tilde{H}^*(SX) \\
& & u,v & \longmapsto & \sigma(a),\sigma(b) \\
\sigma(a * b) & \longmapsto & u \cup v & &
\end{array}$$

<u>Proof.</u> Set $B = X * X$, $D = e \times CX \cup CX \times e \subset B$; the Hopf construction $g = H(m)$ induces maps of exact triads

$$\begin{array}{ccc}
(D; X \times e, e \times X) & \longrightarrow & (CB; X \times CX, CX \times X) \\
\Big\downarrow {\scriptstyle g} & & \Big\downarrow {\scriptstyle \bar{g}} \\
(SX; C^-X, C^+X) & \longrightarrow & (PX; C^-X, C^+X).
\end{array}$$

These maps give rise to the following commutative diagram in which all unlabeled homomorphisms are the obvious ones, δ denotes the boundary operator in the appropriate cohomology sequence and r_1,r_2,i_1,i_2 have been defined before diagram (2.1).

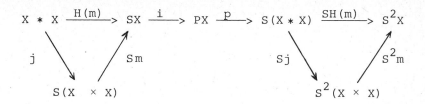

$$
\begin{array}{ccccc}
H^n(SX)\otimes H^m(SX) & \xleftarrow{i^*\otimes i^*} & H^n(PX)\otimes H^m(PX) & \xrightarrow{\cup} & H^{n+m}(PX)\\
\cong\uparrow & & \cong\uparrow & & \uparrow\\
H^n(SX,C^-X)\otimes H^n(SX,C^+X) & \longleftarrow & H^n(PX,C^-X)\otimes H^m(PX,C^+X) & \xrightarrow{\cup} & H^{n+m}(PX,SX)\\
\cong\downarrow g^*\otimes g^* & & \downarrow \bar{g}^*\otimes \bar{g}^* & & \cong\downarrow \bar{g}^*\\
H^n(D,X\times e)\otimes H^m(D,e\times X) & \xleftarrow[\cong]{} & H^n(CB,X\times CX)\otimes H^m(CB,X\times CX) & \xrightarrow{\cup} & H^{n+m}(CB,B)\\
 & & & & \uparrow\delta\\
 & & & & H^{n+m-1}(B)\\
\cong\uparrow\delta\otimes\delta & & \cong\uparrow\delta\otimes\delta & & \uparrow j^*\sigma\\
H^{n-1}(X\times e)\otimes H^{m-1}(e\times X) & \xleftarrow[\cong]{} & H^{n-1}(X\times CX)\otimes H^{m-1}(CX\times X) & \longrightarrow & H^{n+m-2}(X\times X)\\
 & \nwarrow 1\otimes 1 & \cong\uparrow r_1^*\otimes r_2^* & & \uparrow =\\
 & & H^{n-1}(X)\otimes H^{m-1}(X) & \xrightarrow{\times} & H^{n+m-2}(X\times X).
\end{array}
$$

The commutativity of this diagram is trivial, except for the two squares in the lower right hand side corner, whose commutativity is guaranteed by (2.1). Let us focus our attention on the left column of the diagram. The homomorphism $g^* : H^n(SX,C^-X) \to H^n(D,X\times e)$ is induced by

$$g = H(m) : (CX\times e \cup e\times CX, X\times e) \longrightarrow (SX,C^-X);$$

since e is a unit of the multiplication m, the restriction $H(m)|CX\times e$ identifies $CX\times e$ with the upper cone C^+X of SX and hence we infer that the composition

$$
\begin{array}{ccccccc}
H^{n-1}(X\times e) & \xrightarrow{\delta} & H^n(D,X\times e) & \xleftarrow[\cong]{g^*} & H^n(SX,C^-X) & \longrightarrow & H^n(SX)\\
 & \searrow\delta & \cong\downarrow & \nearrow\cong & & & \\
 & & H^n(CX\times e,\ X\times e) & & & &
\end{array}
$$

coincides with the negative suspension isomorphism. Similarly, $H^{m-1}(e\times X) \xrightarrow[\cong]{\delta} H^m(D,e\times X) \xleftarrow[\cong]{g^*} H^m(SX,C^+X) \xrightarrow[\cong]{} H^m(SX)$ is equal to the suspension isomorphism. The composition of the homomorphisms of the right column (start at the bottom) coincides with $p^*(-\sigma)j^*\sigma$. These informations show that the previous large commutative diagram reduces to

$$\begin{array}{ccc}
H^n(PX) \otimes H^m(PX) & \xrightarrow{\ \cup\ } & H^{n+m}(PX) \\
\downarrow{\scriptstyle i^* \otimes i^*} & & \\
H^n(SX) \otimes H^m(SX) & & \uparrow{\scriptstyle -p^*\sigma j^*\sigma} \\
\uparrow{\scriptstyle -\sigma\otimes\sigma} & & \\
H^{n-1}(X) \otimes H^{m-1}(X) & \xrightarrow{\ \times\ } & H^{n+m-2}(X \times X)
\end{array}$$

and the Theorem is proved.

Now suppose that X is a torsion free H-space of type $(2q-1, 2n-1)$, $1 \leq q \leq n$. The integral cohomology of X is an exterior algebra with two generators

$$H^*(X) = \Lambda_{\mathbb{Z}}(x_{2q-1}, x_{2n-1}) ,$$

$\dim x_j = j$ (see [179]). The integral cohomology ring of PX can also be determined without great difficulty (cf. [3], [59]):

<u>Theorem 2.4.</u> *Let* X *be a torsion free* H-*space of type* $(2q-1, 2n-1)$, $1 \leq q \leq n$. *Then, the integral cohomology of the projective plane* PX *is torsion free and splits as*

$$H^*(PX) = A \oplus S ,$$

where A *is a subring of the form*

$$A \cong \mathbb{Z}[a,b]/\langle a^3, a^2b, ab^2, b^3\rangle ,$$

$\dim a = 2q$, $\dim b = 2n$, *and* S *is a free abelian subgroup with generators in dimension* $4q + 2n - 1$, $2q + 4n - 1$ *and* $4q + 4n - 2$.

<u>Proof.</u> The definition of $H(m)$ and its Puppe sequence give rise to the following commutative diagram with exact upper row:

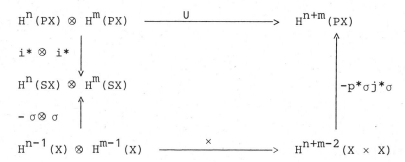

$$\begin{array}{ccccccccc} H^r(S^2X) & \xrightarrow{SH(m)^*} & H^r(S(X*X)) & \xrightarrow{p^*} & H^r(PX) & \xrightarrow{i^*} & H^r(SX) & \xrightarrow{H(m)^*} & H^r(X*X) \\ \uparrow \cong \sigma & & \uparrow \cong \sigma & & & & \searrow (Sm)^* & & \nearrow j^* \\ H^{r-1}(SX) & \xrightarrow{H(m)^*} & H^{r-1}(X*X) & & & & & H^r(S(X\times X)). & \end{array}$$

Set $x_{2q-1} = x_1$, $x_{2n-1} = x_2$ and $x_{2q-1}x_{2n-1} = x_3$ in $H^*(X)$; then, $\{\sigma x_1, \sigma x_2, \sigma x_3\}$ is a basis of $\tilde{H}^*(SX)$ and $\{x_r * x_s \mid r,s = 1,2,3\}$ is a basis of $\tilde{H}^*(X * X)$. For dimensional reasons x_1 and x_2 are *primitive*, i.e., $m^*(x_s) = p_1^*(x_s) + p_2^*(x_s)$, $s = 1,2$; hence, $H(m)^*(\sigma x_s) = 0 (s=1,2)$ and $H(m)^*(\sigma x_3) = x_1 * x_2 + x_2 * x_1$ (just recall that $H(m)^* = j^*\sigma m^*$ and use the remarks at the beginning of the section). It follows that the image of $H(m)^*$ is a direct summand of $H^*(X * X)$ and the group $\tilde{H}^*(PX)$ is therefore free abelian. Let $a_s \in H^*(PX)$ be such that $i^*(a_s) = \sigma x_s$, $s = 1,2$. Then, the elements

(2.5) $a_1, a_2, p^*\sigma(x_1 * x_1), p^*\sigma(x_1 * x_2), p^*\sigma(x_2 * x_2)$

together with

(2.6) $p^*\sigma(x_1 * x_3), p^*\sigma(x_3 * x_1), p^*\sigma(x_2 * x_3), p^*\sigma(x_3 * x_2), p^*\sigma(x_3 * x_3)$

form a basis of $\tilde{H}^*(PX)$. But Theorem 2.3 shows that

$$a_r \cup a_s = p^*\sigma(x_r * x_s),\ r,s = 1,2\ ;$$

hence, $\tilde{H}^*(PX) = A \oplus S$, where A is the free abelian group with base $\{1, a_1, a_2, a_1 \cup a_1, a_1 \cup a_2, a_2 \cup a_2\}$ (see (2.5)) and S is the complement generated by the elements of (2.6). Notice that the generators of S satisfy the dimension conditions claimed in the statement. It remains to prove that $a_1^3 = a_1^2 a_2 = a_1 a_2^2 = a_2^3 = 0$. Since PX is obtained by attaching a cone to a suspension, we can cover it by three open contractible subspaces and so, the three-fold products vanish in $\tilde{H}^*(PX)$. (Alternatively, we also reach this conclusion using Theorem 2.3, since $i^*(a_r \cup a_s) = 0$).

Let QX be the $(4n+1)$ - skeleton of PX, where X

is a torsion free H-space of type $(2q-1,2n-1)$. Because of Theorem 2.4 the CW-complex QX is torsion free and

$$(2.7)\qquad H^{even}(QX) \cong \mathbb{Z}[a,b]/\langle a^3,a^2b,ab^2,b^3\rangle$$

with $\dim a = 2q$, $\dim b = 2n$.

Theorem 2.8. *If* X *is a torsion free* H-*space of type* $(2q-1,2n-1)$, $1 \le q \le n$, *there exists a torsion free finite* CW-*complex* QX *such that*

$$KU(QX) \cong \mathbb{Z}[x,y]/\langle x^3,x^2y,xy^2,y^3\rangle$$

where x *and* y *have exact filtration*(*) 2q *and* 2n, *respectively.*

Proof. Take QX to be, as before, the $(4n+1)$-skeleton of PX ; being torsion free, $H^{even}(QX)$ is isomorphic to the graded ring $\mathcal{G}KU(QX)$ (see (0.10) and (0.11)); because of (2.7),

$$\mathcal{G}KU(X) \cong \mathbb{Z}[a,b]/\langle a^3,a^2b,ab^2,b^3\rangle$$

with $a \in \mathcal{G}^{2q}KU(QX)$, $b \in \mathcal{G}^{2n}KU(QX)$. Let $x \in \mathcal{F}^{2q}KU(QX)$ and $y \in \mathcal{F}^{2n}KU(QX)$ (see (0.11)) be representatives of a and b, respectively. Then $\{x,y,x^2,xy,y^2\}$ is a basis of the free abelian group $KU(QX)$. It remains to show that $x^3 = x^2y = xy^2 = y^3 = 0$. To this end observe that the Chern character injects the ring $\widetilde{KU}(QX)$ into $\widetilde{H}^{even}(QX;\mathbb{Q})$ (see (0.11)). But in the latter ring, three-fold products vanish by (2.7). Hence, the same holds in $\widetilde{KU}(QX)$ and (2.8) is proved.

The following Theorem will be proved in Section 5.

Theorem 2.9. *Let* Q *be a torsion free* CW-*complex such that*

$$KU(Q) \cong \mathbb{Z}[x,y]/\langle x^3,x^2y,xy^2,y^3\rangle$$

where x *and* y *are elements of exact filtration* 2q *and* 2n, *respectively*, $1 \le q \le n$. *Then*, (q,n) *is one of the following pairs*: (1,1), (1,2), (1,4), (2,2), (2,3), (2,4), (4,4).

The reader should now observe that Theorems 2.8 and 2.9 imply immediately the following Type Classification result:

(*) We say that $x \in KU(QX)$ is of *exact filtration* 2q if it is represented by an element in $\mathcal{F}^{2q}(KU(X)) - \mathcal{F}^{2q+2}(KU(X))$.

Theorem 2.10. *The type of a connected, torsion free* H-*space of rank* 2 *is equal to one of the following pairs:* (1,1), (1,3), (1,7), (3,3), (3,5), (3,7) *or* (7,7).

3. Torsion free H - Spaces of Type (3,7)

Let X be a torsion free space of type (3,7); because $H^*(X) = \Lambda_{\mathbb{Z}}(x_3, x_7)$ and $\pi_1(X) = H_1(X) = 0$, X has the homotopy type of a 3-cell complex

$$X = S^3 \cup e^7 \cup e^{10} .$$

Consider the subcomplex $Z = S^3 \cup e^7 \xrightarrow{k} X$ and the composition

$$g : Z * Z \xrightarrow{k * k} X * X \xrightarrow{H(m)} SX ;$$

let $RX = SX \cup_g C(Z * Z)$ be the mapping cone of g which, in this context will be called *reduced projective plane* of X.

Proposition 3.1. (i) $H^*(RX) = \mathbb{Z}[a,b]/\langle a^3, a^2b, ba^2, b^3 \rangle$, *dim* $a = 4$, *dim* $b = 8$.

(ii) RX *contains* $SZ = S^4 \cup e^8$ *as a subcomplex. If* $i : SX \to RX$ *is the inclusion map,* $i^*(a)$ *and* $i^*(b)$ *generate the groups* $H^4(SZ) \cong \mathbb{Z}$ *and* $H^8(SZ) \cong \mathbb{Z}$, *respectively.*

Proof. (i) The commutative diagram

$$\begin{array}{ccccc}
Z * Z & \xrightarrow{g} & SX & \xrightarrow{g_1} & C_g = RX \\
\downarrow{\scriptstyle k * k} & & \downarrow{\scriptstyle =} & & \downarrow{\scriptstyle h} \\
X * X & \xrightarrow{H(m) = f} & SX & \xrightarrow{f_1} & C_f = PX
\end{array}$$

induces a homomorphism of the exact cohomology sequence of f into that of g,

$$(3.2)\qquad \begin{array}{ccccccccc}
\longrightarrow & H^{r-1}(SX) & \xrightarrow{f^*} & H^{r-1}(X * X) & \longrightarrow & H^r(PX) & \xrightarrow{f_1^*} & H^r(SX) & \longrightarrow \\
 & \downarrow{\scriptstyle =} & & \downarrow{\scriptstyle (k * k)^*} & & \downarrow{\scriptstyle h^*} & & \downarrow{\scriptstyle =} & \\
\longrightarrow & H^{r-1}(SX) & \xrightarrow[g^*]{} & H^{r-1}(Z * Z) & \longrightarrow & H^r(RX) & \xrightarrow[g_1^*]{} & H^r(SX) & \longrightarrow .
\end{array}$$

Since the subcomplex Z carries the generators x_3 and x_7 of $H^*(X)$ it follows easily that $(k * k)^*$ is an isomorphism whenever $H^r(Z * Z) \neq 0$. Using (3.2) and (2.4) we readily establish that $h^*|A : A \cong H^*(RX)$, where A is the ring described in Theorem 2.4.

(ii) The composition $SZ \xrightarrow{Sk} SX \xrightarrow{g_1} RX$ embeds SZ into RX. In dimensions 4 and 8 the homomorphisms $g_1^* \circ h^*$, h^* and $(Sk)^*$ are isomorphisms thus, part (ii) follows.

From now on we shall be more specific about the attaching maps of the CW-complex $X = S^3 \cup e^7 \cup e^{10}$; although we shall deliberately confuse maps and homotopy classes (thus, $S^3 \cup_\alpha e^7 \cup_\beta e^{10}$ with $\alpha \in \pi_6(S^3)$, $\beta \in \pi_9(S^3 \cup_\alpha e^7)$ is really a homotopy type).

(3.3) Recall that the homotopy group $\pi_6(S^3)$ is isomorphic to $\mathbb{Z}_{12}$ [242]. We claim that the attaching map ω of the 7-cell in the Lie group $Sp(2) = S^3 \cup_\omega e^7 \cup e^{10}$, $\omega \in \pi_6(S^3)$ *represents a generator* of $\pi_6(S^3)$. This is seen as follows. Take $Z_\omega = S^3 \cup_\omega e^7$ and the sequence

$$\pi_7(Z_\omega, S^3) \xrightarrow{\partial} \pi_6(S^3) \longrightarrow \pi_6(Z_\omega) \quad ,$$

part of the exact homotopy sequence of the pair (Z_ω, S^3) . The characteristic map $\sigma : (D^7, S^6) \longrightarrow (Z_\omega, S^3)$ for the 7-cell in Z_ω generates the cyclic group $\pi_7(Z_\omega, S^3)$ and thus, $\partial\sigma = \omega$ generates the image of ∂ . Since $\pi_6(Z_\omega) \cong \pi_6(Sp(2)) \cong \pi_6(Sp) = 0$ (see (0.6)), ∂ is onto and thus, ω generates $\pi_6(S^3)$.

Our aim is to prove the following result.

<u>Theorem 3.4.</u> *Let* $S^3 \cup_{n\omega} e^7 \cup e^{10}$ *be an* H-*space. Then* $n \not\equiv 2 \pmod 4$.

This theorem shows that of the list $E_0, E_1, E_2, E_3, E_4, E_5$ and E_6 of possible torsion free rank 2 H-spaces of type (3,7), presented in the Introduction, the spaces E_2 and E_6 certainly are not H-spaces.

To achieve our goal we shall work with the functor

$$L(Y) = KO(Y) \oplus KSp(Y)$$

and so, we now put together the basic properties of $L(Y)$ (see

[11] and [218]). Observe that the tensor product of real and quaternionic vector bundles induces a $\mathbb{Z}_2$ - graded ring structure on L(Y). (If ξ is a real vector bundle and η, η' are quaternionic ones, $\xi \otimes \eta$ is a quaternionic and $\eta \otimes \eta'$ is a real vector bundle). Thus, we have an isomorphism of *rings* [47]

$$L(Y) = KO(Y) \oplus KSp(Y) \cong KO^{0}(Y) \oplus KO^{4}(Y)$$

compatible with the usual filtration (see (0.6)). For a based space $(Y,*)$, there is a canonical decomposition

$$L(Y) \cong L(*) \oplus \tilde{L}(Y) .$$

The free abelian "subgroup" $L(*)$ is generated by $1 \in KO(Y)$ and an element $[\epsilon] \in KSp(Y)$, represented by the trivial quaternionic line bundle, satisfying $[\epsilon]^2 = 4 \in KO(Y)$. If v is a generator of $\widetilde{KO}(S^{8k})$ then $\{v,[\epsilon]v\}$ is a basis of $\tilde{L}(S^{8k})$ (i.e., $[\epsilon]v$ generates $\widetilde{KSp}(S^{8k})$). Similarly, if u generates $\widetilde{KSp}(S^{8k+4})$, $\{u,[\epsilon]u\}$ is a basis of $\tilde{L}(S^{8k+4})$. The exterior powers of real and quaternionic vector bundles define a $\mathbb{Z}_2$ - graded λ - ring structure on L(Y). The λ - operations give rise to $\mathbb{Z}_2$ - graded Adams operations ψ^k, $k \in \mathbb{N}$, on the ring L(Y). These coincide on the subring KO(Y) with the usual real Adams operations. For a quaternionic bundle η over Y, the element $\psi^k([\eta])$ *is real if* k *is even* and *quaternionic* if k *is odd*. The operations ψ^k are ring homomorphisms and satisfy

$$\psi^k\psi^m = \psi^{km} = \psi^m\psi^k .$$

Furthermore,

$$\psi^k([\epsilon]) = \begin{cases} 2, & \text{if } k \text{ is even} \\ [\epsilon], & \text{if } k \text{ is odd.} \end{cases}$$

The following two commutative diagrams give the relations between the ψ - operations on $\widetilde{KSp}(Y)$ and the real ψ - operations on $\widetilde{KO}(S^4 \wedge Y)$ (we write B for the Bott isomorphism $\widetilde{KSp}(Y) \cong \widetilde{KO}(S^4 \wedge Y)$).

k even:

$$\begin{array}{ccccc} & & \widetilde{KSp}(Y) & \xrightarrow[\cong]{B} & \widetilde{KO}(S^4 \wedge Y) \\ & \swarrow{\scriptstyle \psi^k} & & & \downarrow{\scriptstyle \psi^k_{\mathbb{R}}} \\ \widetilde{KO}(Y) & & & & \\ & \searrow{\scriptstyle \frac{k^2}{2}[\epsilon]} & & & \\ & & \widetilde{KSp}(Y) & \xrightarrow[\cong]{B} & \widetilde{KO}(S^4 \wedge Y) \end{array}$$

k odd:

$$\begin{array}{ccc} \widetilde{KSp}(Y) & \xrightarrow[\cong]{B} & \widetilde{KO}(S^4 \wedge Y) \\ {\scriptstyle k^2\psi^k}\downarrow & & \downarrow{\scriptstyle \psi^k_{\mathbb{R}}} \\ \widetilde{KSp}(Y) & \xrightarrow[\cong]{B} & \widetilde{KO}(S^4 \wedge Y) \end{array} \quad .$$

Recall now that the reduced projective plane RX_n of an H-space of the form $X_n = S^3 \cup_{n\omega} e^7 \cup e^{10}$ contains the (reduced) suspension $\Sigma Z_n = S^4 \cup_{n\Sigma\omega} e^8$ as a subcomplex. To compute the ring $L(\Sigma Z_n)$ consider the cofibration $S^4 \xrightarrow{j} S^4 \cup_{n\Sigma\omega} e^8 \xrightarrow{p} S^8$, giving rise to exact sequences

(3.5) $\quad 0 \longrightarrow \widetilde{KO}(S^8) \cong \mathbb{Z} \xrightarrow{p^!} \widetilde{KO}(\Sigma Z_n) \xrightarrow{j^!} \widetilde{KO}(S^4) \cong \mathbb{Z} \longrightarrow 0$

(3.6) $\quad 0 \longrightarrow \widetilde{KSp}(S^8) \cong \mathbb{Z} \xrightarrow{p^!} \widetilde{KSp}(\Sigma Z_n) \xrightarrow{j^!} \widetilde{KSp}(S^4) \cong \mathbb{Z} \longrightarrow 0$.

The ring $\widetilde{L}(\Sigma Z_n)$ is therefore torsion free of rank 4.

<u>Theorem 3.7.</u> *Let* $u \in \widetilde{KSp}(\Sigma Z_n)$ *be such that* $j^!(u)$ *is a generator of* $\widetilde{KSp}(S^4)$ *and let* $v \in \widetilde{KO}(\Sigma Z_n)$ *be the* $p^!$*- image of a generator of* $\widetilde{KO}(S^8)$. *Then,*

(i) $\{u, [\epsilon]v\}$ *is a basis for* $\widetilde{KSp}(\Sigma Z_n)$ *and* $\{[\epsilon]u, v\}$ *is a basis of* $\widetilde{KO}(\Sigma Z_n)$.

(ii)

$$\psi^k(u) = \begin{cases} \frac{k^2}{2}[\epsilon]u + q(n)\,\frac{k^2(k^2-1)}{6}\,v, & k \text{ even} \\[2ex] k^2 u + q(n)\,\frac{k^2(k^2-1)}{12}[\epsilon]v, & k \text{ odd} \end{cases}$$

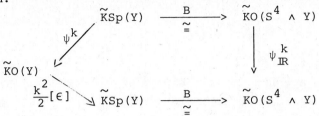

where $q(n)$ *is an integer* of the form $q(n) = nm + 12t$, *with* m *prime to* 12.

(iii) $$\psi^k(v) = k^4 v \; .$$

Proof. Let u and v be as in the statement. Then, if $j^!(u)$ is a generator of $\widetilde{KSp}(S^4)$, the element $[\epsilon]j^!(u) = j^!([\epsilon]u)$ generates $\widetilde{KO}(S^4)$; similarly, $[\epsilon]v$ is the $p^!$-image of a generator of $\widetilde{KSp}(S^8)$. Part (i) now follows from the exact sequences (3.5) and (3.6).

In order to prove (ii), consider the KO - theory of the cofibre sequence $S^8 \xrightarrow{j} S^8 \cup_\alpha e^{12} \xrightarrow{p} S^{12}$, $\alpha \in \pi_{11}(S^8) \cong \mathbb{Z}_{24}$. We have $\widetilde{KO}(S^8 \cup_\alpha e^{12}) \cong \mathbb{Z} \oplus \mathbb{Z}$ and so, choose generators w and z such that $j^!(w)$ generates $\widetilde{KO}(S^8)$ and z is the $p^!$ - image of a generator of $\widetilde{KO}(S^{12})$. According to Adams [4, Part IV], the ψ - operations on w are given by

$$\psi^k_{\mathbb{R}}(w) = k^4 w + \lambda(\alpha) k^4(k^2-1) z$$

where $\lambda(\alpha)$ is a rational number that does not depend on k; in fact, $\lambda(\alpha)$ (mod 1) is the so-called e - invariant of α [4, Part IV, § 3]. The homomorphism

$$e'_{\mathbb{R}} : \pi_{11}(S^8) \longrightarrow \mathbb{Q}/\mathbb{Z}$$

which takes α into $\lambda(\alpha)$ (mod 1) is injective (see [4, Part IV, 7.17]) and the element $\Sigma^5\omega \in \pi_{11}(S^8)$ has order 12, according to [242; (5.5), (13.2) and (13.6)]. We conclude that

$$\lambda(n \cdot \Sigma^5\omega) = n\,\frac{m}{12} + t$$

where $m,t \in \mathbb{Z}$ and m is prime to 12.
Setting $q(n) = n \cdot m + 12t$, we conclude that for the ψ - operations of $\widetilde{KO}\left(S^8 \cup_{n\Sigma^5\omega} e^{12}\right)$ one has

$$\psi^k_{\mathbb{R}}(w) = k^4 w + q(n)\,\frac{k^4(k^2-1)}{12}\, z \; .$$

Consider the Bott isomorphism $B : \widetilde{KSp}(S^4 \cup_{n\Sigma\omega} e^8) \cong \widetilde{KO}(S^4 \wedge (S^4 \cup_{n\Sigma\omega} e^8)) \cong \widetilde{KO}(S^8 \cup_{n\Sigma^5\omega} e^{12})$ and set $w = B(u)$, $z = B([\epsilon]v)$. Recalling the properties of ψ^k listed before, we obtain, *for* k *even*:

$$B\left(\frac{h^2}{2}[\epsilon]\,\psi^k(u)\right) = \psi^k_{\mathbb{R}}(w) = k^4 w + q(n)\,\frac{k^4(k^2-1)}{12}\, z \; =$$

$$= B\left(k^4 u + q(n)\ \frac{k^4(k^2-1)}{12}[\epsilon]v\right) .$$

Hence, $[\epsilon]\psi^k(u) = 2k^2 u + q(n)\ \frac{k^2(k^2-1)}{6}\ [\epsilon]v$ and multiplying by $[\epsilon]$,

$$\psi^k(u) = \frac{k^2}{2}\ [\epsilon]u + q(n)\ \frac{k^2(k^2-1)}{6}\ v .$$

If k is odd we proceed in a similar way.
Finally, part (iii) follows using the ψ - operations on $\widetilde{KO}(S^8)$.

Theorem 3.8. *Let* R_n *be a finite* CW - *complex containing the subcomplex* $S^4 \cup_{n\Sigma\omega} e^8$ *and satisfying conditions* (i) *and* (ii) *of Proposition* 3.1 . *Then,*

(i) $L(R_n) \cong L(*) \otimes \mathbb{Z}[x,y]/\langle x^3, x^3y, xy^2, y^3\rangle$, *where* $x \in \widetilde{KSp}(R_n)$ *and* $y \in \widetilde{KO}(R_n)$. *In particular,* $\{[\epsilon]x, y, x^2, [\epsilon]xy, y^2\}$ *is a basis of* $\widetilde{KO}(R_n)$ *and* $\{x, [\epsilon]y, [\epsilon]x^2, xy, [\epsilon]y^2\}$ *is a basis of* $\widetilde{KSp}(R_n)$.

(ii)
$$\psi^k(x) = \begin{cases} \frac{k^2}{2}[\epsilon]x + q(n)\ \frac{k^2(k^2-1)}{6}\ y + a_k x^2 + b_k[\epsilon]xy + c_k y^2, & k \text{ even}, \\ k^2x^2 + q(n)\ \frac{k^2(k^2-1)}{12}[\epsilon]y + a_k x^2 + b_k xy + c_k[\epsilon]y^2, & k \text{ odd}, \end{cases}$$

where a_k, b_k, c_k *are integers and* $q(n) = n\cdot m + 12t$ *with* $m, t \in \mathbb{Z}$ *and* $(m,12) = 1$.

(iii) $\psi^k(y) = k^4 y + d_k[\epsilon]xy + e_k y^2$, *where* d_k, e_k *are integers.*

Proof. (i) The E_2 - term of the KO - theory spectral sequence of R_n is given by

$$E_2 = H^*(R_n) \otimes KO^*(*) = \mathbb{Z}[a,b]/\langle a^3, a^2b, ab^2, b^3\rangle \otimes KO^*(*) ,$$

dim a = 4, dim b = 8. Using the derivation property of the differentials, it is easy to show that the previous spectral sequence collapses. (In fact, assuming that $d_r = 0$, $r = 2,\dots, m-1$, we obtain for example, $d_m(a \otimes 1) \in H^{4+m}(R_n) \otimes KO^{-m+1}(*)$; but this tensor product is always trivial since $H^{4+m}(R_n) = 0$ if $m \not\equiv 0 \pmod 4$ and $KO^{-m+1}(*) = 0$ if $m \equiv 0 \pmod 4$, and therefore, $d_m(a\otimes 1) = 0$). We conclude that $E_2 \cong E_\infty \cong \mathcal{G}KO^*(R_n)$ and so, for the subring $L(R_n) \subseteq KO^*(R_n)$ we obtain

(3.9) $\mathcal{G}L(R_n) \cong H^*(R_n) \otimes L(*) \cong Z[a,b]/\langle a^3,a^2b,ab^2,b^3\rangle \otimes L(*)$.

It follows that $L(R_n)$ is torsion free. Now choose $x \in \widetilde{KO}^4(R_n) \cong \widetilde{KSp}(R_n)$ and $y \in \widetilde{KO}^0(R_n)$ representing $a\otimes 1$ and $b\otimes 1$, respectively. Because of (3.9) we conclude that the elements $x, y[\epsilon], x^2[\epsilon], xy, y^2[\epsilon] \in \widetilde{KSp}(R_n)$ and $x[\epsilon], y, x^2, xy[\epsilon], y^2 \in \widetilde{KO}(R_n)$ form a basis of $\widetilde{L}(R_n)$.

It remains to prove that $x^3 = x^2y = xy^2 = y^3 = 0$. First observe that the composition

$$\widetilde{KO}(R_n) \xrightarrow{c} \widetilde{KU}(R_n) \xrightarrow{ch} \widetilde{H}^*(R_n;\mathbb{Q})$$

is an injection of rings, since $\widetilde{KO}(R_n)$ has no torsion (0.11). Observe that according to (3.1) all 3-fold products of the latter ring vanish, and so, the same can be said about $\widetilde{KO}(R_n)$. Next, notice that in $L(R_n)$ the product of a non-zero element with $[\epsilon]$ is always non-trivial. We can now prove that $x^3 = 0$ as follows. The element $x[\epsilon] \in \widetilde{KO}(R_n)$ is such that $0 = (x[\epsilon])^3 = 4x^3[\epsilon]$; since $L(R_n)$ is torsion free $x^3[\epsilon] = 0$ and so $x^3 = 0$. The remaining relations are established in a similar way and part (i) of (3.8) is established.

Let us now prove (ii). Consider the inclusion $i : S^4 \cup_{n\Sigma\omega} e^8 \longrightarrow R_n$; from the previous discussion it is clear that the elements

$$u = i^!(x) \in \widetilde{KSp}(S^4 \cup_{n\Sigma\omega} e^8)$$

$$v = i^!(y) \in \widetilde{KO}(S^4 \cup_{n\Sigma\omega} e^8)$$

satisfy the hypothesis of Theorem 3.7 (Observe the homomorphism $E_2(i) : E_2(R_n) \cong \mathcal{G}KO^*(R_n) \longrightarrow E_2(S^4 \cup_{n\Sigma\omega} e^8) \cong \mathcal{G}KO^*(S^4 \cup_{n\Sigma\omega} e^8)$). We obtain the ψ - operations on x using $\psi^k i^! = i^! \psi^k$, (3.7) (ii) and $i^!(x^2) = i^!(xy) = i^!(y^2) = 0$ (the space $S^4 \cup_{n\Sigma\omega} e^8$ is a suspension).

Finally we prove (iii). The complexification homomorphism $c : KO(R_n) \longrightarrow KU(R_n)$ is injective and does not increase filtration because $KO(R_n)$ is torsion free. From

[14, Proprosition 5.6] we infer that

$$\psi^k(c(y)) = k^4 c(y) + \text{ elements of filtration } > 8$$

(this consequence of Atiyah's work will also be heavily exploited in § 5; the reader could therefore consult (5.1) which describes more precisely the property of ψ^k we just used). But c commutes with the Adams operations; thus,

$$\psi^k(y) = k^4 y + \text{ elements of filtration } > 8 .$$

This concludes the proof of the Theorem.

Proof of Theorem 3.4. The L - theory of a CW - complex R_n satisfying (3.1) is given by the previous Theorem; this Theorem, together with the properties of L - theory described after the statement of (3.4), shows that the generators $[\epsilon]$, x, y $\in L(R_n)$ satisfy

$$\psi^2([\epsilon]) = 2$$

$$\psi^2(y) = 2^4 y + d_2[\epsilon]xy + e_2 y^2$$

$$\psi^2(x) = 2[\epsilon]x + 2q(n)y + a_2 x^2 + b_2[\epsilon]xy + c_2 y^2$$

$$\psi^3([\epsilon]) = [\epsilon]$$

$$\psi^3(y) = 3^4 y + d_3[\epsilon]xy + e_3 y^2$$

$$\psi^3(x) = 9x + 6q(n)[\epsilon]y + a_3[\epsilon]x^2 + b_3 xy + c_3[\epsilon]y^2 .$$

Recalling that $[\epsilon]^2 = 4$, we have:

$$\psi^2\psi^3(y) = 2^4 3^4 y + \left(3^4 d_2 + 2^6 d_3\right)[\epsilon]xy + \\ + \left(3^4 e_2 + 2^6 q(n) d_3 + 2^8 e_3\right)y^2 ,$$

$$\psi^3\psi^2(y) = 2^4 3^4 y + \left(2^4 d_3 + 3^6 d_2\right)[\epsilon]xy + \\ + \left(2^4 e_3 + 2^3 3^5 q(n) d_2 + 3^8 e_2\right)y^2 .$$

Since $\psi^2\psi^3 = \psi^3\psi^2$,

$$2d_3 = 3^3 d_2$$

$$30e_3 + 2^3 q(n) d_3 = 10\cdot 3^4\cdot e_2 + 3^5 q(n) d_2 .$$

Eliminating d_2 in the second equation we obtain

$$3e_3 = 3^4 e_2 + q(n) d_3$$

and since $q(n) = n \cdot m + 12t \equiv n \pmod 2$ (recall that $(m,12) = 1$), we conclude

$$e_3 \equiv e_2 + nd_3 \pmod 2 . \tag{3.10}$$

For the element x we have:

$$\psi^2\psi^3(x) = 18[\epsilon]x + 210\, q(n)y + Ax^2 + B[\epsilon]xy + $$
$$+ \left(3^2c_2 + 2^2 3q(n)e_2 + 2^3q(n)^2a_3 + 2^5q(n)b_3 + 2^9c_3\right)y^2$$
$$\psi^3\psi^2(x) = 18[\epsilon]x + 210\, q(n)y + A'x^2 + B'[\epsilon]xy + $$
$$+ \left(2^3c_3 + 2q(n)e_3 + 2^3 3^3 q(n)^2 a_2 + 2^3 3^5 q(n)b_2 + 3^8c_2\right)y^2 ,$$

where A,B,A',B' are expression whose explicit form is not needed in our argument. Equating the coefficients of y^2, modulo 8, we obtain:

$$2^2 3q(n)e_2 \equiv 2q(n)e_3 + (3^8-3^2)c_2 \equiv 2q(n)e_3 \pmod 8 .$$

Thus

$$6q(n)e_2 \equiv q(n)e_3 \pmod 4 ;$$

since either $q(n) \equiv n \pmod 4$ or $q(n) \equiv 3n \pmod 4$, we see that

$$2ne_2 \equiv ne_3 \pmod 4 . \tag{3.11}$$

Now, the relation

$$\psi^2(y) = y^2 - 2\lambda_2(y)$$

holds in the λ - ring $L(R_n)$. Hence,

$$\psi^2(y) \equiv y^2 \pmod 2$$

and thus, $e_2 \equiv 1 \pmod 2$. With this, the congruences (3.10) and (3.11) become

$$e_3 + nd_3 \equiv 1 \pmod 2$$
$$ne_3 \equiv 2n \pmod 4$$

and we infer $$n \not\equiv 2 \pmod 4 .$$

4. The Homotopy Type Classification

In this section we give a complete list of homotopy types of torsion free rank 2 H-spaces. More precisely, we shall prove the following.

Theorem 4.1. *A torsion free finite H-space of rank 2 is homotopy equivalent to one of the following eleven spaces:*

$S^1 \times S^1$, $S^1 \times S^3$, $S^1 \times S^7$, $S^3 \times S^3$, $SU(3)$, $E_o = S^3 \times S^7$,

$E_1 = Sp(2)$, E_3, E_4, E_5 and $S^7 \times S^7$.

All these spaces represent mutually distinct homotopy types.

We begin by recalling, once more, that the integral cohomology of a torsion free H-space X of type $(2q-1, 2n-1)$ is an exterior algebra

$$(4.2) \qquad H^*(X) \cong \Lambda_{\mathbb{Z}}(x_{2q-1}, x_{2n-1}), \ \dim x_j = j \quad [179].$$

We first dispose of the cases $(2q-1, 2n-1) = (1,1)$, $(1,3)$ and $(1,7)$ by means of the following result (recall that the fundamental group of an H-space is abelian).

Theorem 4.3. *Let X be a connected H-space which is a CW-complex. If the fundamental group of X contains a free direct summand of rank k , that is to say, if* $\pi_1(X) \cong \mathbb{Z}^k \oplus \pi$, *then* $(S^1)^k \times X_1$, *where* $\pi_1(X_1) \cong \pi$.

Proof. Assume first that $k = 1$. Let $f : S^1 \to X$ be a representative of a generator for the subgroup $\mathbb{Z} \subseteq \pi_1(X)$ and let $p : X_1 \to X$ be the covering corresponding to the subgroup $\pi \subseteq \pi_1(X)$, so that $\pi_1(X_1) \cong \pi$. The composition

$$S^1 \times X_1 \xrightarrow{f \times p} X \times X \xrightarrow{m} X$$

induces an isomorphism between the homotopy groups of $S^1 \times X_1$ and X , hence $X \simeq S^1 \times X_1$.

The case $k > 1$ follows easily by induction since the space X_1 is again an H-space (cf. Chapter 7, Lemma 2.2).

Corollary 4.4. *If X is a torsion free H-space of type* $(1, 2n-1)$, *then* $X \simeq S^1 \times S^{2n-1}$.

Proof. Follows immediately from (4.2) and (4.3).

We turn next to the case $q = n$.

Theorem 4.5. *If* X *is a torsion free* H-*space of type* $(2n-1, 2n-1)$, *then* $X \simeq S^{2n-1} \times S^{2n-1}$.

Proof. The Theorem holds for $n = 1$; let us suppose $n \geq 2$. The Hurewicz Theorem, the Universal Coefficient Theorem and (4.2) imply that

$$\pi_{2n-1}(X) \cong H_{2n-1}(X) \cong H^{2n-1}(X) \cong \mathbb{Z} \oplus \mathbb{Z} ;$$

thus, let $f_i : S^{2n-1} \to X$, $i = 1,2$, be maps representing a basis of $\pi_{2n-1}(X)$. The composition

$$S^{2n-1} \times S^{2n-1} \xrightarrow{f_1 \times f_2} X \times X \xrightarrow{m} X$$

induces an isomorphism between the integral cohomology groups in dimension $2n - 1$ and hence, it induces an isomorphism between the exterior algebras $H^*(X)$ and $H^*(S^{2n-1} \times S^{2n-1})$. By Whitehead's Theorem, $S^{2n-1} \times S^{2n-1} \simeq X$.

Now we consider the type (3.5).

Theorem 4.6. *A torsion free* H-*space of type* $(3,5)$ *is homotopically equivalent to the Lie group* $SU(3)$.

Proof. An H-space of the kind given in the statement is simply connected and therefore belongs to the homotopy type $S^3 \cup_\alpha e^5 \cup_\beta e^8$, $\alpha \in \pi_4(S^3) \cong \mathbb{Z}_2$, $\beta \in \pi_7(S^3 \cup_\alpha e^5)$ (see (4.2)). We discuss first the case $\alpha = 0$. Consider the space

$$X_o = (S^3 \vee S^5) \cup e^8 ;$$

the inclusions $i_3 : S^3 \to X_o$, $i_5 : S^5 \to X_o$ induce isomorphisms $i_s^* : H^s(S^s) \cong H^s(X_o)$, $s = 3,5$. If X_o were an H-space, as in the proof of (4.5) we see that

$$S^3 \times S^5 \xrightarrow{i_3 \times i_5} X_o \times X_o \longrightarrow X_o$$

would induce an isomorphism between the integral cohomology groups and thus, $S^3 \times S^5 \simeq X_o$. But a retract of an H-space is still an H-space; thus, the last homotopy equivalence would imply that S^5 is an H-space, contradicting Corollary 5.2 of Chapter 1. Hence, a complex of the form $(S^3 \vee S^5) \cup e^8$ cannot carry a Hopf multiplication. So, let now

$$X = S^3 \cup_\eta e^5 \cup_\beta e^8$$

$\eta \in \pi_4(S^3) \cong \mathbb{Z}_2$, $\eta \neq 0$, and $\beta \in \pi_7(S^3 \cup_\eta e^5)$, be an H-space. Now, it is known that the Lie group $SU(3)$ has a cellular structure of the form $S^3 \cup_\eta e^5 \cup_\gamma e^8$ and hence, X and $SU(3)$ have the same 7 - skeleton. Since $\pi_7(SU(3)) = 0$ [243], the inclusion of $S^3 \cup_\eta e^5$ into $SU(3)$ extends to a map $f : X \to SU(3)$, inducing isomorphisms between the integral cohomology groups in dimensions 3 and 5. But $H^*(X)$ and $H^*(SU(3))$ are both exterior algebras with generators in dimensions 3 and 5, thus, f^* is an isomorphism and therefore, $X \simeq SU(3)$.

Finally, we investigate the most interesting case, namely the type (3,7). Classical examples of H-spaces having this type are $S^3 \times S^7$ and the Lie group $Sp(2)$. Both spaces are the total spaces of principal S^3 - bundles over the sphere S^7; we search for other bundles of this type, whose total spaces carry an H-multiplication. Recall that the principal S^3-bundles over S^7 are classified by the elements of $\pi_7(BS^3) \cong \pi_6(S^3) \cong \mathbb{Z}_{12}$ and that $\mathbb{Z}_{12}$ has a canonical generator ω characterized by $Sp(2) = S^3 \cup_\omega e^7 \cup e^{10}$. Let E_n be the total space of the S^3 - principal bundle over S^7 induced from the S^3-bundle $Sp(2) \xrightarrow{p_1} S^7$ by a map $f_n : S^7 \longrightarrow S^7$ of degree n :

$$\begin{array}{ccccc} E_n & \xrightarrow{g_n} & Sp(2) & & \\ {\scriptstyle p_n}\downarrow & & \downarrow{\scriptstyle p_1} & & \\ S^7 & \xrightarrow[f_n]{} & S^7 & \xrightarrow[\omega']{} & BS^3 \end{array} \tag{4.7}$$

(Here ω' corresponds to ω under $\pi_7(BS^3) \cong \pi_6(S^3)$). Choose a base point $* \in S^7$. The S^3 - bundle $E_n \to S^7$ is trivial over $S^7 - \{*\}$ and there is a map of pairs $h : (D^7 \times S^3, S^6 \times S^3) \to (E_n, S^3)$ which is a homeomorphism of $(D^7 - S^6) \times S^3$ onto $p_n^{-1}(S^7 - \{*\})$; (here D^7 is the closed 7 - ball). Hence we obtain a CW- decompositon $E_n = S^3 \cup (e^7 \times (* \cup e^3)) = (S^3 \cup e^7) \cup e^{10}$. The bundle map $g_n : E_n \to Sp(2)$ gives rise to a commutative triangle

$$\begin{array}{ccc} \pi_7(S^7) \cong \pi_7(E_n, S^3) & \xrightarrow{(g_n)_*} & \pi_7(Sp(2), S^3) \cong \pi_7(S^7) \\ & \partial' \searrow \quad \swarrow \partial & \\ & \pi_6(S^3) & . \end{array}$$

Since the homomorphism $(g_n)_*$ is multiplication by n and $\partial(1) = \omega$, we conclude that the attaching map of the 7-cell in E_n is equal to $n\omega$; thus

(4.8) $$E_n = \left(S^3 \cup_{n\omega} e^7\right) \cup e^{10} .$$

<u>Proposition 4.9.</u> *The spaces* E_n *and* E_m, $-6 \le n, m \le 6$ *are homotopy equivalent if, and only if,* $n = \pm m$.

<u>Proof.</u> It is clear that E_n and E_{-n} are homotopy equivalent (even homeomorphic). Conversely, let $h : E_n \to E_m$ be a homotopy equivalence, which we can assume to be cellular. Since S^3 carries the 3-dimensional homology of E_n and E_m, $h' = h|S^3$ must be of degree ± 1 on S^3. A comparison of the homotopy sequences of the pairs (E_n, S^3) and (E_m, S^3) shows that

$$h_* : \pi_7(E_n, S^3) \cong \pi_7(E_m, S^3) ;$$

notice that these groups are cyclic infinite and moreover, the following diagram commutes:

$$\begin{array}{ccc} \pi_7(E_n, S^3) & \xrightarrow[\cong]{h_*} & \pi_7(E_m, S^3) \\ \downarrow \partial_n & & \downarrow \partial_m \\ \pi_6(S^3) & \xrightarrow[\cong]{h'_*} & \pi_6(S^3) \end{array} .$$

In addition, observe that a map of degree -1 on S^3 induces the isomorphism -1 on the homotopy groups of S^3 because S^3 is an H-space. Thus, starting with a generator $1_n \in \pi_7(E_n, S^3) \cong \mathbb{Z}$

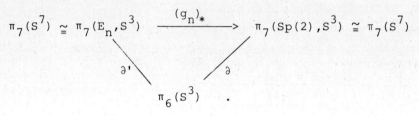

we compute $h_*^!\partial_n(1_n) = h_*^!(n\omega) = \pm n\omega$ and $\partial_m h_*(1_n) = \partial_m(\pm 1_m) = \pm m\omega$ and conclude that $n = \pm m$.

Now we restrict our attention to the seven spaces $E_o, E_1, E_2, E_3, E_4, E_5$ and E_6, which by (4.9) represent mutually distinct homotopy types. Of these, two are not H-spaces, namely E_2 and E_6 (this follows immediately from (3.4) and (4.8)); we claim that the remainder carry an H-multiplication. More precisely,

Theorem 4.10. *The spaces* $E_o = S^3 \times S^7$, $E_1 = Sp(2)$, E_3, E_4 *and* E_5 *are* H-*spaces.*

We shall give a proof of this Theorem relying on Localization Theory of 1 - connected CW-complexes. We shall review here the basic results of Localization Theory needed in the proof of (4.10); the reader who desires to become more familiar with such theory is advised to consult [107], where a proof of Theorem 4.10 can also be found.

We denote the set of all prime numbers by π . For any subset $P \subseteq \pi$, we say that a group G is P - *local* if the function $x \longmapsto x^n$, $x \in G$, is a bijection for all $n \in \pi - P$. For any 1-connected CW-complex X we write X_P for a P - *localization* of X (X_P is a 1-connected CW-complex with P - local homotopy groups and there is a map $e_P : X \to X_P$ inducing isomorphisms $(e_P)_* \otimes \mathbb{Z}_P : \pi_*(X) \otimes \mathbb{Z}_P \xrightarrow{\cong} \pi_*(X_P) \otimes \mathbb{Z}_P$, where $\mathbb{Z}_P \subseteq \mathbb{Q}$ is the ring of integers localized at P). For 1-connected CW-complexes P - localizations always exist and are unique up to homotopy type. A map $f : X \to Y$ gives rise to a map $f_P : X_P \to Y_P$; if $f_* : \pi_*(X) \to \pi_*(Y)$ is a P - isomorphism, then f_P is a homotopy equivalence. The space $X_\emptyset$ corresponding to the empty subset $\emptyset \subset \pi$ is a *rationalization* of X. For any $P \subseteq \pi$ one has $(X_P)_\emptyset = X_\emptyset$ and there is a factorization

$$e_\emptyset : X \xrightarrow[e_P]{} X_P \xrightarrow[e_{P,\emptyset}]{} X_\emptyset \quad ,$$

where $e_{P,\emptyset}$ is the rationalization map $X_P \to X_\emptyset$. A rationalization of the sphere S^{2q+1} is given by the Eilenberg-MacLane space $K(\mathbb{Q},2q+1)$. Finally, we note that if X is an H-space, any localization X_P inherits an H-space structure such that $e_P : X \to X_P$ is an H-map.

Proof of Theorem 4.10. Consider the diagram

$$\begin{array}{ccccc} S^3 \times S^7 \cong E_{12} & \xrightarrow{h} & E_3 & \xrightarrow{g} & E_1 = Sp(2) \\ \downarrow & & \downarrow & & \downarrow \\ S^7 & \xrightarrow{f_4} & S^7 & \xrightarrow{f_3} & S^7 \end{array}$$

(f_r is a map of degree r, $r = 3,4$). The sphere S^7 is an H-space and so, using the left-distributive law, we conclude that $(f_3)_* : \pi_*(S^7) \to \pi_*(S^7)$ is just multiplication by 3. The bundle morphism (g,f_3) induces a morphism between the appropriate exact homotopy sequences; we conclude that g is a $(\pi - \{3\})$ - equivalence and thus, $(E_3)_P \simeq (Sp(2))_P$, where $P = \pi - \{3\}$. Similarly, h is a $(\pi - \{2\})$ - equivalence and so, $(E_3)_{P'} \simeq (S^3 \times S^7)_{P'}$, with $P' = \{3\}$. Now, since $Sp(2)$ and $S^3 \times S^7$ are H-spaces, it follows that $(E_3)_P$ and $(E_3)_{P'}$ are H-spaces. Therefore, the multiplications on the latter two spaces give rise to two H-space structures on $(E_3)_\emptyset \simeq (S^3 \times S^7)_\emptyset \simeq K(\mathbb{Q},3) \times K(\mathbb{Q},7)$. These structures must coincide, since (up to homotopy) there is a unique Hopf-multiplication on the product $K(\mathbb{Q},3) \times K(\mathbb{Q},7)$ of Eilenberg-MacLane spaces. (In fact, two multiplications on an H-space X "differ" by an element of $[X \wedge X, X]$). Now, we turn $e_{P,\emptyset} : (E_3)_P \to (E_3)_\emptyset$ into a fibration and consider the pull-back diagram

$$\begin{array}{ccc} E & \longrightarrow & (E_3)_P \\ \downarrow & & \downarrow e_{P,\emptyset} \\ (E_3)_{P'} & \xrightarrow{e_{P',\emptyset}} & (E_3)_\emptyset \simeq K(\mathbb{Q},3) \times K(\mathbb{Q},7) \end{array} .$$

Since $P' = \pi - P$, the basic pull-back theorem of Localization Theory (see [107, Theorem II.5.1]) shows that E is homotopy equivalent to E_3. Furthermore, $e_{P,\emptyset}$ and $e_{P',\emptyset}$ being H-maps, the space E carries an H-space structure and thus, E_3 is an H-space. Interchanging the roles of 3 and 4 in the proof, we see that E_4 is an H-space. Finally, proceeding as before, but with the diagram

$$\begin{array}{ccccc} Sp(2) \cong E_{-1} \simeq E_{35} & \longrightarrow & E_5 & \longrightarrow & Sp(2) \\ \downarrow & & \downarrow & & \downarrow \\ E^7 & \xrightarrow[f_7]{} & S^7 & \xrightarrow[f_5]{} & S^7 \end{array}$$

one shows that E_5 is an H-space.

Theorem 4.11. *Let* $X_n = S^3 \cup_{n\omega} e^7 \cup_\beta e^{10}$, $-6 < n < 6$, $n \not\equiv 2 \pmod 4$, *be an* H-*space. Then* X_n *is homotopy equivalent to* E_n.

In order to prove the Theorem we need the next three lemmas.

Lemma 4.12. *The group* $\pi_9(E_n)$ *is finite and its order divides the greatest common divisor* $(n,6)$.

Proof. The exact homotopy sequence of the fibration

$$S^3 \xrightarrow[i]{} E_n \xrightarrow[p_n]{} S^7$$

and the finiteness of the groups $\pi_k(S^{2r+1}), k \neq 2r+1$; show that $\pi_k(E_n)$ is finite for $k \neq 3,7$. In addition, since $\pi_9(S^7) \cong \mathbb{Z}_2$ [242] and $\pi_9(S^3) \cong \mathbb{Z}_3$ (generated by $S^9 \xrightarrow{\Sigma^3\omega} S^6 \xrightarrow{\omega} S^3$) [214], the exact homotopy sequence proves the existence of an exact sequence

$$(4.13) \qquad \mathbb{Z}_3 \xrightarrow[i_*]{} \pi_9(E_n) \xrightarrow[(p_n)_*]{} \mathbb{Z}_2 .$$

The left-distributive law valid for the H-space S^3 implies that $n(\omega \circ \Sigma^3\omega) = (n\omega)\circ\Sigma^3\omega$. If $n \not\equiv 0 \pmod 3$ the element $n(\omega \circ \Sigma^3\omega)$ generates $\pi_9(S^3)$ and $i_*(n(\omega\circ\Sigma^3\omega)) = (i\circ n\omega)\circ\Sigma^3\omega = 0$; therefore,

$$(4.14) \qquad 0 \longrightarrow \pi_9(E_n) \longrightarrow \mathbb{Z}_2$$

is exact for $n \not\equiv 0 \pmod 3$. We recall now that $\pi_8(S^3) \cong \mathbb{Z}_2$, generated by $S^8 \xrightarrow{\eta^2} S^6 \xrightarrow{\omega} S^3$ [242, Prop. 5.9]. If n is odd, the left-distributive law shows that $i_*(\omega\eta^2) =$

$i_*(n(\omega\eta^2)) = i_*((n\omega)\circ\eta^2) = (i\circ n\omega)\circ\eta^2 = 0$ and thus,

(4.15) $$\mathbb{Z}_3 \longrightarrow \pi_9(E_n) \longrightarrow 0$$

is exact. The statement follows from (4.13), (4.14) and (4.15).

Lemma 4.16. *The order of the torsion subgroup of* $\pi_9(S^3 \cup_{n\omega} e^7)$ *divides* $(n,6)$.

Proof. Consider the following part of the exact homotopy sequence of the pair $(E_n, S^3 \cup_{n\omega} e^7)$:

$$\pi_{10}(E_n) \longrightarrow \pi_{10}(E_n, S^3 \cup_{n\omega} e^7) \xrightarrow{\partial} \pi_9(S^3 \cup_{n\omega} e^7) \longrightarrow \pi_9(E_n).$$

The group $\pi_{10}(E_n)$ is finite and $\pi_{10}(E_n, S^3 \cup_{n\omega} e^7) \cong \mathbb{Z}$; thus, ∂ is injective and the result follows from (4.12).

Lemma 4.17. *There is a map* $g : S^3 \vee S^7 \to S^3 \cup_{n\omega} e^7$ *which is the canonical inclusion on* S^3 *and such that* $(g|S^7)_* : H_7(S^7) \to H_7(S^3 \cup_{n\omega} e^7)$ *is multiplication by* k, *where* k *is prime to* n.

Proof. Let k be an integer such that $(n,k) = 1$ and $kn \equiv 0 \pmod{12}$. Consider

$$\begin{array}{ccccc} S^3 \times S^7 \simeq E_{nk} & \xrightarrow{h_k} & E_n & \xrightarrow{g_n} & Sp(2) \\ \downarrow & & \downarrow & & \downarrow \\ S^7 & \xrightarrow[f_k]{} & S^7 & \xrightarrow[f_n]{} & S^7 \end{array} ;$$

The restriction $g = h_k|S^3 \vee S^7$ provides a map with the required properties.

Proof of Theorem 4.11. Let $Z_n = S^3 \cup_{n\omega} e^7$ and fix a map $g : S^3 \vee S^7 \to Z_n$ satisfying the properties stated in (4.17). Assume now that $X_n = (S^3 \cup_{n\omega} e^7) \cup_\beta e^{10} = Z_n \cup_\beta e^{10}$, $\beta \in \pi_9(Z_n)$, is an H-space with multiplication m (and strict unit). The composition

$$h : S^3 \times S^7 \xrightarrow{g|S^3 \times g|S^7} X_n \times X_n \xrightarrow{m} X_n$$

in an extension of g. The homomorphism

$$h_* : H^*(X_n) \cong \Lambda_{\mathbb{Z}}(x_3,x_7) \to H^*(S^3 \times S^7) \cong \Lambda_{\mathbb{Z}}(y_3,y_7)$$

is such that $h^*(x_7) = ky_7$ (see (4.17)) and thus, h has degree $\pm k$ on the top cell. Consider the following commutative diagram induced by h :

$$\begin{array}{ccc} \mathbb{Z} \cong \pi_{10}(S^3 \times S^7, S^3 \vee S^7) & \xrightarrow{\partial} & \pi_9(S^3 \vee S^7) \\ \Big\downarrow h_* & & \Big\downarrow (h|S^3 \vee S^7)_* = g_* \\ \mathbb{Z} \cong \pi_{10}(X_n,Z_n) & \xrightarrow[\partial']{} & \pi_9(Z_n) \end{array}$$

and represent the generators of the two infinite cyclic groups on the left by the characteristic maps $\gamma_1 : (D^{10},S^9) \to (S^3 \times S^7, S^3 \vee S^7)$ and $\gamma_2 : (D^{10},S^9) \to (X_n,Z_n)$ (i.e., $\partial\gamma_1 = [\iota_3,\iota_7]$ and $\partial'\gamma_2 = \beta$). Since h has degree $\pm k$ on the top cell, $h_*\gamma_1 = \pm k\gamma_2$ and $g_*\partial(\gamma_1) = \partial' h_*(\gamma_2) = \pm k\partial'(\gamma_2) = \pm k\beta$. For the H-space $E_n = Z_n \cup_{\beta'} e^{10}$, $\beta' \in \pi_9(Z_n)$, the *same* map $g : S^3 \vee S^7 \to Z_n$ implies that $g_*\partial(\gamma_1) = \pm k\beta'$ and thus, $k\beta = k\beta'$ or $k\beta = -k\beta'$, i.e.,

$$k(\beta-\beta') = 0 \quad \text{or} \quad k(\beta+\beta') = 0 \; .$$

Since $(k,n) = 1$, Lemma 4.16 implies that either $\beta = \beta'$ or $\beta = -\beta'$, thus proving that X_n and E_n have the same homotopy type. (It is trivial that $Z_n \cup_\beta e^{10} \simeq Z_n \cup_{-\beta} e^{10}$).

Theorem 4.18. *A torsion free* H*-space of type* (3,7) *is homotopy equivalent to one of the following spaces:*

$$E_o = S^3 \times S^7,\ E_1 = Sp(2),\ E_3,\ E_4 \ \text{ or } \ E_5 \; .$$

All these spaces represent mutually distinct homotopy types.

Proof. An H-space satisfying the hypothesis of (4.18) is homotopy equivalent to a 3-cell complex of the form $S^3 \cup_{n\omega} e^7 \cup e^{10}$ (see § 3). The result now follows from (3.9), (4.9), (4.10) and (4.11).

Proof of Theorem 4.1. Combine (2.10), (4.4), (4.5), (4.6) and (4.18).

5. K - Theoretical Proof of the Type Classification Theorem

We have seen in § 2 that our proof of the Type Classification Theorem for torsion free H-spaces of rank 2 (Theorem 2.10) depends directly on Theorem 2.9. In this section we give a proof of the latter theorem.

(5.1) In what follows we shall make constant use of the following properties of the Adams Operations ψ^k in KU - theory (see (0.12) and [14;(5.6)]) :

(i) The operations ψ^k are ring homomorphisms;

(ii) $\psi^k \psi^m = \psi^m \psi^k$;

(iii) $\psi^2(z) \equiv z^2 \pmod 2$, for every $z \in KU(X)$;

(iv) if Y is a torsion free finite CW-complex

and $z \in KU(Y)$ has exact filtration $2q$ (see footnote for Theorem 2.8) then, for $0 \leq r \leq q$,

$$\psi^k(z) = k^{q-r} y_1 + y_2 \ , \ y_1 \in KU(Y), \ y_2 \in \mathcal{F}^{2q+2r+2} KU(Y) \ ;$$

in addition,

$$\psi^k(z) = k^q z + y_2 \ , \ y_2 \in \mathcal{F}^{2q+2} KU(Y) \ .$$

The following result of elementary number theory (already used in Chapter 1, Theorem 5.1) will also be needed:

<u>Lemma 5.2.</u> *For any integer* k *let* $v_2(k)$ *be the exponent of the highest power of* 2 *dividing* k.(*) *Then*

(i) $$v_2(3^m-1) = \begin{cases} 1 \ , & \textit{if } m \textit{ is odd} \ ; \\ 2 + v_2(m) \ , & \textit{if } m \textit{ is even} \ ; \end{cases}$$

(ii) *if* $m \neq 1,2$ *or* 4, $v_2(3^m-1) < m$;

(iii) *if* $q \leq r$ *then* $v_2(3^{q+r}-1) < r$, *except for the pairs* (q,r) *of the following list* : (1,1), (1,3), (2,2), (3,3), (3,5), (4,4).

<u>Proof.</u> (i) If m is odd, we verify the claim calculating mod 4 : $3^{2n+1} - 1 \equiv 2 \pmod 4$. If m is even, write

(*) We shall also encounter this function in Chapters 4,5 and 6.

$v_2(m) = a$ and so, $m = 2^a \cdot u$, u odd, and $3^m = v^{2^a}$, where $v = 3^u$. Hence,

$$3^m - 1 = v^{2^a} - 1 = \left(v^{2^{a-1}} + 1\right)\left(v^{2^{a-2}} + 1\right) \ldots (v+1)(v-1)$$

and the claim follows from

$$v^{2^s} + 1 \equiv 2 \pmod 4 \quad \text{if} \quad s \geq 1 ,$$

$$v + 1 = 3^u + 1 \equiv 4 \pmod 8 ,$$

and

$$v - 1 \equiv 2 \pmod 4 .$$

Part (ii) follows easily from (i). As for (iii), observe its veracity for $r \leq 7$ (by direct inspection). Assume that $r \geq 8$ and let α be the integer such that $2^{\alpha-1} \leq r < 2^\alpha$. But $\alpha \geq 4$ because $r \geq 8$; using (i),

$$v_2(3^{q+r}-1) \leq 2 + v_2(q+r) \leq 2 + \alpha < 2^{\alpha-1} \leq r .$$

Proof of Theorem 2.9. We shall distinguish two cases, namely, $q \leq n < 2q$ and $n \geq 2q$.

Case 1. $q \leq n < 2q$.

Write $n = q + r$, $0 \leq r < q$ and consider a CW-complex Q satisfying the hypothesis of (2.9). The elements x, y, x^2, xy and y^2 of $KU(Q)$ have filtration $2q$, $2(q+r)$, $4q$, $2(2q+r)$ and $4(q+r)$, respectively. From (5.1) (iv) we conclude that

$$\psi^k(x) = k^q x + k^{q-r} a_k y + b_k x^2 + c_k xy + d_k y^2$$

$$\psi^k(y) = k^{q+r} y + k^{2r} e_k x^2 + k^r f_k xy + g_k y^2 ,$$

where $a_k, b_k, c_k, d_k, e_k, f_k$ and g_k are integers. Hence,

$$\begin{aligned}\psi^m\psi^k(x) = {} & m^q k^q x + \left(m^{q-r}k^q a_m + m^{q+r}k^{q-r}a_k\right)y + \\ & \left(k^q b_m + m^{2r}k^{q-r}e_m a_k + m^{2q}b_k\right)x^2 + \\ & \left(k^q c_m + m^r k^{q-r} f_m a_k + 2m^{2q-r}a_m b_k + m^{2q+r}c_k\right)xy + \\ & \left(k^q d_m + k^{q-r}g_m a_k + m^{2q-2r}a_m^2 b_k + m^{2q}a_m c_k + m^{2q+2r}d_k\right)y^2 ,\end{aligned}$$

$$\psi^m\psi^k(y) = m^{q+r}k^{q+r}y + Bx^2 + Cxy +$$

$$+ \left(k^{q+r}g_m + m^{2q-2r}k^{2r}a_m^2e_k + m^{2q}k^r a_m f_k + m^{2q+2r}g_k\right)y^2 ,$$

where B and C are expressions whose explicit form is not needed in the sequel. Equating the coefficients of y, x^2, xy and y^2 in the equality

$$\psi^2\psi^k(x) = \psi^k\psi^2(x) ,$$

and those of y^2 in

$$\psi^2\psi^k(y) = \psi^k\psi^2(y) ,$$

we obtain the following conditions:

(R1) $\quad 2^r(2^r-1)a_k = k^r(k^r-1)a_2 \quad ;$

(R2) $\quad 2^q(2^q-1)b_k + 2^{2r}k^{q-r}a_k e_2 = k^q(k^q-1)b_2 + 2^{q-r}k^{2r}a_2 e_k \quad ;$

(R3) $\quad 2^q(2^{q+r}-1)c_k + 2^r k^{q-r}a_k f_2 + 2^{2q-r+1}b_k a_2 =$

$$k^q(k^{q+r}-1)c_2 + 2^{q-r}k^r a_2 f_k + 2 \cdot k^{2q-r}b_2 a_k \quad ;$$

(R4) $\quad 2^q(2^{q+2r}-1)d_k + k^{q-r}a_k g_2 + 2^{2q-2r}b_k a_2^2 + 2^{2q}c_k a_2 =$

$$k^q(k^{q+2r}-1)d_2 + 2^{q-r}a_2 g_k + k^{2q-2r}b_2 a_k^2 + k^{2q}c_2 a_k \quad ;$$

(R5) $\quad 2^{q+r}(2^{q+r}-1)g_k + 2^{2q-2r}k^{2r}e_k a_2^2 + 2^{2q}k^r a_2 f_k =$

$$k^{q+r}(k^{q+r}-1)g_2 + 2^{2r}k^{2q-2r}e_2 a_k^2 + 2^r k^{2q}a_k f_2 \ .$$

In addition, using (5.1) (iii), we obtain the congruences

(R6) $\quad c_2 \equiv 0 \pmod 2 \quad , \quad b_2 \equiv 1 \pmod 2$

$\quad\quad\quad d_2 \equiv 0 \pmod 2 \quad , \quad g_2 \equiv 1 \pmod 2 \ .$

Lemma 5.3. *If* $r = 0$ *and* $q \neq 1,2$ *or* 4 $(q > 0)$, *not all the relations* (R1) *to* (R6) *are satisfied.*

Proof. Since $r = 0$, both x and y have exact filtration $2q$, (5.1) (iv) implies that $a_k = 0$. Then, (R2) with $k = 3$ becomes

$$2^q(2^q-1)b_3 = 3^q(3^q-1)b_2 .$$

since b_2 is odd (see (R6)), 2^q divides $3^q - 1$. By (5.2) (ii) this is possible only for $q = 1,2$ or 4.

Lemma 5.4. *If* q *is odd*, $q \geq 3$ *and* $q > r \geq 1$ *not all the relations* (R1) *to* (R6) *are satisfied.*

Proof. Suppose first that $q - r \geq 2$, $q \geq 3$, $r \geq 1$. In this case, (R2) with $k = 3$, implies

$$0 \equiv 3^q(3^q-1)b_2 \pmod 4 .$$

Since q is odd, $3^q - 1 \equiv 2 \pmod 4$, by (5.2) (i), and hence,

$$0 \equiv 2b_2 \pmod 4 ,$$

contradicting (R6).

If $q - r = 1$, $q \geq 3$, then $r \geq 2$ and (R5) implies

$$0 \equiv 3^{2q-1}(3^{2q-1}-1)g_2 \equiv 2g_2 \pmod 4 ,$$

again contradicting (R6).

Lemma 5.5. *If* $q \geq 3$, r *is odd and* $q > r \geq 1$, *not all the relations* (R1) *to* (R6) *are satisfied.*

Proof. By the previous Lemma it suffices to prove the claim for q even.

If $q > r \geq 2$, equation (R5) with $k = 3$, implies:

$$0 \equiv 3^{q+r}(3^{q+r}-1)g_2 \pmod 4 .$$

Since $q + r$ is odd, $3^{q+r} - 1 \equiv 2 \pmod 4$ and the former congruence implies that $0 \equiv 2g_2 \pmod 4$ contradicting (R6).

If $q > r = 1$, we infer from (R4) and $d_2 \equiv 0 \pmod 2$ that

$$k^{q-1}a_kg_2 \equiv k^{2q-2}a_k^2b_2 + k^{2q}a_kc_2 \pmod 4 .$$

Since c_2 is even and both b_2 and g_2 are odd (see (R6)) we conclude that

(5.6) $$a_k \not\equiv 2 \pmod 4 ,\ k \text{ odd.}$$

Relation (R5) with $k = 3$ and reduced mod 4 gives

$$0 \equiv 2g_2u + 2a_3f_2 \pmod 4, \; u \text{ odd.}$$

Since $g_2 \equiv 1 \pmod 2$, we conclude that a_3 must be odd. Consider now (R1) with $r = 1$:

$$2a_k = k(k-1)a_2 .$$

Since a_3 is odd, the same holds for a_2. But this implies that $a_5 \equiv 2 \pmod 4$, contradicting (5.6).

Lemma 5.7. *If* $q \geq 3$ *and* $q > r \geq 1$, *not all the relations* (R1) *to* (R6) *are satisfied.*

Proof. By Lemmas (5.4) and (5.5) it suffices to prove the statement for q and r even, $q > r \geq 2$. Using (5.2) (ii) and (R1) we conclude that

$$(5.8) \qquad v_2(a_3) \begin{cases} = v_2(a_2) + 1 , & \text{if } r = 2 \\ \leq v_2(a_2) & , \text{if } r > 2. \end{cases}$$

Suppose first that $q \geq 6$. From (5.2) (i) and (iii) with q and r interchanged, we have

$$q > v_2(3^{q+r}-1) = 2 + v_2(q+r).$$

Reducing (R3) (with $k = 3$) mod $2^{3+v_2(q+r)}$, and recalling that c_2 is even, we obtain that

$$2^r3^{q-r}f_2a_3 - 2^{q-r}3^rf_3a_2 \equiv$$

$$2 \cdot 3^{2q-r}b_2a_3 \pmod{2^{3+v_2(q+r)}} .$$

Since b_2 is odd, (5.8) shows that the 2-divisibility of the terms appearing on the left handside of the last congruence is strictly greater than the one of the terms on the right handside ($q \geq 6$, $r \geq 2$, $q-r \geq 2$). This implies

$$(5.9) \qquad a_3 \equiv 0 \pmod{2^{2+v_2(q+r)}}$$

and because of (5.8),

$$(5.10) \qquad \begin{aligned} a_2 &\equiv 0 \pmod{2^{1+v_2(q+r)}} , && \text{if } r = 2 \\ a_2 &\equiv 0 \pmod{2^{2+v_2(q+r)}} , && \text{if } r > 2 . \end{aligned}$$

Reducing (R5) mod $2^{3+v_2(q+r)} = 2^{1+v_2(3^{q+r}-1)}$ and using (5.9) and (5.10) $(q \geq 6,\ r \geq 2,\ q-r \geq 2)$, we obtain

$$0 \equiv (3^{q+r}-1)g_2 \pmod{2^{1+v_2(3^{q+r}-1)}} .$$

This contradicts $g_2 \equiv 1 \pmod 2$ and the Lemma is proved for $q \geq 6$.

Finally, consider the case $q = 4,\ r = 2$. By (5.8) the integer a_3 is even. Reducing (R4) mod 4 we obtain

$$3^2 a_3 g_2 \equiv 3^4 b_2 a_3^2 + 3^8 c_2 a_3 \pmod 4 .$$

Since $c_2 \equiv 0 \pmod 2$ and $g_2 \equiv 1 \pmod 2$ we conclude that $a_3 \not\equiv 2 \pmod 4$ thus, $a_3 \equiv 0 \pmod 4$. Reducing (R5) mod 16 we obtain that

$$0 \equiv 3^6(3^6-1)g_2 + 2^2 3^8 a_3 f_2 \equiv 8g_2 \pmod{16}$$

contradicting $g_2 \equiv 1 \pmod 2$.

Remark 5.11. Lemmas 5.3 and 5.7 show that, if $q \leq n < 2q$, then

$$(q,n) = (1,1),\ (2,2),\ (2,3) \text{ or } (4,4) .$$

Case 2. $n \geq 2q$.

Write $n = q + r$, $r \geq q$ and consider a CW-complex Q satisfying the hypothesis of (2.9). The elements $x, x^2, y, xy, y^2 \in KU(Q)$ have filtration $2q, 4q, 2(q+r), 2(2q+r), 4(q+r)$, respectively. For the ψ - operations we have:

$$\psi^k(x) = k^q x + a_k y + b_k x^2 + c_k xy + d_k y^2$$

$$\psi^k(y) = k^{q+r} y + k^r f_k xy + g_k y^2$$

and hence,

$$\psi^m\psi^k(x) = m^q k^q x + \left(k^q b_m + m^{2q} b_k\right)x^2 + \left(k^q a_m + m^{q+r} a_k\right)y + \left(k^q c_m + 2m^q a_m b_k + m^r f_m a_k + m^{2q+r} c_k\right)xy + A'y^2$$

$$\psi^m\psi^k(y) = m^{q+r}k^{q+r}y + \left(m^r k^{q+r} f_m + m^{2q+r} k^r f_k\right)xy + \left(k^{q+r} g_m + m^{q+r} k^r a_m f_k + m^{2q+2r} g_k\right)y^2 .$$

Equating the coefficients of x^2, y and xy in $\psi^2\psi^3(x) = \psi^3\psi^2(x)$ and those of xy and y^2 in $\psi^2\psi^3(y) = \psi^3\psi^2(y)$, we obtain:

(R'1) $2^q(2^q-1)b_3 = 3^q(3^q-1)b_2$;

(R'2) $2^q(2^r-1)a_3 = 3^q(3^r-1)a_2$;

(R'3) $2^q(2^{q+r}-1)c_3 + 2^{q+1}a_2b_3 + 2^r f_2 a_3 =$
$3^q(3^{q+r}-1)c_2 + 2 \cdot 3^q a_3 b_2 + 3^r f_3 a_2$;

(R'4) $2^q(2^q-1)f_3 = 3^q(3^q-1)f_2$;

(R'5) $2^{q+r}(2^{q+r}-1)g_3 + 2^{q+r}3^r a_2 f_3 =$
$3^{q+r}(3^{q+r}-1)g_2 + 2^r 3^{q+r} a_3 f_2$;

(R'6) $a_2 \equiv 0 \pmod 2$, $b_2 \equiv 1 \pmod 2$,
$g_2 \equiv 1 \pmod 2$.

Lemma 5.12. *If* $q = 1$, $r \geq 1$ *and* $r \neq 1,3$, *not all the relations* (R'1) *to* (R'6) *are satisfied.*

Proof. Consider (R'5) for $q = 1$ and reduce modulo 2^r :

$$0 \equiv 3^{1+r}(3^{1+r}-1)g_2 \pmod{2^r} .$$

Since $g_2 \equiv 1 \pmod 2$, it follows that 2^r divides $3^{1+r} - 1$; using Lemma 5.2, (iii), we conclude that $r = 1,3$, contradiction.

Lemma 5.13. *If* $r \geq q \geq 2$ *and* $r > 2$, *not all the relations* (R'1) *to* (R'6) *are satisfied.*

Proof. Reducing (R'5) mod 2^r one gets

$$0 \equiv (3^{q+r}-1)g_2 \pmod{2^r} .$$

According to the hypothesis and (5.2) (iii), the previous congruence is possible only if $(q,r) = (3,3)$, $(3,5)$ or $(4,4)$. The first two cases can be excluded working with (R'1), which, for $q = 3$ gives $(3^3-1)b_2 \equiv 0 \pmod 8$, and thus, contradicting $b_2 \equiv 1 \pmod 2$. Now we investigate the case $q = r = 4$.

With these values of q and r, (R'2) and (R'4) become

$$a_3 = 27a_2 \quad \text{and} \quad f_3 = 27f_2 ,$$

respectively. Hence

$$a_3f_2 = a_2f_3 .$$

Since $a_2 \equiv 0 \pmod 2$, it also follows that

$$a_3 \equiv 0 \pmod 2 .$$

Reduce (R'3) modulo 4 to obtain

$$0 \equiv 2a_3b_2 + a_2f_3 \equiv a_2f_3 \equiv a_3f_2 \pmod 4 .$$

Now consider (R'5) mod 2^6 and get

$$0 \equiv 3^8(3^8-1)g_2 + 2^4 3^8 a_3f_2 \equiv 2^5g_2 \pmod{2^6} ,$$

contradicting $g_2 \equiv 1 \pmod 2$.

Remark 5.14. Lemmas 5.12 and 5.13 show that if $2q \leq n$, $(q,n) = (1,2)$, $(1,4)$ or $(2,4)$.

Remarks 5.11 and 5.14 conclude the proof of Theorem 2.9.

CHAPTER 3

HOMOTOPY AND STABLY COMPLEX STRUCTURE

In this chapter we discuss examples of manifolds constructed by W.A.Sutherland, showing that the property of admitting a stably complex structure is not homotopy invariant.

1. The Question of Complex Structure.

Let M be a complex manifold of complex dimension n. The admissible charts of M define on the same underlying topological space the structure of a real differentiable manifold of dimension $2n$. It is well known that this manifold, which is called realification of the original complex manifold, is orientable. Conversely, given a real orientable differentiable manifold of even dimension $2n$, the question arises whether it admits a complex structure compatible with the given real manifold structure. If $n = 1$, that is to say if we are dealing with orientable surfaces, such a complex structure always exists. The situation is different when $n \geq 2$; for example it was shown by Hopf [125] and Ehresmann [88] that the sphere S^4 cannot be given a complex structure.

Examples of complex manifolds are the complex projective spaces $\mathbb{C}P^n$ and non-singular algebraic varieties. The latter objects are compact complex manifolds which can be analytically embedded into some complex projective space (see [66]). The notion of complex manifold is more general than the notion of algebraic manifold. In [125] Hopf exhibited a complex structure on $S^1 \times S^3$; this example was generalized by Calabi and Eckmann [61] giving a complex structure on $S^{2p+1} \times S^{2q+1}$.

The study of complex manifolds has been pursued in different ways. One line of development was concentrating on special families of manifolds, satisfying certain additional properties. In this context we mention the so called *Kähler Manifolds*. These are complex hermitian manifolds the metric $ds^2 = \Sigma g_{ik}(z,\bar{z})dz^j dz^k$ of which is kählerian, i.e. such that the associated exterior differential form $\omega = \Sigma g_{ik}(z,\bar{z})dz^j \wedge dz^k$ is closed. Kähler manifolds have been investigated extensively (Hodge theory) and have interesting properties (see [249] and [116]). Non-singular algebraic varieties are kählerian; but there are Kähler manifolds which are not algebraic. Furthermore, not any complex manifold admits a Kähler metric; we quote once again the examples of Calabi and Eckmann.

Another approach in the study of complex structures consists in viewing a weaker situation: the almost complex structure. The following observation leads to the definition of this notion. The tangent bundle of a complex manifold M admits a complex vector bundle structure. Multiplication by $\sqrt{-1}$ defines at each point $x \in M$ a linear transformation $\mathcal{J}_x$ of the tangent space TM_x with $\mathcal{J}_x \circ \mathcal{J}_x = -Id(TM_x)$ and such that $\mathcal{J}_x$ depends continously on x. If on a real differentiable manifold M a continuous field $\mathcal{J}_x$ of linear transformations with $\mathcal{J}_x^2 = -\text{Identity}$ is given (without reference to any complex structure), M is called an *almost complex manifold* with *almost complex structure* $\mathcal{J}$. It is easy to see that an almost complex structure defines a canonical orientation on the manifold.

Some necessary conditions of existence of almost complex structures on M can be stated in terms of characteristic classes. Historically, several results on 4-dimensional manifolds have been obtained as early as 1948, by Wentsün Wu [262]. Since then several papers had been written following the same lines of thought; in particular, we wish to bring the attention of the reader to the paper of W.S.Massey [172]. Using different methods, Kirchhoff has shown that, if the sphere S^{2n} admits an almost complex structure, then S^{2n+1} is parallelizable [151]. At the end of chapter 1, we have already pointed out that Borel and Serre proved back in 1953 that the only spheres carrying an almost complex structure are S^2 and S^6.

The study and construction of almost complex manifolds have revealed some interesting phenomena. For instance, the property of admitting an almost complex structure is not a homotopy invariant (cf. [125]). Also, this property depends on the orientation of the manifold, in a sense explained by the following example. The complex structure on $\mathbb{C}P^2$ induces a canonical orientation on this manifold. It can be proved that there is no almost complex structure on $\mathbb{C}P^2$ which induces the opposite orientation (see [145]).

2. Almost Complex Manifolds and Stably Complex Manifolds.

Let M be a $2n$-dimensional orientable manifold and let $\tau(M)$ be its tangent bundle. After reduction, we may suppose that the structure group of the $2n$-dimensional real vector bundle $\tau(M)$ is $SO(2n)$; let us denote by $\tau'(M)$ the associated bundle with fibre $\Gamma_n = SO(2n)/U(n)$.

Definition 2.1. *The manifold* M *is said to admit an almost complex structure if the* Γ_n*-bundle* $\tau'(M)$ *has a cross-section.*

Hence, the existence of an almost complex structure on M amounts to the possibility of reducing the structure group $SO(2n)$ to $U(n)$. It is well known that such a reduction requires, among other conditions, the nullity of each odd-dimensional Stiefel-Whitney class [231; cor. 4.1.9]. Note that the preceding situation may be viewed as a particular case within the more general context of X-structures (see [176; p.18]).

Before we formaly establish the notion of stably complex structure on a manifold, let us describe a particularly interesting(*) standard construction of an almost complex manifold. Let X be a closed differentiable manifold (i.e., compact and without boundary) of dimension n. Then, *the total space* $M = T^*X$ *of the cotangent bundle* $\tau^*(X) = (T^*X, \pi, X)$ *has a natural structure of almost complex manifold.* In order to prove this statement we have to determine clearly the na-

(*) For its usefulness, see section 4 of chapter 8.

ture of the tangent bundle $\tau^*(M) = \tau(T^*X)$, whose base space is, of course, a 2n-manifold. First of all, let us observe that for every (differentiable) real vector bundle $\xi = (E,p,X)$, the differential

$$dp : TE \longrightarrow TX$$

induces a bundle epimorphism

$$(dp)_* : \tau(E) \longrightarrow p^*\tau(X)$$

whose kernel is the bundle of vectors tangent to the fibres of ξ. We also observe that $\ker(dp)_*$ is canonically isomorphic to $p^*\xi$ since, for every $x \in X$, the space tangent to ξ_x at any point of that fibre is isomorphic to ξ_x. It follows that the sequence

$$(2.2) \quad 0 \longrightarrow p^*\xi \longrightarrow \tau(E) \xrightarrow{(dp)_*} p^*\tau(X) \longrightarrow 0$$

is exact. Taking $\tau^*(X)$ for ξ, we have

$$0 \longrightarrow \pi^*\tau^*(X) \longrightarrow \tau(T^*X) \xrightarrow{(d\pi)_*} \pi^*\tau(X) \longrightarrow 0$$

is exact. Now, since the manifold X is riemannian, $\tau(X)$ and $\tau^*(X)$ are isomorphic and moreover, the last sequence splits. Hence,

$$\begin{aligned}\tau(M) = \tau(T^*X) &\cong \pi^*\tau^*(X) \oplus \pi^*\tau(X)\\ &\cong \pi^*(\tau(X) \oplus \tau(X)) .\end{aligned}$$

But $\tau(X) \oplus \tau(X)$ is just the real vector bundle subjacent to the complexification $\tau_{\mathbb{C}}(X) = \tau(X) \otimes \mathbb{C}$ of $\tau(X)$. This shows that $\tau(M)$ inherits the complex structure defined(**) by $\alpha = \pi^*(\tau_{\mathbb{C}}(X))$. In other words, for any $v \in M = T^*X$, let us write the elements $u \in T_v(M)$ in the form $u = (a,b) \in \pi^*(T_v(X) \oplus T_v(X))$. Then, the automorphism

$$\mathcal{J}_v : \tau_v(M) \longrightarrow \tau_v(M)$$

is given by $\mathcal{J}_v(a,b) = (-b,a)$. Finally, let us give clearly the orientation of the manifold M. Let $x_1,\ldots,x_n$ be the local coordinates of X; then, every $v \in T^*_x(X)$ (where $x \in X$) is of the form $\sum_{k=1}^{n} v_k dx_k$: the 2n-tuple

---

(**) This α-structure depends on the riemannian structure of X

$(x_1, v_1, \ldots, x_n, v_n)$ defines the orientation of $M = T^*X$ induced by α. In order to avoid confusion, we shall call this orientation the α-*orientation of* M. If X is itself oriented (in the preceding construction we did not make any requirements about the orientability or not of X), this orientation induces a natural orientation on T^*X. With the previous notations, one can see that the natural orientation corresponds to that given by the 2n-tuple $(x_1, \ldots, x_n, v_1, \ldots, v_n)$. Hence, the α-orientation of M differs from the natural orientation by the signature of the permutation $(x_1, \ldots, x_n, v_1, \ldots, v_n) \to (x_1, v_1, \ldots, x_n, v_n)$, that is to say, by a factor $(-1)^{(n-1)n/2}$.

The notion of almost stably complex manifold consists in replacing the space Γ_n (see Definition 2.1) by the homogeneous space $\Gamma = SO/U$. More precisely, let $\theta(M)$ be the Γ-bundle associated to $\tau(M)$; then, we give the following:

Definition 2.3. M *admits a stably complex structure if* Γ-*bundle* $\theta(M)$ *has a cross-section.*

For the manifold M, to be stably complex means that any stabilization of $\tau(M)$ with even-dimensional fibre may be endowed with a complex vector bundle structure. This is possible if and only if the element $[\tau(M)] - \dim M$ of $\widetilde{KO}(M)$ belongs to the image of the realification homomorphism $r : \widetilde{KU}(M) \to \widetilde{KO}(M)$. (If ξ is a stably complex structure on a connected and closed manifold M, we can show that it reduces to an almost complex structure if and only if its nth Chern class $c_n(\xi)$ is the Euler characteristic of M : a proof, using K-theory, is given in [145; p.347].)

The π-manifolds provide an interesting family of stably complex manifolds. By definition, the Whitney sum of the trivial line bundle and the tangent bundle of such a manifold is trivial. This property enables us to associate, in a standard way, a stably complex manifold to any orientable k-manifold X. Let f be a differentiable embedding of X into R^d (this is always possible when d is large enough) and let us write $\nu = (E, p, X)$ for the normal bundle of this embedding. We have the following split exact sequence of vector bundles (see chapter 6)

$$0 \longrightarrow \tau(X) \longrightarrow \varepsilon_d(X) \longrightarrow \nu \longrightarrow 0 ,$$

where $\varepsilon_d(X)$ is the trivial d-dimensional bundle over X.

Let $S(\nu)$ be the $(d-k-1)$-sphere bundle associated to ν. We shall denote its total space by M and its projection by π, i.e.

$$S(\nu) = (M,\pi,X) .$$

Since the submanifold M of E admits a field of non-trivial normal vectors in E, we get

$$\tau(M) \oplus \varepsilon \cong \tau(E)|M .$$

Restricting the vector bundles of

$$0 \longrightarrow p^*\nu \longrightarrow \tau(E) \longrightarrow p^*\tau(X) \longrightarrow 0$$

(see 2.2) to M, we obtain the following exact sequence of bundles over M

$$0 \longrightarrow \pi^*\nu \longrightarrow \tau(E)|M \longrightarrow \pi^*\tau(X) \longrightarrow 0 .$$

The previous remarks imply that

$$\tau(M) \oplus \varepsilon \cong \pi^*\nu \oplus \pi^*\tau(X) \cong \pi^*(\nu \oplus \tau(X)) \cong \pi^*\varepsilon_d(X) ,$$

in other words, the total space M of $S(\nu)$ is a $(d-1)$-dimensional π-manifold.

The procedure just described was used by Sutherland to build his examples of manifolds mentioned in the introduction. More precisely, in the following let M be *the* π*-manifold obtained from an embedding of* the manifold $X = S^1 \times \Phi P^2$ *into* R^{n+17}, where we suppose that $n > 18$. Here, ΦP^2 denotes the *Cayley projective plane* with its canonical CW-structure, obtained by attaching a 16-cell to S^8 via the Hopf map $\sigma : S^{15} \to S^8$, i.e.

$$\Phi P^2 = S^8 \cup_\sigma e^{16} .$$

Let $g : X = S^1 \times \Phi P^2 \to S^{17}$ be the canonical projection map collapsing the 16-skeleton of X, and let α be a *non-trivial* vector bundle over S^{17} of dimension $n + 17$, where the integer n is the same as before. Since $n > \dim X$ there is an n-dimensional differentiable vector bundle

$\xi = (\tilde{E},\tilde{p},X)$ over X, such that

$$\xi \oplus \tau(X) \cong g^*\alpha .$$

Let $\tilde{M}$ be the *total space of the sphere bundle*

$$S(\xi) = (\tilde{M},\tilde{\pi},X) ,$$

associated to ξ. As for the bundle $S(\nu)$ there is the following exact sequence over $\tilde{M}$:

$$0 \longrightarrow \tilde{\pi}^*\xi \longrightarrow \tau(\tilde{E})|\tilde{M} \longrightarrow \tilde{\pi}^*\tau(X) \longrightarrow 0 .$$

Since again $\tau(\tilde{M}) \oplus \varepsilon \cong \tau(\tilde{E})|\tilde{M}$, we obtain

(2.4) $$\tau(\tilde{M}) \oplus \varepsilon \cong \tilde{\pi}^*(\xi \oplus \tau(X)) \cong \tilde{\pi}^*g^*\alpha .$$

Thus, the bundles $\tau(\tilde{M})$ and $\tilde{\pi}^*g^*\alpha$ are stably equivalent. With this stable equivalence we will prove in § 4 that $\tilde{M}$ does not admit any stably complex structure.

3. The Homotopy Type of M and $\tilde{M}$.

To compare the homotopy types of M and $\tilde{M}$ we will work with the group $\widetilde{KH}(Y)$ of stable fibre homotopy classes of sphere bundles over the space Y. This is no surprise for the reader, since both M and $\tilde{M}$ are defined as total spaces of certain sphere bundles. Let us briefly recall the definition of $\widetilde{KH}$. The semigroup $H(n)$ of all homotopy equivalences of S^{n-1}, endowed with the compact - open topology, is an associative H - space; let $BH(n)$ be its classifying space. According to Dold-Lashoff [80], the set of fibre homotopy classes of S^{n-1}-fibrations over a finite CW - complex Y is in one-to-one correspondence with the set $[Y,BH(n)]$. Under suspension we have a natural inclusion $H(n) \to H(n+1)$ which gives rise to a map $BH(n) \to BH(n+1)$. We put $BH = \varinjlim BH(n)$; this space admits an H-space structure. For a finite CW - complex Y we define

$$\widetilde{KH}(Y) = [Y,BH] .$$

If $n > 1 + \dim Y$, the canonical map $BH(n) \to BH$ induces a bi-jection $[Y,BH(n)] \cong [Y,BH]$. (For further details, see Dold-Lashoff [80] and Stasheff [227].)

Restriction of orthogonal maps of $\mathbb{R}^n$ to S^{n-1} provides an inclusion $O(n) \to H(n)$, which induces a map $BO(n) \to BH(n)$. These maps give rise to an H - map $J : BO \to BH$. The image of the induced homomorphism $\tilde{J} : \widetilde{KO}(Y) = [Y,BO] \to \widetilde{KH}(Y) = [Y,BH]$ may be identified with the group $\tilde{J}(Y)$ defined in (0.14) (see Atiyah [12]). If Y is the r-sphere we have

$$\widetilde{KH}(S^r) = \pi_r(BH) \cong \pi_r(BH(n)) \cong \pi_{r-1}(H(n)), \quad r \leq n-2 .$$

According to [12], the latter group is for $2 \leq r \leq n-2$ isomorphic to $\pi_{r-2+n}(S^{n-1})$; hence

$$\widetilde{KH}(S^r) \cong \pi^S_{r-1} = \lim_{k\to\infty} \pi_{r-1+k}(S^k) , \qquad r \geq 2 ,$$

(π^S_{r-1} is the so called stable $(r-1)$-stem). In the following we will identify $\widetilde{KH}(S^r)$ and π^S_{r-1}. The homomorphism $\tilde{J} : \widetilde{KO}(S^r) \to \widetilde{KH}(S^r)$ may now be interpreted as a map

$$\pi_r(BO) \cong \pi_{r-1}(O) \xrightarrow{\tilde{J}} \pi^S_{r-1} .$$

It can be proved that $\tilde{J}$ coincides with the classical stable J-homomorphism of G.W.Whitehead (see [134, 15.6]).

After this brief sketch on the relations among the functors $\widetilde{KO}$, $\widetilde{KH}$ and $\tilde{J}$, we provide the following lemma which is the key to all our investigation concerning the manifolds M and $\tilde{M}$.

<u>Lemma 3.1.</u> *Let* $\Sigma^2\sigma : S^{17} \to S^{10}$ *be the double-suspension of the Hopf map* σ . *The non-trivial element* $a \in \widetilde{KO}(S^{17}) \cong \tilde{J}(S^{17}) \cong \mathbb{Z}_2$ *satisfies:*

(i) $a \notin \operatorname{im}\left(\widetilde{KO}(S^{10}) \xrightarrow{(\Sigma^2\sigma)^!} \widetilde{KO}(S^{17})\right)$

(ii) $J(a) \in \operatorname{im}\left(\widetilde{KH}(S^{10}) \xrightarrow{(\Sigma^2\sigma)^*} \widetilde{KH}(S^{17})\right)$.

<u>Proof.</u> Let $\eta \in \pi^S_1$ be the non-trivial element. According to [242, prop. 3.1], the 8-fold suspension of the diagram

$$\begin{array}{ccc} S^{17} & \xrightarrow{\Sigma^2\sigma} & S^{10} \\ \eta\downarrow & & \downarrow\eta \\ S^{16} & \xrightarrow{\Sigma\sigma} & S^{9} \end{array}$$

commutes (up to sign), thus giving rise to a commutative diagram in $\widetilde{KO}$ - theory. Since $\eta^! : \widetilde{KO}(S^9) \cong \widetilde{KO}(S^{10})$ (see [4,IV, 1.2])(*) and because $(\Sigma\sigma)^!$ is obviously zero, we conclude that $(\Sigma^2\sigma)^!$ is trivial and (i) is established.

To prove (ii) we first note that the map $S^{17} \overset{\eta}{\to} S^{16}$ induces an epimorphism in $\widetilde{KO}$ - theory (see [4,IV, 1.2]) and therefore also an epimorphism for the $\tilde{J}$ - groups; hence $\tilde{J}(S^{17}) \subseteq \text{im}\,(\widetilde{KH}(S^{16}) \xrightarrow{\eta^*} \widetilde{KH}(S^{17}))$. The only non-trivial element in the latter sub-group of $\widetilde{KH}(S^{17})$ is $\eta \circ \rho$, as one may check, consulting Toda's tables [242, p.189 and Th.14.1] (we are now following Toda's notation). One concludes that

$$J(a) = \eta \circ \rho .$$

According to [242], the group $\widetilde{KH}(S^{10})$ is isomorphic to $\mathbb{Z}_2 \oplus \mathbb{Z}_2 \oplus \mathbb{Z}_2$ generated by ν^3, μ and $\eta \circ \varepsilon$. Again invoking [242, Th. 14.1], we obtain

$$(\Sigma^2\sigma)^*(\mu) = \mu \circ \sigma = \sigma \circ \mu = \eta \circ \rho$$

and (ii) is proved. (In the latter relation we use the same symbol for the map $\sigma : S^{15} \to S^8$ and the class represented by this map in π_7^S.)

<u>Corollary 3.2.</u> *Let* $g : S^1 \times \mathbb{O}P^2 \to S^{17}$ *be the canonical projection. For the non-trivial element* $a \in \widetilde{KO}(S^{17})$ *one has*

(i) $g^!(a) \neq 0$ *in* $\widetilde{KO}(S^1 \times \mathbb{O}P^2)$;

(ii) $g^*J(a)) = 0$ *in* $\widetilde{KH}(S^1 \times \mathbb{O}P^2)$.

<u>Proof.</u> The map g factors through the smash product $S^1 \wedge \mathbb{O}P^2$,

$$g : S^1 \times \mathbb{O}P^2 \xrightarrow{p} S^9 \cup_{\Sigma\sigma} e^{17} \xrightarrow{q} S^{17} .$$

It is well known that the projection p onto the smash induces a monomorphism in any cohomology theory. Thus, the corollary will be proved if we can show that

$$q^!(a) \neq 0 \quad \text{and} \quad q^*(J(a)) = 0 .$$

These two relations follow readily from Lemma 3.1 and the

(*) The composition $S^{10} \overset{\eta}{\to} S^9 \overset{\eta}{\to} S^8$ induces an epimorphism of $\widetilde{KO}$ - groups.

exact sequences in $\widetilde{KO}$- and $\widetilde{KH}$-theory associated to the relevant part of the Puppe sequence of σ, namely to

$$S^9 \cup_{\Sigma\sigma} e^{17} \xrightarrow{q} S^{17} \xrightarrow{\Sigma^2\sigma} S^{10} .$$

We are now ready to prove the main result of this section.

<u>Theorem 3.3.</u> *The manifolds* M *and* $\widetilde{M}$ *are homotopically equivalent.*

<u>Proof.</u> According to § 2, M is the total space of the sphere bundle associated to the normal bundle ν of an embedding $X \hookrightarrow R^{n+17}$, $n > 18$, whereas $\widetilde{M}$ arises as total space of a sphere bundle $S(\xi)$, with ξ satisfying

$$\xi \oplus \tau(X) \cong g^*\alpha \ , \ \dim \xi = \dim \nu = n \ .$$

Here, α is a (non-trivial) vector bundle over S^{17}.

Part (ii) of corollary 3.2 implies that $\xi \oplus \tau(X)$ is J-trivial, and because $\nu \oplus \tau(X)$ is a trivial bundle we infer that ν and ξ are J-equivalent. Since $\dim \nu = \dim \xi > \dim X + 1$ the bundles $S(\nu)$ and $S(\xi)$ are fibre homotopy equivalent and the theorem is proved.

<u>4. The manifold $\widetilde{M}$ is not stably complex.</u>

In this last section of chapter 3 we show that the manifold $\widetilde{M}$ does not admit a stably complex structure. By (2.4) the tangent bundle $\tau(\widetilde{M})$ is stably equivalent to $\widetilde{\pi}^*g^*\alpha$ and it would therefore be sufficient to prove that the latter bundle is not stably complex. With this in mind we first establish the following result.

<u>Proposition 4.1.</u> *Let* $g : X = S^1 \times \mathbb{O}P^2 \to S^{17}$ *be the canonical projection. Then, for any vector bundle* α *over* S^{17} *which is stably non-trivial, the bundle* $g^*\alpha$ *does not admit a stably complex structure.*

<u>Proof.</u> Our hypothesis on α implies that in $KO(S^{17})$ we have $[\alpha] = a + \dim \alpha$, with $a \neq 0$. According to the remarks following (2.8) we must show that $g^!(a)$ does not belong to $r(\widetilde{KU}(X))$.

The KU-theory of X is torsion free and $r : \widetilde{KU}(X) \to \widetilde{KO}(X)$ is injective. (The projection $X \to \Phi P^2$ induces an isomorphism $\widetilde{KU}(\Phi P^2) \cong \widetilde{KU}(X)$; for the two-cell complex $\Phi P^2 = S^8 \cup e^{16}$ it is easy to see that the realification is injective).

According to (3.2) (i), the element $g^!(a)$ is non-zero of order 2 and hence $g^!(a) \notin r(\widetilde{KU}(X))$, proving the proposition.

<u>Theorem 4.2.</u> *The manifold* $\tilde{M}$ *does not admit a stably complex structure.*

<u>Proof.</u> The (n-1)-sphere bundle $S(\xi) = (\tilde{M}, \tilde{\pi}, X)$ has a cross-section $s : X \to \tilde{M}$, since by hypothesis $\dim X < n - 1$. In view of (2.4) we obtain that

$$(4.3) \qquad s^*(\tau(\tilde{M}) \oplus \varepsilon) \cong g^*\alpha \ .$$

By Proposition 4.1, the bundle $g^*\alpha$ does not admit a stably complex structure and we conclude with (4.3) that the same holds for $\tau(\tilde{M})$.

CHAPTER 4

VECTOR FIELDS ON SPHERES

1. Introduction

Let S^{n-1} be the unit sphere in the euclidean n-space $\mathbb{R}^n$. A *(non-zero) vector field* on S^{n-1} is a continuous function

$$v : S^{n-1} \longrightarrow \mathbb{R}^n - \{0\}$$

which associates to each $x \in S^{n-1}$ a vector $v(x)$ orthogonal to x, that is to say, a vector "tangent" to S^{n-1} at x. For example, if n is even, say $n = 2m$, then

$$v(x) = (-x_2, x_1, -x_4, x_3, \ldots, -x_{2m}, x_{2m-1})$$

defines a non-zero vector field on S^{2m-1}. It was Poincaré who, in one of his celebrated papers inspired by what he called "le problème astronomique de la stabilité du système solaire" [197], first observed that the sphere S^2 does not have a vector field. Later on, Brouwer proved that the spheres S^{2n}, $n \geq 1$, do not have a vector field [55]. Actually, the non-existence of vector fields on even dimensional spheres is a very simple question from nowadays standpoint. In fact, suppose that a sphere S^{n-1} possesses a vector field v and let w be the normalized field given by

$$w(x) = \frac{1}{\|v(x)\|} v(x) ,$$

for every $x \in S^{n-1}$. Then,

$$h(t,x) = (\cos \pi t)x + (\sin \pi t)w(x) , \qquad 0 \leq t \leq 1 ,$$

defines a homotopy between the identity of S^{n-1} and the antipodal map. Since the degree of the former map is 1 and that of the latter is $(-1)^n$, one concludes that n must be even.

The natural question now is to determine which spheres

admit several vector fields that are linearly independent at each point. More generally, we shall study the following problem. Let $\mathbb{F}$ be one of the classical division algebras over $\mathbb{R}$, that is, $\mathbb{F} = \mathbb{R}$, $\mathbb{C}$ or $\mathbb{H}$, and let $S\,\mathbb{F}^n$ be the unit sphere of $\mathbb{F}^n$ with respect to the standard inner product (|). A non-zero $\mathbb{F}$ - *vector field* on $S\,\mathbb{F}^n$ is a map

$$v : S\,\mathbb{F}^n \longrightarrow \mathbb{F}^n - \{0\}$$

which assigns to each $x \in S\,\mathbb{F}^n$ a vector $v(x)$ with $(x|v(x)) = 0$. We say that r such fields $v^{(1)},\dots,v^{(r)}$ are *linearly independent* when, for every $x \in S\,\mathbb{F}^n$, the vectors

$$v^{(1)}(x),\dots,v^{(r)}(x)$$

are linearly independent *over* $\mathbb{F}$. The "vector field problem" (for the spheres) then reads as follows: *find the maximal number of linearly independent* $\mathbb{F}$ - *vector fields on the sphere* $S\,\mathbb{F}^n$.

There are two reasonable ways of attacking this problem. On the one hand, one might try to exhibit as many vector fields as possible on a given sphere. This, of course, should be done in the most simple way and is the sort of problem one tries to solve by giving explicit constructions using linear algebra methods. On the other hand, one might try to show that a given sphere does not admit a certain number of linearly independent vector fields. This is the kind of problem one usually approaches with Algebraic Topology methods. As it turned out, the vector field problem for the reals was solved precisely along the previous lines. The strongest positive result derives from the works of Hurwitz [133], Radon [200] and Eckmann [85] in Linear Algebra and Representation Theory of Finite Groups (the first two papers were already published in 1923). It may be stated as follows. Write $n = 2^{4d+c}(2a+1)$, where a,d,c are integers with $0 \le c \le 3$ and let $\rho(n)$ be the integer defined by

$$\rho(n) = 2^c + 8d \ ;$$

then, the sphere S^{n-1} admits $\rho(n) - 1$ linearly independent vector fields. The negative part of the vector field problem over $\mathbb{R}$ was solved by Adams who, inventing powerful methods in Topological K - theory, was able to prove that the number of vector fields given by Hurwitz and Radon cannot be improved. This result is contained in Adams' famous paper "Vector Fields on Spheres" [2], published about 40 years after the papers of Hurwitz and Radon.

The situation is quite different in the complex and quaternionic cases. Indeed, no explicit construction giving two or more linearly independent vector fields on $S\mathbb{C}^n$ is known, up to the present moment. As for the quaternions, one does not even have a single non-zero vector field on $S\mathbb{H}^n$ which is given by a formula "qui peut être montrée en public". Nevertheless, it is a remarkable fact that the vector field problem for $\mathbb{C}$ and $\mathbb{H}$ could be "solved" by the following result.

The sphere $S\mathbb{C}^n$ (resp. $S\mathbb{H}^n$) admits $k \geq 1$ linearly independent complex (resp. quaternionic) vector fields if, and only if, n is a multiple of an explicitly given integer b_k (resp. c_k), called the kth complex (resp. quaternionic) *James Number* (see Theorems 5.9 and 6.6).(*) It should be noted that these numbers can be explicitly computed.

We now enlarge upon the historical remarks already made, to give a short history of the complete solution of the Vector Field Problem. The first important result after those obtained by Poincaré, Brouwer, Hurwitz and Radon was the following Theorem of Alexandroff and Hopf: "A differentiable manifold M has a (continuous) vector field if, and only if, its Euler characteristic is zero" [10]. As a trivial consequence, we note that all odd dimensional differentiable manifolds have a vector field. Actually, Alexandroff-Hopf's Theorem is contained in the results of Stiefel [233], who had addressed himself to the more general question of finding the maximal number of linearly independent vector fields on a differentiable manifold; it is on that work that Stiefel introduced his widely known class of differentiable manifolds, called the *Stiefel manifolds*.

The next key result was obtained by Eckmann [83] (and independently by G.Whitehead [250]), who proved that on

(*) Our indexing of these numbers differs by a shift of 1 from that one adopted in [139], [12] and [8] .

a sphere S^{4n+1} there is no pair of linearly independent vector fields. The importance of this paper lies not only in the fact that it exhibited a class of spheres for which the number of linearly independent vector fields would not go over $\rho(n) - 1$ (actually, it should be noted that Eckmann had shown before that S^5 cannot have two linearly independent vector fields [82]), but also, because the powerful ideas and techniques of homotopy groups were used for the first time. In this same order of ideas, the next important theorem we should quote is that announced by Steenrod and J.H.C.Whitehead in 1951 [232] : "Let n be an integer and let 2^k be the highest power of 2 dividing $n + 1$; then S^n does not have 2^k linearly independent vector fields." Their proof relies on cohomology properties of the real projective spaces (Steenrod Operations). Around that time there was also a lot of work done on the question of parallelizability of the spheres; the reader is invited to consult Chapter 1 for the relevant results.

The next generation of results came about with the papers of Ioan James on Stiefel manifolds [138], [139] and [140]. In them James proved that for each positive integer k, the sphere $S\mathbb{F}^n$, with $\mathbb{F} = \mathbb{C}$ or $\mathbb{H}$, admits k linearly independent vector fields if, and only if, n is a multiple of a certain number b_k (if $\mathbb{F} = \mathbb{C}$) or c_k (if $\mathbb{F} = \mathbb{H}$), without being able to determine explicitly these numbers. Making use of his notions of S - reducibility and S - coreducibility, he then reduced the vector field problem to a problem concerning truncated projective spaces. Later on, Atiyah proved (see [12]) that the numbers b_k and c_k are the J - orders of the canonical line bundle over the projective space $\mathbb{F}P^k$. Next, Atiyah and Todd showed [37], with the aid of complex K - theory, that if $S\mathbb{C}^n$ has k linearly independent vector fields, then n must be a multiple of an explicitly determined number m_k. This condition was shown to be sufficient by Adams and Walker in 1965 [8], who proved that $b_k = m_k$. In the meantime, as we observed before, Adams had solved the vector field problem for $S\mathbb{R}^n$ using real and complex K - theory, together with James' result on S - coreducibility of truncated projective spaces.

Finally, in 1973, Sigrist and Suter published a paper in which they solved the vector field problem for the quaternionic sphere $S\mathbb{H}^n$ [219]. The method used was similar to that of [8], but it required the use of the "Adams Conjecture", then already proved by Quillen.

The Chapter is organized as follows. In § 2 we first formulate the vector field problem in terms of cross-sections of Stiefel fibrations. We then relate the cross-section problem to a question concerning the J - order of the canonical Hopf bundle over an appropriate projective space. In § 3 we compute the KO and KU - theory of the projective spaces. The vector field problem for the reals is treated in section 4. We follow the method of [49] and also sketch a proof of the Hurwitz - Radon - Eckmann Theorem. In § 5 we solve the cross-section problem for complex Stiefel fibrations following and idea of K.Lam [160] and working primarily with the functor J' of [4]. Section 6 is devoted to the quaternionic case. We close the chapter with a comparison between the complex and quaternionic cases.

2. Vector Fields and Sphere Bundles over Projective Spaces.

In this section we first interpret vector fields on spheres as cross-sections of the so called Stiefel Fibrings. Following Woodward [261] and James [141], [143] we then relate such cross-sections to homotopy properties of certain sphere bundles over a projective space $\mathbb{F}P^{k-1}$, reducing the vector field problem to the computation of the J - order of the canonical line bundle over $\mathbb{F}P^{k-1}$.

(2.1) Let $\mathbb{F}$ be one of the classical real division algebras $\mathbb{R}, \mathbb{C}$ or $\mathbb{H}$, with the usual norm and conjugation operation. Let $\mathbb{F}^n$ be the euclidean n - space corresponding to $\mathbb{F}$. We regard $\mathbb{F}^n$ as a right $\mathbb{F}$ - module and consider the standard inner product $(\ |\ )$ so that if $x = (x_1,\dots,x_n)$ and $y = (y_1,\dots,y_n)$ are elements of $\mathbb{F}^n$, then

$$(x|y) = \bar{x}_1 y_1 + \dots + \bar{x}_n y_n\ ,\ \|x\| = (\bar{x}_1 x_1 + \dots + \bar{x}_n x_n)^{1/2}.$$

An ordered set $\{x^{(1)},\dots,x^{(k)}\}$ of k vectors in

$\mathbb{F}^n$ is called an *orthonormal* k-*frame* if $(x^{(i)}|x^{(j)}) = \delta_{ij}$ $(i,j=1,\dots,k)$. Such a frame can be identified to the $n \times k$ - matrix having the orthonormal columns $x^{(1)},\dots,x^{(k)}$, moreover, with the linear norm preserving map $\mathbb{F}^k \to \mathbb{F}^n$ given by $(y_1,\dots,y_k) \to x^{(1)}y_1 + \dots + x^{(k)}y_k$. The set of all orthonormal k - frames in $\mathbb{F}^n$,

$$\mathbb{F}_{n,k} = \left\{\{x^{(1)},\dots,x^{(k)}\} | x^{(i)} \in \mathbb{F}^n, (x^{(i)}|x^{(j)}) = \delta_{ij}, i,j=1,\dots,k\right\}$$

$(n \geq k)$, corresponds to a subset of the space $\mathbb{F}^n \times \dots \times \mathbb{F}^n$ (k factors) and hence, it inherits a canonical topology. Actually, $\mathbb{F}_{n,k}$ is a smooth (real) manifold, the so called *Stiefel Manifold*. We shall frequently identify $\mathbb{F}_{n,k}$ with the space of all $\mathbb{F}$ - linear norm preserving maps from $\mathbb{F}^k$ to $\mathbb{F}^n$.

The space $\mathbb{F}_{n,1}$ coincides with the sphere $S\mathbb{F}^n$ formed by the vectors of norm 1 in $\mathbb{F}^n$ ($S\mathbb{F}^n$ is a sphere of dimension $dn - 1$, where $d = \dim_{\mathbb{R}}(\mathbb{F})$). The projection map

$$p : \mathbb{F}_{n,k} \longrightarrow S\mathbb{F}^n$$

defined by $p(\{x^{(1)},\dots,x^{(k)}\}) = x^{(1)}$, is a fibration with fibre $\mathbb{F}_{n-1,k-1}$ (see 231; § 7.8). Clearly, a cross-section of p gives rise to $k - 1$ orthonormal $\mathbb{F}$ - vector fields on the sphere $S\mathbb{F}^n$; introducing some notation, if s is a cross-section of p and $x \in S\mathbb{F}^n$, $s(x)$ is an orthonormal k - frame $\{x,s^1(x),\dots,s^{k-1}(x)\}$ and the vectors $s^1(x),\dots,s^{k-1}(x)$ are tangent to $S\mathbb{F}^n$ and perpendicular to each other. Conversely, suppose that we are given $k - 1$ linearly independent $\mathbb{F}$ - vector fields on $S\mathbb{F}^n$. Then the Gram-Schmidt orthogonalization process provides us with $k - 1$ orthonormal vector fields $v^1(x),\dots,v^{k-1}(x)$ (the process is continuous) and one gets a cross-section of p defined by $s(x) = \{x,v^1(x),\dots,v^{k-1}(x)\}$. Hence:

(2.2) *The sphere* $S\mathbb{F}^n$ *admits* $k - 1$ *linearly independent* $\mathbb{F}$-*vector fields if, and only if, the Stiefel fibration* $p : \mathbb{F}_{n,k} \to S\mathbb{F}^n$ *has a cross-section.*

The vector field problem formulated in the introduction is therefore equivalent to the following problem: *for what pairs* (n,k) *does the fibration* $\mathbb{F}_{n,k} \to S\mathbb{F}^n$ *admit a cross-*

section?

(2.3) The sphere $S\mathbb{F}$ is a Lie subgroup of the multiplicative group $\mathbb{F}^* = \mathbb{F} - \{0\}$ (that is, $S\mathbb{R} = S^0$, $S\mathbb{C} = S^1$ and $S\mathbb{H} = S^3$). We consider the canonical right action of $S\mathbb{F}$ on the sphere $S\mathbb{F}^m$; factoring out this action, one obtains the projective space $\mathbb{F}P^{m-1}$ associated to $\mathbb{F}^m$. Let ξ be the canonical $\mathbb{F}$-line bundle associated to the principal $S\mathbb{F}$-bundle $S\mathbb{F}^m \to \mathbb{F}P^{m-1}$. The total space E_ξ of this line bundle is given by

(2.4) $$E_\xi = (S\mathbb{F}^m \times \mathbb{F})/(v,y) \sim (vz,yz) \ , \qquad (z \in S\mathbb{F}) \ .$$

It follows that $S(n\xi)$, the sphere bundle associated to the vector bundle $n\xi$, is obtained by factoring out the diagonal action of $S\mathbb{F}$ in $S\mathbb{F}^m \times S\mathbb{F}^n$, i.e.,

$$S(n\xi) = (S\mathbb{F}^m \times S\mathbb{F}^n)/(v,x) \sim (vz,xz)$$

$(v \in S\mathbb{F}^m,\ x \in S\mathbb{F}^n,\ z \in S\mathbb{F})$.

Consider now a cross-section $s : S\mathbb{F}^n \to \mathbb{F}_{n,k}$. We view the element $s(x)$ as a norm-preserving $\mathbb{F}$- linear transformation $s_x : \mathbb{F}^k \to \mathbb{F}^n$ given by $s_x(v_1,\ldots,v_k) = x \cdot v_1 + \sum_{i=1}^{k-1} s^i(x) \cdot v_{i+1}$, for every k - tuple $(v_1,\ldots,v_k) \in \mathbb{F}^k$. Hence, we get a map $s_x : S\mathbb{F}^k \to S\mathbb{F}^n$ compatible with the $S\mathbb{F}$-action, i.e.,

$$s_x(v \cdot z) = s_x(v) \cdot z \ ,$$

for every $v \in S\mathbb{F}^k$ and every $z \in S\mathbb{F}$. Clearly, this map depends continuously on x. The self-map

$$\sigma : S\mathbb{F}^k \times S\mathbb{F}^n \longrightarrow S\mathbb{F}^k \times S\mathbb{F}^n$$

defined by $\sigma(v,x) = (v,s_x(v))$, is continuous and has the following properties:

(i) $\sigma(v \cdot z,x) = \sigma(v,x) \cdot z$, for every $v \in S\mathbb{F}^k$, $x \in S\mathbb{F}^n$ and $z \in S\mathbb{F}$;

(ii) $\sigma(e^{(1)},x) = (e^{(1)},x)$, where $e^{(1)} = (1,0,\ldots,0)$ and $x \in S\mathbb{F}^n$.

Because of (i), σ gives rise to a map $\overline{\sigma}$ making the following diagram commutative:

$$\begin{array}{ccc} (S\mathbb{F}^k/(v\sim vz))\times S\mathbb{F}^n & \xrightarrow{\ \overline{\sigma}\ } & (S\mathbb{F}^k\times S\mathbb{F}^n)/(v,x)\sim(vz,xz) \\ & \searrow \quad \swarrow & \\ & S\mathbb{F}^k/v\sim vz = \mathbb{F}P^{k-1} & \end{array}$$

that is to say, $\overline{\sigma}$ is a map between sphere bundles over $\mathbb{F}P^{k-1}$, $\overline{\sigma} : S(n\varepsilon) \to S(n\xi)$ (here ε is the trivial $\mathbb{F}$ - line bundle). On the other hand, (ii) shows that $\overline{\sigma}$ is the identity on the fibre over $[e^{(1)}] \in \mathbb{F}P^{k-1}$ and so, it follows from a well-known theorem of Dold [77] that $\overline{\sigma}$ is a fibre homotopy equivalence.

At this point we introduce a notation which will be used in the remainder of the Chapter: if α is a real vector bundle over X, we denote by $\tilde{J}(\alpha)$ the element of $\tilde{J}(X)$ represented by the stable class of α, that is to say, $\tilde{J}(\alpha) = \tilde{J}(x)$, where $x = [\alpha] - \dim \alpha \in \widetilde{KO}(X)$.

With this and the preceeding observations, we can state the following

<u>Theorem 2.5.</u> *If the fibration* $\mathbb{F}_{n,k} \to S\mathbb{F}^n$ *has a cross-section, then the sphere bundle* $S(n\xi)$ *over* $\mathbb{F}P^{k-1}$ *is fibre homotopically trivial. In particular,* $n \cdot \tilde{J}(\xi) = 0$ *in the group* $\tilde{J}(\mathbb{F}P^{k-1})$, *that is to say,* n *is a multiple of the* $\tilde{J}$ *- order of* ξ.

The next Theorem is a partial converse of (2.5) and will be proved in (2.12).

<u>Theorem 2.6.</u> *Suppose that* $n > 2k$. *If the sphere bundle* $S(n\xi)$ *over* $\mathbb{F}P^{k-1}$ *has stably trivial fibre homotopy type, that is to say, if* n *is a multiple of the order of* $\tilde{J}(\xi) \in \tilde{J}(\mathbb{F}P^{k-1})$, *then the fibration* $\mathbb{F}_{n,k} \to S\mathbb{F}^n$ *has a cross-section.*

(2.7) In an attempt to prove Theorem 2.6., we try to reverse the considerations preceeding (2.5). Let $\overline{\sigma}$ be a fibre homotopy equivalence between $S(n\varepsilon)$ and $S(n\xi)$. It follows from elementary principles in bundle theory that $\overline{\sigma}$ can be lifted to a map σ over $S\mathbb{F}^k$, that is, to a map

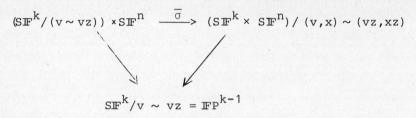

$$\begin{aligned} \sigma : S\mathbb{F}^k \times S\mathbb{F}^n &\longrightarrow S\mathbb{F}^k \times S\mathbb{F}^n \\ (v,x) &\longmapsto (v,\sigma_1(v,x)), \end{aligned}$$

which is compatible with the appropriate $S\mathbb{F}$ - action, i.e., such that, for every $v \in S\mathbb{F}^k$, $x \in S\mathbb{F}^n$ and $z \in S\mathbb{F}$, $\sigma(v \cdot z,x) = \sigma(v,x) \cdot z$. Now we focus on the map

$$\sigma_1 : S\mathbb{F}^k \times S\mathbb{F}^n \longrightarrow S\mathbb{F}^n$$

obtained by composing σ with the projection onto the second factor. The map

$$\begin{aligned} \sigma_1(\ ,x) : S\mathbb{F}^k &\longrightarrow S\mathbb{F}^n \\ v &\longmapsto \sigma_1(v,x) , \end{aligned}$$

in general, will not be the restriction of a linear, norm preserving map from $\mathbb{F}^k$ into $\mathbb{F}^n$, but is compatible with the $S\mathbb{F}$-action on the two spheres, i.e., $\sigma_1(v \cdot z,x) = \sigma_1(v,x) \cdot z$. The adjoint of σ_1 provides, therefore, a continuous map

$$s : S\mathbb{F}^n \longrightarrow X_{n,k}(\mathbb{F})$$

defined by $s(x) = \sigma_1(\ ,x)$, for every $x \in S\mathbb{F}^n$, where $X_{n,k}(\mathbb{F})$ is the *space of all* $S\mathbb{F}$ - *equivariant maps from* $S\mathbb{F}^k$ *into* $S\mathbb{F}^n$, endowed with the compact-open topology. We choose $e^{(1)} = (1,0,\dots,0)$ as base point of $S\mathbb{F}^k$. Evaluation at $e^{(1)}$ defines a continuous map

$$e : X_{n,k}(\mathbb{F}) \longrightarrow S\mathbb{F}^n ;$$

we shall prove in (2.11) that e is a Hurewicz fibration. Observe that the map

$$\sigma_1(e^{(1)}, \ ) : S\mathbb{F}^n \longrightarrow S\mathbb{F}^n$$

has degree one, since we may presuppose that, the fibre homotopy equivalence $\bar{\sigma}$ restricted to fibres is of degree one. Hence, the composition $e \circ s$ is homotopic to the identity map $1_{S\mathbb{F}^n}$, and because e is a Hurewicz fibration (as we shall see), s can be deformed to a cross-section of e. So far, we have proved the following result.

<u>Proposition 2.8.</u> *If the sphere bundle* $S(n\xi)$ *over* $\mathbb{F}P^{k-1}$ *is fibre homotopically trivial, the projection* $e : X_{n,k}(\mathbb{F}) \to S\mathbb{F}^n$

admits a cross-section.

(2.9) The Stiefel manifold $\mathbb{F}_{n,k}$ may be identified with a subspace of $X_{n,k}(\mathbb{F})$, namely, with the subspace of those $S\mathbb{F}$-equivariant maps that are given by linear norm preserving maps from $\mathbb{F}^k$ into $\mathbb{F}^n$. We have a commutative diagram

$$\begin{array}{ccc} \mathbb{F}_{n,k} & \xrightarrow{\quad i_k \quad} & X_{n,k}(\mathbb{F}) \\ & {\scriptstyle p}\searrow \quad \swarrow{\scriptstyle e} & \\ & S\mathbb{F}^n & \end{array} \tag{2.10}$$

The following Lemma implies that a cross-section of e can be deformed into a cross-section of p, provided n is sufficiently large when compared to k (see [102] and [242]).

Lemma 2.11. *Let* d *be the dimension of* $\mathbb{F}$ *as a real vector space. Then, the homomorphism*

$$(i_k)_* : \pi_j(\mathbb{F}_{n,k}) \longrightarrow \pi_j(X_{n,k}(\mathbb{F}))$$

induced by the inclusion $i_k : \mathbb{F}_{n,k} \to X_{n,k}(\mathbb{F})$ *is an isomorphism if* $j < 2d(n-k+1)-3$.

Proof. Choose the standard inclusion $f_k : S\mathbb{F}^k \to S\mathbb{F}^n$, which takes any k-tuple $(x_1,\dots,x_k)$ into $(x_1,\dots,x_k, 0,\dots,0)$ as the base-point of $F_{n,k}$ and $X_{n,k}(\mathbb{F})$.

We prove the Lemma by induction on k. If $k = 1$, we have $\mathbb{F}_{n,1} = X_{n,1}(\mathbb{F}) = S\mathbb{F}^n$, and the result is true. Let then $k > 1$ and assume by induction that $(i_{k-1})_*$ is an isomorphism for $j < 2d(n-k+2)-3$. Let us identify $S\mathbb{F}^{k-1}$ with the subspace $\{x = (x_1,\dots,x_{k-1},0) \mid \|x\| = 1\}$ of $S\mathbb{F}^k$. Restriction of linear and $S\mathbb{F}$-equivariant maps $S\mathbb{F}^k \to S\mathbb{F}^n$ to $S\mathbb{F}^{k-1}$ defines maps

$$q : \mathbb{F}_{n,k} \longrightarrow \mathbb{F}_{n,k-1}$$

and

$$\bar{q} : X_{n,k}(\mathbb{F}) \longrightarrow X_{n,k-1}(\mathbb{F}) ,$$

respectively. Notice that $\bar{q} \circ i_k = i_{k-1} \circ q$. It is well known that q is the projection of a fibre bundle with fibre homeomorphic to $S\mathbb{F}^{n-k+1}$ (see [231; § 7.8]). Furthermore, we have the following two facts:

(A) the map $\bar{q}$ is a Hurewicz fibration with fibres homotopically equivalent to the iterated loop space $\Omega^{d(k-1)}S\mathbb{F}^n$;

(B) let $j_k = i_k|q^{-1}(f_{k-1})$ be the inclusion of the fibre $S\mathbb{F}^{n-k+1}$ into the fibre $Y_{n,k} = \bar{q}^{-1}(f_{k-1}) \sim \Omega^{d(k-1)}S\mathbb{F}^n$. Then the induced homomorphism

$$(j_k)_* : \pi_j(S\mathbb{F}^{n-k+1}) \longrightarrow \pi_j(Y_{n,k}) \simeq \pi_{j+d(k-1)}(S\mathbb{F}^n)$$

is the $d(k-1)$ - fold suspension.

Statements (A) and (B) will be proved shortly.

Consider now the following homomorphism of exact homotopy sequences:

$$\begin{array}{ccccccccc}
\to \pi_{j+1}(\mathbb{F}_{n,k-1}) & \longrightarrow & \pi_j(S\mathbb{F}^{n-k+1}) & \longrightarrow & \pi_j(\mathbb{F}_{n,k}) & \xrightarrow{q_*} & \pi_j(\mathbb{F}_{n,k-1}) & \longrightarrow & \cdots \\
\downarrow (i_{k-1})_* & & \downarrow (j_k)_* & & \downarrow (i_k)_* & & \downarrow (i_{k-1})_* & & \\
\to \pi_{j+1}(X_{n,k-1}(\mathbb{F})) & \longrightarrow & \pi_j(Y_{n,k}) & \longrightarrow & \pi_j(X_{n,k}(\mathbb{F})) & \xrightarrow{\bar{q}_*} & \pi_j(X_{n,k-1}(\mathbb{F})) & \to & \cdots .
\end{array}$$

Assertion (B) and the Freudenthal Suspension Theorem show that the above suspension is an isomorphism for $j < 2d(n-k+1)-3$; it follows inductively that $(i_k)_*$ is an isomorphism in the relevant range.

For the sake of completeness we give next the proofs of (A) and (B).

<u>Proof of (A).</u> Consider the subspace

$$E^{k-1} = \left\{(x_1,\dots,x_{k-1},(1-\|x\|^2)^{1/2}) \,\middle|\, \sum_{i=1}^{k-1} \bar{x}_i x_i = \|x\|^2 = 1\right\}$$

of $S\mathbb{F}^k$. This subspace is homeomorphic to a $d(k-1)$ - ball and has the canonically embedded sphere $S\mathbb{F}^{k-1}$ as its boundary. The following two facts are readily checked:

(A_1) - any $S\mathbb{F}$ - equivariant map defined on $S\mathbb{F}^k$ is uniquely determined by its restriction to the subspace E^{k-1};

(A_2) - any continuous map $E^{k-1} \to S\mathbb{F}^n$ which is $S\mathbb{F}$ -

equivariant on the boundary $\partial E^{k-1} = S\mathbb{F}^{k-1}$ can be extended continuously (and uniquely) to an $S\mathbb{F}$-equivariant map $S\mathbb{F}^k \to S\mathbb{F}^n$.

Observe that $S\mathbb{F}^{k-1} = \partial E^{k-1} \longrightarrow E^{k-1}$ is a cofibration and that both spaces are compact; hence, by [225; Theorem 2.8.2], the map

$$\phi : \mathrm{Map}(E^{k-1}, S\mathbb{F}^n) \longrightarrow \mathrm{Map}(S\mathbb{F}^{k-1}, S\mathbb{F}^n)$$

defined by $\phi(f : E^{k-1} \to S\mathbb{F}^n) = f|S\mathbb{F}^{k-1}$ is a Hurewicz fibration. By Dold's Theorem, the fibres are of the homotopy type of $\phi^{-1}(f_o) \simeq \Omega^{d(k-1)}S\mathbb{F}^n$, where $f_o \in \mathrm{Map}(S\mathbb{F}^{k-1}, S\mathbb{F}^n)$ is the constant map onto $(0,\ldots,0,1) \in S\mathbb{F}^n$. Now form the commutative diagram

$$\begin{array}{ccc} X_{n,k}(\mathbb{F}) & \xrightarrow{\ j\ } & \mathrm{Map}(E^{k-1}, S\mathbb{F}^n) \\ \Big\downarrow \bar{q} & & \Big\downarrow \phi \\ X_{n,k-1}(\mathbb{F}) & \hookrightarrow & \mathrm{Map}(S\mathbb{F}^{k-1}, S\mathbb{F}^n) \end{array}$$

where j is defined by restriction on E^{k-1}; as the reader can easily check, it turns out that this is a pull-back and hence, claim (A) is verified.

Remark. The statement that

$$e : X_{n,k}(\mathbb{F}) \longrightarrow S\mathbb{F}^n = X_{n,1}(\mathbb{F})$$

is a Hurewicz fibration, made earlier, follows trivially from (A).

Proof of (B). Direct inspection shows that the inclusion

$$j_k : S\mathbb{F}^{n-k+1} \longrightarrow Y_{n,k} = \phi^{-1}(f_{k-1})$$

is the adjoint of the map

$$g : B\mathbb{F}^{k-1} \times S\mathbb{F}^{n-k+1} \longrightarrow S\mathbb{F}^n$$

given by $g(x,y) = (x_1,\ldots,x_{k-1},(1-\|x\|^2)^{1/2}\cdot y_1,\ldots$ $\ldots,(1-\|x\|^2)^{1/2}\cdot y_{n-k+1})$, for every $x \in B\mathbb{F}^{k-1}$ and $y \in S\mathbb{F}^{n-k+1}$; here

$B\mathbb{F}^{k-1}$ is the closed unit ball of $\mathbb{F}^{k-1}$. Write $(x_1,\ldots,x_{k-1}, u_1,\ldots,u_{n-k+1}) = (x,u) \in S\mathbb{F}^n$ and consider the following homotopy of the identity map $1_{S\mathbb{F}^n}$:

$$h_t(x,u) = \frac{[(1-t\|x\|)(x,u) + (0,\ldots,0,t)]}{\|(1-t\|x\|)(x,u) + (0,\ldots,0,t)\|} .$$

The map h_1 gives rise to a map from $Y_{n,k}$ into $\phi^{-1}(f_o) \cong \Omega^{d(k-1)}S\mathbb{F}^n$, which in turn, induces an isomorphism between the corresponding homotopy groups. The reader can now check that the composition

$$h_1 \circ j_k : SF^{n-k+1} \longrightarrow Y_{n,k} \longrightarrow \Omega^{d(k-1)}SF^n$$

is the adjoint to the map

$$BF^{k-1}/SF^{k-1} \times SF^{n-k+1} \longrightarrow SF^n$$

which sends an arbitrary element $(x,y) \in BF^{k-1}/SF^{k-1} \times SF^{n-k+1}$ into

$$\frac{[(1-\|x\|)(x,(1-\|x\|^2)^{1/2}\cdot y) + (0,\ldots,0,1)]}{\|(1-\|x\|)(x,(1-\|x\|^2)^{1/2}\cdot y) + (0,\ldots,0,1)\|} .$$

The latter map is readily seen to be homotopic to the canonical map

$$BF^{k-1}/SF^{k-1} \times SF^{n-k+1} \longrightarrow BF^{k-1}/SF^{k-1} \wedge SF^{n-k+1} \cong SF^n ,$$

whose adjoint is the $d(k-1)$ - fold suspension [225; § 8.5]. It follows that $h_1 \circ j_k$ is homotopic to this suspension; this concludes the proof of (B).

(2.12) <u>Proof of Theorem 2.6.</u> Suppose that $n > 2k$ and that $n\tilde{J}(\xi) = 0$ in $\tilde{J}(\mathbb{F}P^{k-1})$. Since the dimension of the CW-complex $\mathbb{F}P^{k-1}$ is smaller than the dimension of the fibre of the sphere bundle $S(n\xi)$, it follows that $S(n\xi)$ is fibre homotopically trivial (*). Hence (2.8) implies that the fi-

---

(*) According to Atiyah [12], in the stable range $(m-2 \geq \dim X)$ we may identify $\tilde{J}(X)$ with the set of fibre homotopy types of orthogonal S^{m-1} - bundles over X .

bration $e : X_{n,k}(F) \to SF^n$ admits a cross-section s . Since $n > 2k$, it follows from (2.11 that $(i_k)_* : \pi_{dn-1}(\mathbb{F}_{n,k}) \cong \pi_{dn-1}(X_{n,k}(\mathbb{F}))$ and so, there exists $s' : S\mathbb{F}^n \to \mathbb{F}_{n,k}$ such that $(i_k)_*[s'] = [s]$. Now (2.10) shows that s' is a cross-section of p.

Remark 2.13. In sections 4 and 5 we shall prove that if $n\tilde{J}(\xi) = 0$ in $\tilde{J}(\mathbb{F}P^{k-1})$, then $n > 2k$, except for a small number of cases (see proofs of (4.2), (5.11) and (6.7)). From (2.5), (2.6), (4.2), (5.11) and (6.7) we shall in fact conclude that *the fibration* $\mathbb{F}_{n,k} \to S\mathbb{F}^n$ *has a cross-section if, and only if,* n *is a multiple of the order of the element* $\tilde{J}(\xi)$ *in the group* $\tilde{J}(\mathbb{F}P^{k-1})$.

3. The K - Theory of the projective Spaces.

In this section we compute the K - theory of the projective spaces $\mathbb{F}P^m$. The results will be used later to determine the $\tilde{J}$ - order of the Hopf line bundle $\xi = \xi_m(\mathbb{F})$ over $\mathbb{F}P^m$. The K - theory of the various projective spaces was first discussed in [2], [8], [37] and [202].

Proposition 3.1. *Consider* $v = [\xi_m(\mathbb{C})] - 1 \in \widetilde{KU}(\mathbb{C}P^m)$. *Then*

(i) the ring $KU(\mathbb{C}P^m)$ *is a truncated polynomial ring (over the integers) generated by* v *, i.e.,*

$$KU(\mathbb{C}P^m) \cong \mathbb{Z}[v]/\langle v^{m+1}\rangle ;$$

the group $KU^1(\mathbb{C}P^m)$ *is zero;*

(ii) the operations ψ^p *are given by*

$$\psi^p(v) = (1+v)^p - 1 .$$

Proof. The integral cohomology of $\mathbb{C}P^m$ is a truncated polynomial ring

$$H^*(\mathbb{C}P^m;\mathbb{Z}) \cong \mathbb{Z}[a]/\langle a^{m+1}\rangle$$

with generator $a \in H^2(\mathbb{C}P^m;\mathbb{Z})$. With (0.10) we infer that

$$H^*(\mathbb{C}P^m;\mathbb{Z}) = H^{even}(\mathbb{C}P^m;\mathbb{Z}) \cong \mathcal{G}KU(\mathbb{C}P^m).$$

The first Chern class of ξ_m being equal to a , the element $v = \xi_m - 1$ thus represents $a \in \mathcal{G}KU(\mathbb{C}P^m)$ (see (0.11)) and part (i) follows.

Since $\xi = \xi_m(\mathbb{C})$ is a line bundle, $\psi^p(\xi) = \xi^p$ (see (0.12)) and so, part (ii) holds.

We now turn to the real K - theory of $\mathbb{C}P^m$.

<u>Proposition 3.2.</u> *Let* $y = r[\xi_m(\mathbb{C})] - 2 \in \widetilde{KO}(CP^m)$. *Then,*

(i) $KO(\mathbb{C}P^m)$ *is a truncated polynomial ring (over the integers) generated by* y, *with the following relations:*

$$y^{t+1} = 0, \text{ if } \quad m = 2t \quad (t \geq 0)$$

$$2y^{2s+1} = 0,\ y^{2s+2} = 0, \text{ if } \quad m = 4s+1 \quad (s \geq 0)$$

$$y^{2s+2} = 0, \text{ if } \quad m = 4s+3 \quad (s \geq 0)\ ;$$

(ii) *the complexification* $c : KO(\mathbb{C}P^m) \to KU(\mathbb{C}P^m)$, *given by* $c(y) = v + \overline{v} = \xi_m(\mathbb{C}) + \xi_m^{-1}(\mathbb{C}) - 2$, *is a monomorphism if* $m \not\equiv 1$ *(mod. 4);*

(iii) *the operations* ψ^p *are given by*

$$\psi^p(y) = T_p(y)\ ,$$

where T_p *is the unique polynomial of degree* p *with integral coefficients, such that*

$$T_p(z+z^{-1}-2) = z^p + z^{-p} - 2\ .$$

(3.3) In order to prove (3.2) we shall first determine the K - theory of the spaces

$$Y_t = \mathbb{C}P^{2t}/\mathbb{C}P^{2t-2} = S^{4t-2} \cup_\eta e^{4t}\ ,$$

where η is the non-trivial element of $\pi_2^S = \mathbb{Z}_2$. Working with the exact sequence associated to the cofibration

$$\mathbb{C}P^{2t-2} \xrightarrow{\ i\ } \mathbb{C}P^{2t} \xrightarrow{\ q\ } Y$$

and using (3.1), we infer that $\widetilde{KU}(Y_t)$ is free abelian with two generators a_{2t-1} and a_{2t}, determined by $q^!(a_j) = v^j$, $j = 2t - 1, 2t$. The homomorphism $q^!$ is indeed a monomorphism, since $KU^1(\mathbb{C}P^{2t-2}) = 0$ (see (3.1)); furthermore, the complex conjugation on KU coincides with the operation ψ^{-1}. These remarks, plus (3.1) (ii), show that

$$(3.4)\qquad \begin{aligned} \bar{a}_{2t-1} &= -a_{2t-1} + (2t-1)a_{2t} \\ \bar{a}_{2t} &= a_{2t}. \end{aligned}$$

The following Lemma gives the KO - Theory of Y_t.

<u>Lemma 3.5</u> (i) *The group* $\widetilde{KO}(Y_t)$ *is isomorphic to* $\mathbb{Z}$ *and is generated by*

$$a = r\left(-a_{2t-1} + ta_{2t}\right);$$

(ii) *the complexification* $c : \widetilde{KO}(Y_t) \to \widetilde{KU}(Y_t)$ *is injective and is given by*

$$c(a) = a_{2t}.$$

<u>Proof.</u> Since $cr(a_j) = a_j + \bar{a}_j$ we obtain from (3.4) that $c(a) = a_{2t}$. Hence, a generates an infinite cyclic summand in $\widetilde{KO}(Y_t)$. It remains to show that this is all of $\widetilde{KO}(Y_t)$.

We consider the following exact sequence associated to $S^{4t-2} \xrightarrow[i]{} Y_t \xrightarrow[j]{} S^{4t}$:

$$\begin{array}{ccccccccc} \widetilde{KO}^{-1}(S^{4t-2}) & \longrightarrow & \widetilde{KO}(S^{4t}) & \longrightarrow & \widetilde{KO}(Y_t) & \longrightarrow & \widetilde{KO}(S^{4t-2}) & \longrightarrow & \widetilde{KO}^{1}(S^{4t}) \\ \| & & \cong & & & & \cong & & \| \\ 0 & & \mathbb{Z} & & & & 0 \text{ or } \mathbb{Z}_2 & & 0\,. \end{array}$$

If $t \equiv 0 \pmod{2}$ we have $\widetilde{KO}(S^{4t-2}) = 0$, hence

$$\widetilde{KO}(Y_t) \cong \widetilde{KO}(S^{4t}) \cong \mathbb{Z}\,.$$

If $t \equiv 1 \pmod{2}$, then $\widetilde{KO}(S^{4t-2}) \cong \mathbb{Z}_2$; in this case, we map the previous sequence by complexification into the corresponding KU - theory sequence and obtain a commutative diagram with exact rows

$$\begin{array}{ccccccccc} 0 & \longrightarrow & \mathbb{Z} & \xrightarrow{j^!_{\mathbb{R}}} & \widetilde{KO}(Y_t) & \xrightarrow{i^!_{\mathbb{R}}} & \mathbb{Z}_2 & \longrightarrow & 0 \\ & & \downarrow{\scriptstyle c_{S^{4t}}} & & \downarrow{\scriptstyle c_{Y_t}} & & \downarrow{\scriptstyle c_{S^{4t-2}}} & & \\ 0 & \longrightarrow & \mathbb{Z} & \xrightarrow[j^!_{\mathbb{C}}]{} & \widetilde{KU}(Y_t) & \xrightarrow[i^!_{\mathbb{C}}]{} & \mathbb{Z} & \longrightarrow & 0\,. \end{array}$$

Since $c_{S^{4t-2}}$ is trivial, the image of c_{Y_t} is contained in $\ker i^!_{\mathbb{C}}$, which is generated by $a_{2t} = c_{Y_t}(a)$. Hence,

$$\text{image } c_{Y_t} = \ker i^!_{\mathbb{C}} = \text{image } j^!_{\mathbb{C}} \cong \mathbb{Z} .$$

Now, the homomorphism $c_{S^{4t}}$ is multiplication by 2 (see (0.6)) and we conclude that the upper row does not split. This implies that $\widetilde{KO}(Y_t) \cong \mathbb{Z}$, and so, (3.5) is proved.

Proof of Proposition 3.2. We first establish parts (i) and (ii) for m even, say, $m = 2t$; this is done by induction on t , starting with the trivial case $\mathbb{C}P^0$ = point.

Consider the following commutative diagram with exact rows, induced by

$$\mathbb{C}P^{2t-2} \xrightarrow[i]{} \mathbb{C}P^{2t} \xrightarrow[q]{} Y_t :$$

$$\begin{array}{ccccccccc}
 & & \widetilde{KO}(Y_t) & \xrightarrow[q^!_{\mathbb{R}}]{} & \widetilde{KO}(\mathbb{C}P^{2t}) & \xrightarrow[i^!_{\mathbb{R}}]{} & \widetilde{KO}(\mathbb{C}P^{2t-2}) & & \\
 & & \downarrow{\scriptstyle c'} & & \downarrow{\scriptstyle c} & & \downarrow{\scriptstyle c''} & & \\
0 & \longrightarrow & \widetilde{KU}(Y_t) & \xrightarrow[q^!_{\mathbb{C}}]{} & \widetilde{KU}(\mathbb{C}P^{2t}) & \xrightarrow[i^!_{\mathbb{C}}]{} & \widetilde{KU}(\mathbb{C}P^{2t-2}) & \longrightarrow & 0 .
\end{array}$$

Here, the vertical homomorphism are given by complexification. Since $i^*_{\mathbb{R}}\xi_{2t} = \xi_{2t-2}$, the map $i^!_{\mathbb{R}}$ is an epimorphism by induction (the element $y = r\xi_{2t-2} - 2$ is a generator of $\widetilde{KO}(\mathbb{C}P^{2t-2})$). On the other hand, using (3.5) (ii), we infer that $q^!_{\mathbb{R}}$ is injective. Hence, $\widetilde{KO}(\mathbb{C}P^{2t})$ is free abelian with base

$$\{y, y^2, \ldots, y^{t-1}, q^!_{\mathbb{R}}(a)\}$$

and therefore, c is injective. Observing that

$$c(y) = v + \overline{v} = v^2 + \text{higher terms},$$

$(c \circ r(\xi) = \xi + \overline{\xi})$ and because of (3.1), (3.3) and (3.5), we obtain that

$$c(y^t) = v^{2t} = q^!_{\mathbb{C}} \circ c'(a) = c \circ q^!_{\mathbb{R}}(a) .$$

We conclude that $q^!_{\mathbb{R}}(a) = y^t$ and that $y^{t+1} = 0$. So far, we have proved (i) and (ii) of (3.2) for m even.

Suppose now that $m = 4s + 3$ $(s \geq 0)$. Working with the cofibration

$$\mathbb{C}P^{4s+2} \xrightarrow[i]{} \mathbb{C}P^{4s+3} \longrightarrow S^{8s+6}$$

we get readily that

$$i^!_{\mathbb{R}} : \widetilde{KO}(\mathbb{C}P^{4s+3}) \cong \widetilde{KO}(\mathbb{C}P^{4s+2}) \cong Z[y]/\langle y^{2s+2}\rangle .$$

Finally, let $m = 4s + 1$ $(s \geq 0)$. The cofibration

$$\mathbb{C}P^{4s+1} \xrightarrow[i]{} \mathbb{C}P^{4s+2} \xrightarrow[j]{} S^{8s+4}$$

gives rise to the following commutative diagram with exact rows:

$$\begin{array}{ccccccc}
\mathbb{Z} \cong \widetilde{KO}(S^{8s+4}) & \xrightarrow{j^!_{\mathbb{R}}} & \widetilde{KO}(\mathbb{C}P^{4s+2}) & \xrightarrow{i^!_{\mathbb{R}}} & \widetilde{KO}(\mathbb{C}P^{4s+1}) & \longrightarrow & \widetilde{KO}^1(S^{8s+4}) = \\
\downarrow c' & & \downarrow c & & \downarrow c'' & & \downarrow \\
\mathbb{Z} \cong \widetilde{KU}(S^{8s+4}) & \xrightarrow[j^!_{\mathbb{C}}]{} & \widetilde{KU}(\mathbb{C}P^{4s+2}) & \xrightarrow[i^!_{\mathbb{C}}]{} & \widetilde{KU}(\mathbb{C}P^{4s+1}) & \longrightarrow & 0
\end{array}$$

(the vertical arrows are given by the complexification). The map c is injective and the map c' is just multiplication by 2 (see (0.6)); the homomorphism $j^!_{\mathbb{C}}$ maps a generator of $\widetilde{KU}(S^{8s+4})$ onto $v^{4s+2} = c(y^{2s+1})$. It follows that the image of $j^!_{\mathbb{R}}$ is generated by $2y^{2s+1}$. Since $i^!_{\mathbb{R}}$ is an epimorphism, we conclude that

$$KO(\mathbb{C}P^{4s+1}) \cong KO(\mathbb{C}P^{4s+2})/\langle 2y^{2s+1}\rangle \cong \mathbb{Z}[y]/\langle 2y^{2s+1}, y^{2s+2}\rangle .$$

Parts (i) and (ii) of (3.2) are now established for all cases.

In order to determine the ψ - operations on y, we again map $KO(\mathbb{C}P^m)$ into $KU(\mathbb{C}P^m)$ by complexification. We have

$$c(y) = c \circ r(\xi) - 2 = \xi + \xi^{-1} - 2 \in KU(\mathbb{C}P^m) ,$$

(here $\xi = \xi_m(\mathbb{C})$) and because complexification is a ψ - ring homomorphism,

$$c \circ \psi^p_R(y) = \psi^p_C \circ c(y) = \psi^p_C(\xi + \xi^{-1} - 2) = \xi^p + \xi^{-p} - 2$$

$$= T_p(\xi + \xi^{-1} - 2) = T_p(c(y)) = c \circ T_p(y).$$

If $m \not\equiv 1$ (mod.4), then c is injective and we infer that $\psi^p_{\mathbb{R}}(y) = T_p(y)$. For $m \equiv 1$ (mod.4) we get the same result via the epimorphism $KO(\mathbb{C}P^{4s+2}) \to KO(\mathbb{C}P^{4s+1})$ This completes the proof of (3.2).

In § 4 we shall make use of the following properties of the polynomials $T_p(x)$, that is, of $\psi^p(y)$.

<u>Lemma 3.6.</u> (i) $T_2(x) = 4x + x^2$;

(ii) *for any odd integer* $p = 2q + 1$,

$$T_p(x) = x \cdot \left\{ \sum_{j=0}^{q} \frac{2q+1}{2j+1} \binom{q+j}{2j} x^j \right\}^2 .$$

<u>Proof.</u> We shall prove formally only (ii), since the first part of the Lemma is easily verified. In the ring $\mathbb{Z}[z,z^{-1}]$ we compute

$$\begin{aligned} T_{2q+1}(z+z^{-1}-2) &= z^{2q+1} + z^{-2q-1} - 2 = (z^{2q+1}-1)^2 \, z^{-2q-1} \\ &= \left\{(z^{2q} + z^{2q-1} + \ldots + z+1)\, z^{-q}\right\}^2 (z-1)^2 \, z^{-1} = \\ &= \left\{ \sum_{k=-q}^{q} z^k \right\}^2 (z+z^{-1}-2) . \end{aligned}$$

Using the identity

$$\sum_{k=-q}^{q} z^k = (z+z^{-1}) \cdot \sum_{k=-q+1}^{q-1} z^k - \sum_{k=-q+2}^{q-2} z^k , \quad q \geq 2 ,$$

one proves inductively that there exists a unique polynomial $S_q(x)$ of degree q , with integral coefficients, such that

$$\sum_{k=-q}^{q} z^k = S_q(z+z^{-1}-2) , \quad q = 0,1,\ldots .$$

(In fact, one has the recursion formula

$$S_q(x) = (x+2)\, S_{q-1}(x) - S_{q-2}(x), \quad q \geq 2 ,$$

with

$$S_0(x) = 1 , \; S_1(x) = 3 + x ;$$

notice also that these formulas imply, by induction, that the coefficient c_0 of $S_q(x)$ is equal to $2q + 1$). Hence,

$$T_{2q+1}(x) = \left\{S_q(x)\right\}^2 \cdot x .$$

In order to determine the coefficients of

$$S_q(x) = \sum_{j=o}^{q} c_j x^j ,$$

we write down the relation

$$z^{-q}(z^{2q+1}-1) = (z-1)S_q(z+z^{-1}-2) = (z-1) \sum_{j=o}^{q} c_j(z+z^{-1}-2)^j$$

and in it, replace z by e^{2u}, that is to say, replace $z + z^{-1} - 2$ by $(e^u-e^{-u})^2$. This implies

$$e^u(e^{(2q+1)u} - e^{-(2q+1)u}) = e^u(e^u-e^{-u}) \sum_{j=o}^{q} c_j(e^u-e^{-u})^{2j}$$

and hence, that

$$\sinh((2q+1)u) = \sum_{j=o}^{q} 2^{2j} c_j \sinh^{2j+1} u .$$

By differentiating twice both sides of the last relation we obtain (set $c_{q+1} = 0$):

$$(2q+1)^2 \cdot \sinh((2q+1)u) =$$
$$\sum_{j=o}^{q} 2^{2j} c_j (2j+1) \left\{2j \cdot \sinh^{2j-1} u + (2j+1) \sinh^{2j+1} u\right\} =$$
$$\sum_{j=o}^{q} \left\{2^{2j}(2j+1)^2 c_j + 2^{2j+2}(2j+2)(2j+3)c_{j+1}\right\} \cdot \sinh^{2j+1} u ;$$

replacing the left hand side by $(2q+1)^2$ times the expression of $\sinh((2q+1)u)$ obtained before, we conclude that

$$(2q+1)^2\, 2^{2j} c_j = 2^{2j}(2j+1)^2 c_j + 2^{2j+2}(2j+2)(2j+3)c_{j+1} .$$

A simple induction procedure now shows that

$$c_j = (2q+1) \prod_{k=o}^{j-1} \frac{(2q+1)^2 - (2k+1)^2}{2^2(2k+2)(2k+3)} = \frac{2q+1}{2j+1} \binom{q+j}{2j} ,$$

and so, the proof of Lemma 3.6 is complete.

Next we investigate the K - theory of the spaces $\mathbb{H}P^m$. The canonical projection $S^{4m+3} \to \mathbb{H}P^m$ factors through $\mathbb{C}P^{2m+1}$ giving rise to an S^2 - fibre bundle

$$S^2 \longrightarrow \mathbb{C}P^{2m+1} \xrightarrow{\ g\ } \mathbb{H}P^m .$$

The map g induces injections in ordinary cohomology as well as in K - theory. So we might regard $KU(\mathbb{H}P^m)$ as subring of the corresponding ring of $\mathbb{C}P^{2m+1}$.

Let $\xi_m^{\mathbb{C}}(\mathbb{H})$ be the 2 - dimensional complex vector bundle underlying the quaternionic line bundle $\xi_m(\mathbb{H})$. Using (2.4) we readily establish the following bundle isomorphism over $\mathbb{C}P^{2m+1}$:

(3.7) $$g^*\xi_m^{\mathbb{C}}(H) \cong \xi_{2m+1}(\mathbb{C}) \oplus \bar{\xi}_{2m+1}(\mathbb{C})$$

(see also [44; 9.6]).

<u>Proposition 3.8.</u> *Let* $w = \left[\xi_m^{\mathbb{C}}(\mathbb{H})\right] - 2 \in \widetilde{KU}(\mathbb{H}P^m)$.

(i) *The ring* $KU(\mathbb{H}P^m)$ *is generated by* w *subject to the relation* $w^{m+1} = 0$ *, that is to say,*

$$KU(\mathbb{H}P^m) = \mathbb{Z}[w]/\langle w^{m+1}\rangle ;$$

(ii) *the homomorphism* $g^! : KU(\mathbb{H}P^m) \to KU(\mathbb{C}P^{2m+1})$ *is injective and given by* $g^!(w) = v + \bar{v}$ *, where* $v = \xi_{2m+1}(\mathbb{C}) - 1$.

<u>Proof.</u> The cofibration sequence $\mathbb{H}P^{m-1} \to \mathbb{H}P^m \to S^{4m}$ shows, by induction, that $KU(\mathbb{H}P^m)$ is a free abelian group of rank $m + 1$. By (3.7) we have

$$g^!(w) = v + \bar{v} + \text{higher powers of } v ;$$

the elements $1, g^!(w), \ldots, g^!(w)^m$ therefore generate a direct summand of rank m in $KU(\mathbb{C}P^{2m+1}) \cong \mathbb{Z}[v]/\langle v^{2m+2}\rangle$ and the Proposition is proved.

<u>Proposition 3.9.</u> (i) *The complexification homomorphism* $c : KO(\mathbb{H}P^m) \to KU(\mathbb{H}P^m)$ *is injective;*

(ii) *the subring* $c(KO(\mathbb{H}P^m))$ *of* $KU(\mathbb{H}P^m)$ *is generated (as a free abelian group) by* 1 *and the elements* $e_j w^j$, $j = 1, \ldots, m$, *where* e_j *is equal to* 1 *if* j *is even and equal to* 2 *if* j *is odd.*

<u>Proof.</u> Again, we make use of the cofibration $\mathbb{H}P^{m-1} \to \mathbb{H}P^m \to S^{4m}$ and proceed by induction on m, starting with the case $\mathbb{H}P^0$ = point. Consider the exact sequence

$$\begin{array}{ccccccccc}
\widetilde{KO}^{-1}(S^{4m}) & \longrightarrow & \widetilde{KO}^{-1}(\mathbb{H}P^m) & \longrightarrow & \widetilde{KO}^{-1}(\mathbb{H}P^{m-1}) & \longrightarrow & \widetilde{KO}(S^{4m}) & \longrightarrow \\
\cong & & & & & & \cong & \\
0 \text{ or } \mathbb{Z}_2 & & & & & & \mathbb{Z} &
\end{array}$$

$$\widetilde{KO}(\mathbb{H}P^m) \longrightarrow \widetilde{KO}(\mathbb{H}P^{m-1}) \longrightarrow \widetilde{KO}^1(S^{4m}) = 0 ;$$

inductively we conclude that $\widetilde{KO}^{-1}(\mathbb{H}P^m)$ is finite. This gives the short exact sequence

$$0 \longrightarrow KO(S^{4m}) \longrightarrow KO(\mathbb{H}P^m) \longrightarrow KO(\mathbb{H}^{m-1}) \longrightarrow 0$$

and by induction one proves that $KO(\mathbb{H}P^m)$ is *torsion free*. Thus, the relation $r \circ c = 2$ (see (0.6)) implies that c is injective.

To prove part (ii) we first observe that for every integer $q \geq 1$, $2w$ and w^2 are elements of $c(\widetilde{KO}(\mathbb{H}P^q))$. This is seen as follows. Write $\xi = \xi_q(\mathbb{H})$ and $\xi^{\mathbb{C}} = \xi^{\mathbb{C}}_q(\mathbb{H})$. The bundle $\xi^{\mathbb{C}}$ is self-conjugate, that is to say, $\overline{\xi}^{\mathbb{C}} = \xi^{\mathbb{C}}$, and hence $c \circ r(\xi^{\mathbb{C}}) = 2\xi^{\mathbb{C}}$; this shows that $2w = c \circ r(w) \in c(\widetilde{KO}(\mathbb{H}P^q))$. As for w^2, notice that by definition $\xi^{\mathbb{C}}$ is equal to $c'\xi$, where c' stands for the map which associates to a quaternionic vector bundle its underlying complex vector bundle. The tensor product $\xi \otimes_{\mathbb{H}} \xi$ is a real (4-dimensional) vector bundle and we have $c(\xi \otimes_{\mathbb{H}} \xi) = c'\xi \otimes_{\mathbb{C}} c'\xi$. This gives $w^2 \in c(\widetilde{KO}(\mathbb{H}P^q))$. (We quote [47] as a reference to the preceeding remarks).

The second part of (3.9) now follows easily by induction on m from the commutative diagram

$$\begin{array}{ccccccccc}
0 & \longrightarrow & \widetilde{KO}(S^{4m}) & \longrightarrow & KO(\mathbb{H}P^m) & \longrightarrow & KO(\mathbb{H}P^{m-1}) & \longrightarrow & 0 \\
& & \downarrow c & & \downarrow c & & \downarrow c & & \\
0 & \longrightarrow & \widetilde{KU}(S^{4m}) & \longrightarrow & KU(\mathbb{H}P^m) & \longrightarrow & KU(\mathbb{H}P^{m-1}) & \longrightarrow & 0
\end{array}$$

and noting that $c : \widetilde{KO}(S^{4m}) \to \widetilde{KU}(S^{4m})$ is an isomorphism if m is even and multiplication by 2 if m is odd (see (0.6)).

The canonical projection $g : \mathbb{C}P^{2m+1} \to \mathbb{H}P^m$ induces, by restriction, a map

$$g|\mathbb{C}P^{2m} = h : \mathbb{C}P^{2m} \longrightarrow \mathbb{H}P^m .$$

Proposition 3.10. (i) *The homomorphism*

$$h^! : KO(\mathbb{H}P^m) \longrightarrow KO(\mathbb{C}P^{2m}) \cong \mathbb{Z}[y]/\langle y^{m+1}\rangle$$

is a monomorphism. Its image is the free abelian group with base $\{1, 2y, y^2, \ldots, e_j y^j, \ldots, e_m y^m\}$, *where*

$$e_j = \begin{cases} 1, & m \ \textit{even}, \\ 2, & m \ \textit{odd}. \end{cases}$$

(ii) *Let* $\xi_m^{\mathbb{R}}(\mathbb{H})$ *be the real* 4 - *dimensional bundle underlying the cononical quaternionic line bundle and let*

$$z = \xi_m^{\mathbb{R}}(\mathbb{H}) - 4 \in \widetilde{KO}(\mathbb{H}P^m) \ ; \textit{ then}$$

$$h^!(z) = 2y.$$

Proof: Part (i) follows from (3.2),(3.8),(3.9) and the following commutative diagram:

$$\begin{array}{ccccc}
KO(\mathbb{H}P^m) & \xrightarrow{g^!} & KO(\mathbb{C}P^{2m+1}) & \longrightarrow & KO(\mathbb{C}P^{2m}) \\
\downarrow c & & \downarrow c & & \downarrow c \\
KU(\mathbb{H}P^m) & \xrightarrow{g^!} & KU(\mathbb{C}P^{2m+1}) & \longrightarrow & KU(\mathbb{C}P^{2m})
\end{array} .$$

(In particular, (3.2) (ii) shows that $c : KO(\mathbb{C}P^{2m}) \to KU(\mathbb{C}P^{2m})$ is monic). The second part of (3.10) stems from (3.7).

We finally compute the K - theory of the real projective spaces $\mathbb{R}P^m$.

Proposition 3.11. *Let* $u = c\xi_m(\mathbb{R}) - 1 \in \widetilde{KU}(\mathbb{R}P^m)$.

(i) *The ring* $\widetilde{KU}(\mathbb{R}P^m)$ *is generated by* u *subject to the relations*

$$u^2 + 2u = 0 \ \textit{ and } \ u^{[m/2]+1} = 0 \ ;$$

in particular, $\widetilde{KU}(\mathbb{R}P^{2k}) \cong \widetilde{KU}(\mathbb{R}P^{2k+1})$ *is a cyclic group of order* 2^k

(ii) *The operations* ψ^p *are given by*

$$\psi^p(u) = \begin{cases} 0, & \textit{if } p \textit{ is even}, \\ u, & \textit{if } p \textit{ is odd}. \end{cases}$$

Remark. If the relation $u^2 + 2u = 0$ holds, then $u^{k+1} = 0$ if, and only if, $2^k u = 0$.

Proof of (3.11). We begin by observing that $\xi_m(\mathbb{R}) \otimes \xi_m(\mathbb{R}) = \varepsilon_{\mathbb{R}}$. In fact, for any CW-complex X the multiplicative group of line bundles over X is isomorphic to $H^1(X;\mathbb{Z}_2)$.

Since complexification is compatible with tensor products we obtain $(1+u)^2 = 1$ and hence, $2u + u^2 = 0$. Next we show that u is a non-zero element of filtration 2 in $\widetilde{KU}(\mathbb{R}P^m)$, $m \geq 2$, that is to say, if $i : \mathbb{R}P^2 \to \mathbb{R}P^m$ is the inclusion map, then $i^!(u) \neq 0$. The total Stiefel-Whitney class of $\xi_m(\mathbb{R})$ is given by

$$w(\xi_m(\mathbb{R})) = 1 + b ,$$

where $b \in H^1(\mathbb{R}P^m;\mathbb{Z}_2)$ is the generator of $H^*(\mathbb{R}P^m);\mathbb{Z}_2) = \mathbb{Z}_2[b]/\langle b^{m+1}\rangle$ [79; VII, 9.4]. Hence,

$$w(r \circ c(\xi_m(\mathbb{R}))) = w(2\xi_m(\mathbb{R})) = (1+b)^2 = 1 + b^2 \neq 1 ,$$

if $m \geq 2$. It follows that $c(\xi_m(\mathbb{R}))$ and its restriction to $\mathbb{R}P^2$ are stably non-trivial, proving that $u = c(\xi_m(\mathbb{R})) - 1$ is a non-zero element of filtration 2 $(m \geq 2)$.

The ring $H^*(\mathbb{R}P^{2k};\mathbb{Z})$ is generated by an element $a \in H^2(\mathbb{R}P^{2k};\mathbb{Z})$ subject to the relations

$$2a = 0 \quad \text{and} \quad a^{k+1} = 0$$

(see [79]). Since there is no odd-dimensional integral cohomology, the Atiyah-Hirzebruch spectral sequence of $\mathbb{R}P^{2k}$ collapses and we infer that the associated graded ring $\mathcal{G}\widetilde{KU}(\mathbb{R}P^{2k})$ is isomorphic to $H^*(\mathbb{R}P^{2k};\mathbb{Z})$ (see (0.10)). By the above remarks, the element u represents $a \in \mathcal{G}^2\widetilde{KU}(\mathbb{R}P^{2k})$ and we conclude that $u^j \neq 0$, $j = 1,\dots,k$, and $u^{k+1} = 0$.

Part (i) of (3.11) is now proved for $m = 2k \geq 0$ (the case $k = 0$ being trivial). With the cofibration

$$\mathbb{R}P^{2k} \xrightarrow{\ i\ } \mathbb{R}P^{2k+1} \longrightarrow S^{2k+1}$$

one gets readily that $i^! : \widetilde{KU}(\mathbb{R}P^{2k+1}) \cong \widetilde{KU}(\mathbb{R}P^{2k})$.

Part (ii) follows from

$$\psi_R^p(\xi_m(\mathbb{R})) = \{\xi_m(\mathbb{R})\}^p = \begin{cases} 1, & \text{if } p \text{ is even} \\ \xi_m(\mathbb{R}), & \text{if } p \text{ is odd,} \end{cases}$$

and the fact that complexification commutes with the ψ - operations.

<u>Proposition 3.12.</u> *Let* $x = \xi_m(\mathbb{R}) - 1 \in \widetilde{KO}(\mathbb{R}P^m)$.

(i) *The ring* $\widetilde{KO}(\mathbb{R}P^m)$ *is generated by* x *subject to the relations*

$$x^2 + 2x = 0 \quad \textit{and} \quad x^{f(m)+1} = 0 ,$$

where $f(m)$ *is the number of integers* q *with* $q \equiv 0,1,2$ *or* 4 *(mod. 8) and* $0 < q \leq m$. *In particular, the group* $\widetilde{KO}(\mathbb{R}P^m)$ *is cyclic of order* $2^{f(m)}$.

(ii) *The operations* ψ^p *are given by*

$$\psi^p(x) = \begin{cases} 0, & \textit{if } p \textit{ even}, \\ x, & \textit{if } p \textit{ odd.} \end{cases}$$

<u>Remark 3.13.</u> The integer valued function $f(m)$ is given by $f(m+8) = f(m) + 4$ and

m	1	2	3	4	5	6	7	8
f(m)	1	2	2	3	3	3	3	4

In particular, $f(8q) = 4q$.

<u>Proof of (3.12).</u> We begin by showing that the order $|\widetilde{KO}(RP^m)|$ of $\widetilde{KO}(RP^m)$ divides $|\widetilde{KO}(RP^{m-1})| \cdot e_m$, where

$$(3.14) \qquad e_m = \begin{cases} 2, & \text{if } m \equiv 0,1,2,4 \quad (\text{mod.} 8) \\ 1, & \text{if } m \equiv 3,5,6,7 \quad (\text{mod.} 8). \end{cases}$$

Recall that $\mathbb{R}P^m$ is homeomorphic to the mapping cone of the projection

$$p : S^{m-1} \longrightarrow \mathbb{R}P^{m-1}$$

and so, the Puppe sequence of p gives rise to the following exact sequence

$$\widetilde{KO}(S^{m-1}) \xleftarrow{p^!} \widetilde{KO}(\mathbb{R}P^{m-1}) \longleftarrow \widetilde{KO}(\mathbb{R}P^m) \longleftarrow \widetilde{KO}(\Sigma S^{m-1}) \xleftarrow{(\Sigma p)^!} \widetilde{KO}(\Sigma \mathbb{R}P^{m-1}).$$

Since $\widetilde{KO}(\Sigma S^{m-1}) \cong \mathbb{Z}, \mathbb{Z}_2, \mathbb{Z}_2, 0, \mathbb{Z}, 0, 0, 0$ if $m \equiv 0,1,2,3,4,5,6,7 \pmod{8}$ respectively, (3.14) follows immediately from the exact sequence for the case in which $m \not\equiv 0 \pmod{4}$. To treat the case $m \equiv 0 \pmod{4}$, we note that the composition

$$S^{m-1} \xrightarrow{p} \mathbb{R}P^{m-1} \longrightarrow \mathbb{R}P^{m-1}/\mathbb{R}P^{m-2} \cong S^{m-1}$$

(and hence, also its suspension) has degree 2 if m is even [79; V,6.13]. Thus, the image of $(\Sigma p)^!$ contains $2 \cdot \widetilde{KO}(\Sigma S^{m-1}) \cong 2\mathbb{Z}$; the cokernel of $(\Sigma p)^!$ is therefore at most $\mathbb{Z}_2$ and (3.14) follows also for $m \equiv 0 \pmod{4}$.

Using (3.14) we conclude, by induction, that the order $\widetilde{KO}(\mathbb{R}P^m)$ is at most $2^{f(m)}$. Moreover, if $|\widetilde{KO}(\mathbb{R}P^m)| = 2^{f(m)}$ for some m, the same holds for all $n \leq m$ and the homomorphism $\widetilde{KO}(\mathbb{R}P^m) \to \widetilde{KO}(\mathbb{R}P^n)$ induced by the inclusion $\mathbb{R}P^n \longrightarrow \mathbb{R}P^m$ must be onto.

Next we prove that for every integer $q > 0$, $\widetilde{KO}(\mathbb{R}P^{8q})$ is cyclic of order $4q = f(8q)$ and is generated by x, satisfying the relations $x^2 + 2x = 0$ and $x^{4q+1} = 0$. By the preceeding remark this will imply part (i) of (3.12) for all m. In fact, the complexification homomorphism $c : \widetilde{KO}(\mathbb{R}P^m) \to \widetilde{KU}(\mathbb{R}P^m)$ is an epimorphism, since by (3.11) the element $c(x) = u$ is a generator. For $m = 8q$ we therefore have the following inequalities

$$2^{4q} = 2^{f(8q)} \geq |\widetilde{KO}(\mathbb{R}P^{8q})| \geq |\widetilde{KU}(\mathbb{R}P^{8q})| = 2^{4q} ,$$

implying that $c : \widetilde{KO}(\mathbb{R}P^{8q}) \cong \widetilde{KU}(\mathbb{R}P^{8q})$. Now (3.11) shows that $\widetilde{KO}(\mathbb{R}P^{8q})$ is as claimed and part (i) of (3.12) is established.

Part (ii) follows from

$$\psi^p(\xi_m(\mathbb{R})) = \begin{cases} 1 , & \text{if } p \text{ is even,} \\ \xi_m(\mathbb{R}), & \text{if } p \text{ is odd.} \end{cases}$$

<u>Remark 3.15.</u> The projection $S^{2m+1} \to \mathbb{C}P^m$ factors through $\mathbb{R}P^{2m+1}$, giving rise to an S^1 - principal bundle

$$S^1 \longrightarrow \mathbb{R}P^{2m+1} \xrightarrow{g} \mathbb{C}P^m .$$

Using (2.4) we readily establish the following isomorphism of vector bundles over $\mathbb{R}P^{2m+1}$

$$g^*(r\xi_m(\mathbb{C})) \cong \xi_{2m+1}(\mathbb{R}) \oplus \xi_{2m+1}(\mathbb{R}) .$$

(The previous isomorphism of vector bundles can also be established showing that the first Chern classes of $g^*(\xi_m(\mathbb{C}))$ and $c\xi_{2m+1}(\mathbb{R})$ coincide - complex line bundles are classified by their first Chern class.) Hence, in $KO(\mathbb{R}P^{2m+1})$ we have that $g^!(y) = 2x$, with $y = r\xi_m(\mathbb{C}) - 2$.

4. Real Vector Fields on Spheres.

In this section we give a solution of the vector field problem over the reals. As already mentioned in the introduction, this problem was first solved by J.F.Adams in his celebrated paper "Vector Fields on Spheres" [2].

According to (2.5) and (2.6) we have to determine the J - order of the canonical line bundle $\xi_m(\mathbb{R})$, that is to say, the order of $\tilde{J}(x)$ in $\tilde{J}(\mathbb{R}P^m)$, where $x = \xi_m(\mathbb{R}) - 1 \in \widetilde{KO}(\mathbb{R}P^m)$.

Theorem 4.1. *The group* $\tilde{J}(\mathbb{R}P^m)$ *is isomorphic to* $\widetilde{KO}(\mathbb{R}P^m)$. *In particular, the* $\tilde{J}$ *- order of the canonical line bundle* $\xi_m(\mathbb{R})$ *over* $\mathbb{R}P^m$ *is equal to* $2^{f(m)}$ *, where* $f(m)$ *is the function defined in* (3.13).

Proof. Since the groups $\tilde{J}(X)$ and $\tilde{J}''(X)$ are isomorphic for any finite CW - complex X (see (0.14)), it will suffice to show that $\widetilde{KO}(\mathbb{R}P^m) \cong \tilde{J}''(\mathbb{R}P^m)$. The reader should now review the definition of the subgroup $W(\mathbb{R}P^m)$ of $KO(\mathbb{R}P^m)$ and consider the constant function e_o on $\mathbb{Z} \times \mathbb{K}O(\mathbb{R}P^m)$ given by $e_o(k,y) = 2^{f(m)}$. From (3.12) we infer that for every $k \in \mathbb{Z}$ and every $y \in KO(\mathbb{R}P^m)$,

$$k^{e_o(k,y)}(\psi^k(y) - y) = 0 ;$$

hence, $W(\mathbb{R}P^m) = 0$ and $\widetilde{KO}(\mathbb{R}P^m) = \tilde{J}''(\mathbb{R}P^m)$.

It follows from (3.12) that the order of $\tilde{J}''(x)$ is equal to $2^{f(m)}$.

Theorem 4.2. *The sphere* S^{n-1} *admits* m *linearly independent real vector fields if, and only if,* n *is a multiple of* $2^{f(m)}$.

Proof. We set $m = k - 1$ and follow the notation employed in section 2. According to (2.2), (2.5) and (4.1), the condition $n \equiv 0 \pmod{2^{f(k-1)}}$ is necessary for the existence of $k - 1$ linearly independent vector fields on S^{n-1}; we shall show that it is also sufficient.

Let n be a multiple of $2^{f(k-1)}$. Then $n > 2k$, except for a small number of cases, namely for $n = 2^{f(k-1)}$, with $k = 1,\ldots,8$ (see (3.13)). Hence, for all but the latter cases, it follows from (2.2) and (2.6), that S^{n-1} admits $k - 1$ linearly independent vector fields. For the exceptional cases one can explicitely construct the vector fields (see (4.4)). We illustrate this for $k - 1 = 8$, $n = 2^{f(8)} = 16$. Let $\mathbb{H}(2)$ be the $\mathbb{R}$-algebra of the 2×2-matrices over the quaternions and let i,j,k be the usual generators of $\mathbb{H}$. We set in $\mathbb{H}(2)$

$$e_1 = \begin{pmatrix} i & 0 \\ 0 & i \end{pmatrix},\ e_2 = \begin{pmatrix} j & 0 \\ 0 & j \end{pmatrix},\ e_3 = \begin{pmatrix} k & 0 \\ 0 & -k \end{pmatrix},\ e_4 = \begin{pmatrix} 0 & k \\ k & 0 \end{pmatrix} .$$

One checks readily that $e_p e_q = -e_q e_p$, if $p \neq q$, and $e_p^2 = -1_{\mathbb{H}(2)}$ $(p=1,\ldots,4)$. The $\mathbb{R}$-algebra $A = \mathbb{H}(2) \otimes_{\mathbb{R}} \mathbb{H}(2)$ operates on $\mathbb{R}^{16} \cong \mathbb{H}(2)$ by linear transformations as follows:

$$(x_1 \otimes x_2)(x) = x_1 \cdot x \cdot x_2$$

(in fact $A \cong \mathbb{R}(16)$). For the elements $a_1,\ldots,a_8 \in A$ defined by

$$a_p = e_p \otimes 1,\ a_{p+4} = e_p \otimes e_1 e_2 e_3 \qquad (p=1,2,3,4)$$

we show that

$$a_p a_q = -a_q a_p \quad \text{if} \quad p \neq q \quad \text{and} \quad a_p^2 = -1_A$$

$(p,q = 1,\ldots,8)$.

We choose an inner product $(\ |\ )$ in $\mathbb{R}^{16}$ which is invariant under the action of the finite (multiplicative) group generated by $a_1,\ldots,a_8$ and compute

$$(x \mid a_p(x)) = (a_p(x) \mid a_p^2(x)) = (a_p(x) \mid -x) = -(x \mid a_p(x)).$$

Thus $(x|a_p(x)) = 0$. Similarly, one obtains $(a_p(x)|a_q(x)) = 0$, if $p \neq q$. Hence, the frame

$$\{x, a_1(x), \ldots, a_8(x)\},$$

$x \in S^{15} \subseteq \mathbb{R}^{16}$, is orthonormal, providing 8 linearly independent vector fields on S^{15}.

Working with quaternions and Cayley numbers we exhibit in the same way $k-1$ vector fields on the sphere of dimension $2^{f(k-1)} - 1$, $k = 3,\ldots,8$. The case $k = 2$ was treated in the Introduction. Theorem (4.2) is now proved.

Corollary 4.3. *The maximum number of linearly independent vector fields on* S^{n-1} *is equal to* $\rho(n) - 1$, *where* $\rho(n)$ *is defined as follows: write* $n = (2\cdot a(n)+1)2^{b(n)}$, $b(n) = c(n) + 4\cdot d(n)$, *where* $a(n), b(n), c(n), d(n)$ *are integers and* $0 \leq c(n) \leq 3$ *; then,*

$$\rho(n) = 2^{c(n)} + 8d(n).$$

Proof. According to (4.2) we have to find for a given n the greatest m such that $2^{f(m)}$ divides n. Hence, we must look for the greatest m such that $f(m) \leq b(n)$. By (3.13) this is obviously $\rho(n) - 1$.

Remark. As for $k - 1 = 8$, $n = 16$, it is in fact possible to construct explicitely the vector fields in all cases. We shall sketch a proof of the following so-called Hurwitz-Radon-Eckmann Theorem (see [85], [133] and [200]).

Theorem 4.4. *On the sphere* S^{n-1} *there exist* $\rho(n) - 1$ *linearly independent vector fields*

$$v^{(q)}(x) = (v_1^{(q)}(x), \ldots, v_n^{(q)}(x))$$

$q = 1,\ldots,\rho(n) - 1$, *which are given by functions* $v_i^{(q)}(x)$, *linear in* $x = (x_1,\ldots,x_n)$.

Proof. We shall show that S^{n-1} admits m vector fields which are linear in x if n is a multiple of $2^{f(m)}$. This is equivalent to (4.4) as the proof of (4.3) shows. To this end we shall apply the theory of real Clifford algebras (see [26]). Let us recall that a Clifford algebra C_m is an associative algebra with a unit element 1 over $\mathbb{R}$, generated (as an $\mathbb{R}$-algebra) by elements $e_1, e_2, \ldots, e_m$ subject to the relations

$$e_i^2 = -1, \quad e_i e_j = -e_j e_i \quad \text{if } i \neq j \quad (i,j=1,\ldots,m).$$

Let V be a real vector space of dimension n which is a left C_m-module. Thus, C_m acts on V by linear transformations. If $a \in C_m$ and $x \in V$, we write $a(x)$ for the action of a on x. Choose now an inner product $(\ |\ )$ on V which is invariant under the action of the multiplicative group generated by $e_1, e_2, \ldots, e_m$. Let $S(V)$ be the unit sphere in V with respect to this inner product. For any $x \in S(V)$ the elements

$$x, e_1(x), \ \ldots \ , e_m(x)$$

form an orthonormal $(m+1)$-frame in V, providing us with m linearly independent vector fields on the $(n-1)$-sphere $S(V)$. (Using the fact that the selected inner product is invariant under the action of the multiplicative group described, it follows that if $i \neq j$, $(e_i(v)|e_j(v)) = (e_i^2(v)|e_ie_j(v)) = -(v|e_ie_j(v)) = (v|e_je_i(v)) = -(e_j(v)|e_i(v))$ and hence, $(e_i(v)|e_j(v)) = 0$. Similarly, $(v|e_i(v)) = 0$ and $(e_i(v)|e_i(v))=1$).

In view of the preceeding remarks, (4.4) will be proved if $\mathbb{R}^n$ can be given the structure of a C_m-module, whenever n is a multiple of $2^{f(m)}$.

Let $\mathbb{F}(q)$ denote the $\mathbb{R}$-algebra of $q \times q$-matrices with entries in $\mathbb{F}$ ($\mathbb{F} = \mathbb{R}, \mathbb{C}$ or $\mathbb{H}$). The algebras C_m, $m=1,\ldots,8$, can be identified with the following $\mathbb{R}$-algebras:

m	1	2	3	4	5	6	7	8
C_m	$\mathbb{C}$	$\mathbb{H}$	$\mathbb{H} \oplus \mathbb{H}$	$\mathbb{H}(2)$	$\mathbb{C}(4)$	$\mathbb{R}(8)$	$\mathbb{R}(8) \oplus \mathbb{R}(8)$	$\mathbb{R}(16)$

Furthermore, $C_{m+8} \cong C_m \otimes_{\mathbb{R}} \mathbb{R}(16)$. From these statements, we see that C_m, $m = 1,\ldots,8$, acts by linear transformations on $\mathbb{R}^{a_m}$ where a_m is given in the following table:

(4.5)

m	1	2	3	4	5	6	7	8
a_m	2^1	2^2	2^2	2^3	2^3	2^3	2^3	2^4

(The algebra $\mathbb{H} \oplus \mathbb{H}$ acts on $\mathbb{R}^4$ via the algebra homomorphism $\mathbb{H} \oplus \mathbb{H} \to \mathbb{H}$, defined by the projection on the first factor; similarly for $\mathbb{R}(8) \oplus \mathbb{R}(8)$). Notice that if $\mathbb{R}^s$ is a C_m - module, then $\mathbb{R}^s \otimes \mathbb{R}^{16} = \mathbb{R}^{s \cdot 2^4}$ is a C_{m+8} - module; this, together with (4.5) implies that $\mathbb{R}^{2^{f(m)}}$ is a C_m - module (see (3.13)). Hence, for any multiple of $2^{f(m)}$, say n, the vector space $\mathbb{R}^n$ admits a C_m - action and (4.4) is established.

5. Cross-Sections of Complex Stiefel Fibrations.

The cross-section problem for the fibrations

$$\mathbb{C}_{n,k} \longrightarrow S(\mathbb{C}^n) = S^{2n-1}$$

was solved by Atiyah and Todd [37], and by Adams and Walker [8]; the former authors gave necessary conditions, the latter proved that these conditions were also sufficient.

Referring to (2.5) and (2.6) the task is to compute the $\widetilde{J}$ - order of the real bundle underlying the canonical line bundle $\xi_{k-1} = \xi_{k-1}(\mathbb{C})$ over $\mathbb{C}P^{k-1}$. We shall work with the functor $\widetilde{J}'$ defined in Chapter 0. The groups $\widetilde{J}(X)$ and $\widetilde{J}'(X)$ are naturally isomorphic for any finite CW - complex X; we also notice that Adams and Walker use a different J' in [8], defined in terms of the classes bh and sh. Using a method invented by K.Lam [160], we first determine explicitely the order of $\widetilde{J}'(y_m)$, $y_m = r(\xi_m) - 2$, for even m; then, we show that y_{2t+1} and y_{2t} have the same $\widetilde{J}$ - order ($t \geq 1$).

In the sequel we shall write ξ and y for ξ_m and y_m respectively. According to the definition of the functor J' the order b_{2t} of $J'(y) \in \widetilde{J}'(\mathbb{C}P^{2t})$ is the smallest positive integer such that there exists an element $w \in \widetilde{KO}(\mathbb{C}P^{2t})$ with

$$\Theta_p(b_{2t}y) = \frac{\psi^p(1+w)}{1+w} , \tag{5.1}$$

for all primes p (see (0.13) and (0.14)). For the ring

$$KO(\mathbb{C}P^{2t}) \cong \mathbb{Z}[y]/\langle y^{t+1}\rangle$$

(see (3.2)), we have the canonical embeddings

$$KO(\mathbb{C}P^{2t}) \subset KO(\mathbb{C}P^{2t}) \otimes \mathbb{Q}_p \subset KO(\mathbb{C}P^{2t}) \otimes \mathbb{Q} = \mathbb{Q}[y]/\langle y^{t+1}\rangle$$

and we extend the ψ - operations to the latter rings in the obvious way, namely, by taking $\psi^p \otimes 1_{\mathbb{Q}_p}$ and $\psi^p \otimes 1_{\mathbb{Q}}$. The multiplicative groups of polynomials with constant term 1 are denoted by

$$1 + \tilde{Z}[y]/\langle y^{t+1}\rangle \quad \text{and} \quad 1 + \tilde{\mathbb{Q}}[y]/\langle y^{t+1}\rangle ,$$

respectively.

The next result is readily established using the formula

$$\psi^p(y^m) = p^{2m}\, y^m + \text{higher terms},$$

proved in (3.6).

(5.2) *Let* $1 + u$ *be an element of* $1 + \tilde{\mathbb{Q}}[y]/\langle y^{t+1}\rangle$. *Then, the polynomial* $\dfrac{\psi^p(1+u)}{1+u}$ *has integral coefficients for all primes* p *if, and only if,* $u \in \tilde{\mathbb{Z}}[y]/\langle y^{t+1}\rangle$.

We also note that every element $1 + u \in 1 + \tilde{\mathbb{Q}}[y]/\langle y^{t+1}\rangle$ has an nth - root given by

$$(1+u)^{1/n} = 1 + \sum_{j=1}^{t} \binom{1/n}{j} u^j \quad .$$

Since ψ^p is a ring homomorphism, $\psi^p((1+u)^{1/n}) = (\psi^p(1+u))^{1/n}$.

Now, for any integer b , the element $\theta_p(by) = (\theta_p(y))^b$ corresponds to a polynomial in $1 + \mathbb{Q}[y]/\langle y^{t+1}\rangle$. According to (5.1) and (5.2) this polynomial has integral coefficients for all primes p , if b is a multiple of the order of $\tilde{J}'(y)$. We show that the converse also holds.

<u>Lemma 5.3.</u> *The order* b_{2t} *of* $\tilde{J}'(y) \in \tilde{J}'(\mathbb{C}P^{2t})$ *is the smallest positive integer* b, *such that for all primes* p *the element* $(\theta_p(x))^b \in 1 + \mathbb{Q}[y]/\langle y^{t+1}\rangle$ *is a polynomial with integral coefficients.*

<u>Proof.</u> There exists and integer n (for example, $n = b_{2t}$) and an element $z \in \tilde{\mathbb{Z}}[y]/\langle y^{t+1}\rangle$ such that for all primes p one has $(\theta_p(y))^n = \dfrac{\psi^p(1+z)}{1+z}$. Hence, for every prime

p, $\theta_p(y) = \dfrac{\psi^p(1+u)}{1+u}$, where $1 + u = (1+z)^{1/n}$. For any positive integer b we obtain

$$(\theta_p(y))^b = \left(\frac{\psi^p(1+u)}{1+u}\right)^b = \frac{\psi^p(1+w)}{1+w} \quad .$$

with $1 + w = (1+u)^b$. Thus, if for all primes p the polynomial $(\theta_p(y))^b$ has integral coefficients, from (5.2) we conclude that $w \in \tilde{\mathbb{Z}}[y]/\langle y^{t+1}\rangle$ and hence, $b \cdot \tilde{J}'(y) = 0$, proving the Lemma.

We now determine the polynomial $\theta_p(y) \in 1 + \tilde{\mathbb{Q}}[y]/\langle y^{t+1}\rangle$.

Lemma 5.4. (i) $\theta_2(y) = (1+y/4)^{1/2}$.

(ii) *For any odd integer* $p = 2q + 1$ *one has*

$$\theta_p(y) = \frac{1}{2q+1} \sum_{j=0}^{q} \frac{2q+1}{2j+1}\binom{q+j}{2j} y^j =$$

$$1 + \sum_{j=1}^{q-1} \frac{1}{2j+1}\binom{q+j}{2j} y^j + \frac{1}{2q+1} y^q \ .$$

(iii) *If* $p = 2q + 1$ *is a prime, the coefficients*

$$\frac{1}{2j+1}\binom{q+j}{2j},\ j = 0,\dots,q-1 \ \textit{are integers.}$$

Proof. Complexification injects the torsion free ring $KO(\mathbb{C}P^{2t}) \cong \mathbb{Z}[y]/\langle y^{t+1}\rangle$ into the ring $KU(\mathbb{C}P^{2t}) \cong \mathbb{Z}[v]/\langle v^{2t+1}\rangle$. Referring to (3.2 (ii)) this embedding is given by $c(y) = v + \bar{v} = \xi + \xi^{-1} - 2$ (note that $\bar{\xi} = \xi^{-1}$). Consider now the complex 4-dimensional bundle $\eta = 2\xi + 2\xi^{-1}$. One has $\Lambda^4_{\mathbb{C}}(\eta) = 1$ and $r(\eta) = 4r(\xi) = 8 + 4y$. From a well-known property of the Bott operations (see (0.13)) we have

$$c \circ \theta^{\mathbb{R}}_p(4r(\xi)) = \theta^{\mathbb{C}}_p(2(\xi+\xi^{-1})) = \left(\frac{\xi^p - 1}{\xi - 1} \cdot \frac{\xi^{-p} - 1}{\xi^{-1} - 1}\right)^2 =$$

$$= \left(\frac{\xi^p + \xi^{-p} - 2}{\xi + \xi^{-1} - 2}\right)^2 = c\left(\frac{\psi^p(y)}{y}\right)^2 \quad .$$

Hence,

$$\theta_p(8+4y) = p^4(\theta_p(y))^4 = \left(\frac{\psi^p(y)}{y}\right)^2 ,$$

implying that

$$\theta_p(y) = \left(\frac{\psi^p(y)}{p^2 y}\right)^{1/2} = \left(\frac{T_p(y)}{p^2 y}\right)^{1/2}$$

(see (3.2 (iii))). The proof of the Lemma is now concluded using the explicit formula for $T_p(y)$ given in (3.6). (If $2q + 1$ is a prime, then the *integers*

$$\frac{2q+1}{2j+1}\binom{q+j}{2j} , \quad j = 0,1,\ldots,q-1 ,$$

are obviously divisible by $2q + 1$.)

We now recall that given any integer $n \neq 0$ and for any prime number p, the p-*adic valuation* $v_p(n)$ of n is the exponent of p in the prime power decomposition of n; we set $v_p(0) = 0$ for any prime p.(*) Moreover, the p-adic valuation function can be extended trivially to rational numbers.

<u>Lemma 5.5.</u> *Let* b *be a positive integer and* p *be a prime. The following three conditions are equivalent.*

$(i)_p$: *The polynomial* $(\theta_p(y))^b \in \mathbb{Q}[y]/\langle y^{t+1}\rangle$ *has integral coefficients.*

$(ii)_p$: *For all* s *with* $0 \le s \le \left[\frac{2t}{p-1}\right]$ *the power* p^s *divides the binomial coefficient* $\binom{b}{s}$.

$(iii)_p$: $v_p(b) \ge \max\{s + v_p(s) \mid 0 \le s \le \left[\frac{2t}{p-1}\right]\}$.

<u>Proof.</u> (A) $(i)_p \Leftrightarrow (ii)_p$ for odd primes $p = 2q + 1$. According to (5.4) the polynomial $\theta_p(y)$ is of the form

$$\theta_{2q+1}(y) = 1 + w + \frac{1}{p} y^q ,$$

$w \in \mathbb{Z}[y]/\langle y^{t+1}\rangle$. Hence,

$$(\theta_p(y))^b = \sum_{s=0}^{b} \binom{b}{s} (1+w)^{b-s} \frac{y^{qs}}{p^s} .$$

(*) This definition is convenient for our purpose.

Since the elements $(1+w)^{b-s}\, y^{qs}$, $0 \le s \le \left[\frac{t}{q}\right]$, form a partial basis for the abelian group $\mathbb{Z}[y]/\langle y^{t+1}\rangle$ (note that $y^{qs} = 0$ if $qs > t$), we conclude that $(\theta_p(y))^b$ is integral if, and only if, p^s divides $\binom{b}{s}$, $0 \le s \le \left[\frac{t}{q}\right] = \left[\frac{2t}{p-1}\right]$.

(B) $(ii)_p \Leftrightarrow (iii)_p$ for all primes p.

Let m be a non-negative integer. We show that p^s divides $\binom{b}{s}$ for all s with $0 \le s \le m$ if, and only if,

$$v_p(b) \ge \max\left\{s + v_p(s) \mid 0 \le s \le m\right\} .$$

This claim holds trivially for $m = 0$ and $m = 1$. For $m \ge 2$, we write

$$\binom{b}{m} = \frac{b}{m} \cdot \frac{b-1}{1} \cdot \frac{b-2}{2} \cdots \frac{b-(m-1)}{m-1}$$

and observe that if $v_p(b) \ge m - 1$, then

$$v_p\binom{b}{m} = v_p(b) - v_p(m) ;$$

this remark implies that, in case $v_p(b) \ge m - 1$, one has

$$v_p\binom{b}{m} \ge m \Leftrightarrow v_p(b) \ge m + v_p(m) .$$

The claim now follows by induction on m.

(C) $(i)_2 \Leftrightarrow (iii)_2$.

Both conditions are satisfied only if b is an even integer, say $b = 2b'$. The polynomial

$$(1+y/4)^{(1/2)b} = (1+y/4)^{b'} \in \mathbb{Q}[y]/\langle y^{t+1}\rangle$$

has integral coefficients if, and only if, $(1/2^{2s'})\cdot\binom{b'}{s'} \in \mathbb{Z}$, $s' = 0,\dots,t$. As in (B) one shows that the latter holds if, and only if,

$$v_2(b') \ge \max\{2s' + v_2(s') \mid s' = 0,\dots,t\} .$$

Since $v_2(b) = 1 + v_2(b')$ and

$$1 + \max\{2s' + v_2(s') \mid s' = 0,\dots,t\}$$

$$= \max\{2s' + v_2(s') + 1 \mid s' = 0,\dots,t\}$$

$$= \max\{s + v_2(s) \mid s = 0,\dots,2t\} \quad ,$$

the proof is complete.

<u>Proposition 5.6.</u> *The order* b_{2t} *of* $\tilde{J}(y)$ *in* $\tilde{J}(\mathbb{C}P^{2t})$ *is given by*

$$v_p(b_{2t}) = \max\left\{s + v_p(s) \mid 0 \le s \le \left[\frac{2t}{p-1}\right]\right\} \quad ,$$

for every prime p .

<u>Proof.</u> Since $\tilde{J}(y)$ and $\tilde{J}'(y)$ have the same order, the Proposition follows from (5.3) and (5.7).

We show next that the $\tilde{J}$-order of the canonical line bundles over $\mathbb{C}P^{2t+1}$ and $\mathbb{C}P^{2t}$ are equal.

<u>Proposition 5.7.</u> *For* $t \ge 1$ *the order* b_{2t+1} *of* $\tilde{J}(y)$ *in* $\tilde{J}(\mathbb{C}P^{2t+1})$ *is equal to* b_{2t} .

<u>Proof.</u> The homomorphism $i^!$ induced in KO-theory by the inclusion $i : \mathbb{C}P^{2t} \to \mathbb{C}P^{2t+1}$ maps ξ_{2t+1} onto ξ_{2t}. Hence, b_{2t} divides b_{2t+1}. If $t \equiv 1$ (mod.2) the map $i^!$ is an isomorphism. Thus, $\tilde{J}(\mathbb{C}P^{2t+1}) \cong \tilde{J}(\mathbb{C}P^{2t})$ and $b_{2t+1} = b_{2t}$. Now suppose that $t \equiv 0$ (mod.2), say, $t = 2q$. The exact sequence

$$0 \longrightarrow \widetilde{KO}(S^{8q+2}) \longrightarrow \widetilde{KO}(\mathbb{C}P^{4q+1}) \xrightarrow{i^!} \widetilde{KO}(\mathbb{C}P^{4q}) \longrightarrow 0$$

(see (3.2)) shows that $\ker(i^!) \cong \mathbb{Z}_2$ and is generated by y^{2q+1}. Because of the final remarks in (0.14), we conclude that the kernel of $J(i) : \tilde{J}(\mathbb{C}P^{4q+1}) \to \tilde{J}(\mathbb{C}P^{4q})$ is at most $\mathbb{Z}_2$ and is generated by $\tilde{J}(y^{2q+1})$. The homomorphism $J(i)$ maps the element $b_{4q}\tilde{J}(y) \in \tilde{J}(\mathbb{C}P^{4q+1})$ onto zero, that is to say

$$\tau = b_{4q}\tilde{J}(y) \in \ker J(i) \cong \mathbb{Z}_2 \ \text{ or } \ 0 \ .$$

We are going to show that $\tau \neq \tilde{J}(y^{2q+1})$; this implies that $\tau = 0$ and the proposition will be proved. Consider the canonical projection

$$g : \mathbb{R}P^{8q+3} \longrightarrow \mathbb{C}P^{4q+1} \quad .$$

By (3.15) the homomorphism $J(g)$ maps $\tilde{J}(y)$ onto $2\tilde{J}(x)$, hence $\tilde{J}(g)\tau = 2b_{4q}\tilde{J}(x) \in \tilde{J}(\mathbb{R}P^{8q+3})$. Using (5.6), (4.1) and (3.13) it is easy to see that

$$v_2(b_{4q}) \geq 4q + v_2(4q) \geq 4q + 2 = |\tilde{J}(\mathbb{R}P^{8q+3})| \quad ,$$

which implies that $\tilde{J}(g)\tau = 0$. On the other hand, from (3.15) and (4.1) we conclude that

$$\tilde{J}(g)(\tilde{J}(y^{2q+1})) = \tilde{J}(g^! y^{2q+1}) = \tilde{J}((2x)^{2q+1})$$

$$= 2^{4q+1}\,\tilde{J}(x) \neq 0 \; .$$

Thus $\tau \neq \tilde{J}(y^{2q+1})$ and (5.7) is proved.

<u>Theorem 5.8.</u> *The* $\tilde{J}$ - *order of the canonical line bundle over* $\mathbb{C}P^m$ *is equal to the integer* b_m *given by*

$$v_p(b_m) = \max\left\{s + v_p(s) \,\middle|\, 0 \leq s \leq \left[\frac{m}{p-1}\right]\right\}$$

for every prime p . *In particular,* $v_p(b_m) = 0$ *if* $p > m + 1$ *and* $v_p(b_m) \geq 1$ *if* $p \leq m + 1$.

<u>Proof.</u> For any positive integer m and any prime p , let

$$N(m,p) = \max\left\{s + v_p(s) \,\middle|\, 0 \leq s \leq \left[\frac{m}{p-1}\right]\right\} \; .$$

Observing that for $t \geq 1$ and for every prime p one has

$$N(2t,p) = N(2t+1,p) \; ,$$

the theorem follows for $m \geq 2$ from (5.6) and (5.7).

Let us assume that $m = 1$. Then, $\tilde{J}''(\mathbb{C}P^1) = \tilde{J}''(S^2) = \mathbb{Z}_2$, generated by $\tilde{J}''(y)$ (see (3.2) and (3.6)). Hence $|\tilde{J}(y)| = 2$. Since $N(1,2) = 1$ and $N(1,p) = 0$ for p odd, the theorem holds for $m = 1$.

<u>Remark.</u> The first few values of b_m are $b_1 = 2$, $b_2 = b_3 = 24$, $b_4 = b_5 = 2880$, $b_6 = b_7 = 362\,880$.

Theorem 5.9. *The complex Stiefel fibration* $\mathbb{C}_{n,k} \to S^{2n-1}$, $k \geq 2$, *has a cross-section if, and only if,* n *is a multiple of the integer* b_{k-1} , *called Complex James number, defined in* (5.8).

Proof. According to (2.5) and (5.8), if $\mathbb{C}_{n,k} \to S^{2n-1}$ has a cross-section, n is a multiple of b_{k-1}.

Conversely, let n be a multiple of b_{k-1}. Then

$$v_2(n) \geq v_2(b_{k-1}) \geq k - 1 ,$$

that is to say, $n \geq 2^{k-1}$. If $k \geq 5$ then, $n > 2k$; if $k = 3$ or 4, then $b_{k-1} = 24$ and again, $n > 2k$. Hence, for $k \geq 3$, Theorem 2.6 implies the existence of a cross-section of $\mathbb{C}_{n,k} \to S^{2n-1}$.

If $k = 2$ we have that $b_{k-1} = b_1 = 2$ and hence, n is even, say $n = 2q$. In this case, we give an explicit cross-section s of the Stiefel fibration: for every $x = (x_1,\dots,x_{2q}) \in S^{2n-1}$, $s(x) = \{x, s^1(x)\}$, with $s^1(x) = (-\bar{x}_2, \bar{x}_1, \dots, -\bar{x}_{2q}, \bar{x}_{2q-1})$.

Remark 5.10. (1) In view of (2.2), Theorem 5.9 is equivalent to the following statement.

> *"The sphere* S^{2n-1} *admits* m *linearly independent complex vector fields if, and only if,* n *is a multiple of* b_m*".*

(2) Contrary to the situation in the real case there are no known formulas which give explicitly the complex vector fields on the spheres (except for $m = 1$).

6. Cross-Sections of Quaternionic Stiefel Fibrations.

In this last section of the Chapter we shall determine the $\tilde{J}$-order of the canonical line bundle over the quaternionic projective spaces. We shall use methods, results and notation of the preceeding section.

Let $\xi_m^{\mathbb{R}}(\mathbb{H})$ be the real 4-dimensional vector bundle underlying the quaternionic line bundle $\xi_m(\mathbb{H})$ and let $z = \xi_m^{\mathbb{R}}(\mathbb{H}) - 4 \in \widetilde{KO}(\mathbb{H}P^m)$. By definition, the $\widetilde{J}$-order of $\xi_m(\mathbb{H})$ is the order of $\widetilde{J}(z)$ in $\widetilde{J}(\mathbb{H}P^m)$. Once more we use the functor J' to determine this order.

According to (3.10) the homomorphism

$$h^! : KO(\mathbb{H}P^m) \longrightarrow KO(\mathbb{C}P^{2m}) \cong \mathbb{Z}[y]/\langle y^{m+1}\rangle$$

induced by the canonical map $h : \mathbb{C}P^{2m} \to \mathbb{H}P^m$ is injective and identifies the ring $\widetilde{KO}(\mathbb{H}P^m)$ with the ψ-subring $A_m(y)$ generated (as a ring) by $2y$ and y^2. Furthermore, $h^!(z) = 2y$. Since the classes θ_p are natural, we infer that the order of $\widetilde{J}'(z) \in \widetilde{J}'(\mathbb{H}P^m)$ is the smallest positive integer c_m such that there exists an element $u \in A_m(y)$ with

$$\theta_p(c_m \cdot 2y) = \frac{\psi^p(1+u)}{1+u} \tag{6.1}$$

for all primes p. It is easy to see that a polynomial $u \in \widetilde{\mathbb{Q}}[y]/\langle y^{m+1}\rangle$ belongs to $A_m(y)$ if, and only if, the element $\dfrac{\psi^p(1+u)}{1+u}$ belongs to $1 + A_m(y)$, for all primes p. (If $\psi^p(1+u)/(1+u) = 1 + u_p$ lies in $1 + A_m(y)$ for all primes p, then according to (5.2) we have that $u \in \widetilde{\mathbb{Z}}[y]/\langle y^{m+1}\rangle$. Since $\psi^2(y) = 4y + y^2$ is in $A_m(y)$, one obtains that $\psi^2(1+u) \in 1 + A_m(y)$ for every $u \in \widetilde{\mathbb{Z}}[y]/\langle y^{m+1}\rangle$. This implies that $1 + u = (1+u_2)/\psi^2(1+u) \in 1 + A_m(y)$).

With the preceeding remark, the following Lemma is now proved in exactly the same way as (5.3).

<u>Lemma 6.2.</u> *The order of* $\widetilde{J}'(z)$ *in* $\widetilde{J}'(\mathbb{H}P^m)$ *is the smallest positive integer* c_m *such that the polynomial* $(\theta_p(y))^{2c_m} \in 1 + \widetilde{\mathbb{Q}}[y]/\langle y^{m+1}\rangle$ *belongs to* $1 + A_m(y)$, *for all prime numbers* p.

In the following two Lemmas we determine the p-adic valuation of the integer c_m.

<u>Lemma 6.3.</u> *Let* d *and* m *be integers* ≥ 1 *and let* p *be an <u>odd</u> prime. The following are equivalent.*

(i) : *The polynomial* $(\theta_p(y))^{2d}$ *belongs to* $1 + A_m(y)$.

(ii) : *The polynomial* $(\theta_p(y))^d$ *belongs to* $1+\tilde{\mathbb{Z}}[y]/\langle y^{m+1}\rangle$.

(iii) : $v_p(d) \geq \max\left\{s+v_p(s) \mid 0 \leq s \leq \left[\frac{2m}{p-1}\right]\right\}$.

<u>Proof.</u> (i) $\Leftrightarrow$ (ii) : If $(\theta_p(y))^d$ is integral, then $(\theta_p(y))^{2d} \in 1 + A_m(y)$, since the subring $\mathbb{Z} \oplus A_m(y)$ generated by $1, 2y, y^2$ contains all squares of $\mathbb{Z}[y]/\langle y^{m+1}\rangle$.

To prove the converse, we note that by definition the polynomial $\theta_p(y)$ has its coefficients in Q_p (see Chapter 0 and (5.4)). Therefore, if $(\theta_p(y))^d = 1 + w$ does not belong to $1 + \tilde{\mathbb{Z}}[y]/\langle y^{m+1}\rangle$ one of its coefficients has a power of p in its denominator and we conclude that the same must be true for the square $(\theta_p(y))^{2d}$.

(ii) $\Leftrightarrow$ (iii) : This follows from (5.5).

<u>Lemma 6.4.</u> *Let* d *and* m *be integers* ≥ 1. *The following are equivalent.*

(i) *The polynomial* $(\theta_2(y))^{2d}$ *belongs to* $1 + A_m(y)$.

(ii) *For all integers* j *with* $0 \leq j \leq m$, *the binomial coefficient* $\binom{d}{j}$ *is divisible by* $e_j 4^j$, *where* $e_j = 1$ *if* j *is even and* $e_j = 2$ *if* j *is odd.*

(iii) $v_2(d) \geq \max\{2j+v_2(j),\ 2m+1 \mid 0 \leq j \leq m\}$.

<u>Proof.</u> According to (5.4) we have $\theta_2(y) = (1+y/4)^{1/2}$ and hence,

$$(\theta_2(y))^{2d} = \sum_{j=0}^{m} \frac{1}{4^j}\binom{d}{j} y^j .$$

Since $1, 2y, y^2, \ldots, e_j y^j, \ldots, e_m y^m$ form a basis of the free abelian group $\mathbb{Z} \oplus A_m(y)$, we conclude that (i) and (ii) are equivalent.

Note that condition (ii) can be reformulated as follows:

$$\text{(ii)}' : v_2\binom{d}{j} \geq 2j + e_j - 1,\ 0 \leq j \leq m .$$

As in the proof of (5.5) (B) we show that if $v_2(d) \geq 2(m-1) \geq m - 1$, then

$$v_2\binom{d}{m} \geq 2m + e_m - 1 \Leftrightarrow v_2(d) \geq 2m + v_2(m) + e_m - 1 .$$

Starting with the case $m = 1$ we conclude now by induction on m that (ii)' holds if, and only if,

$$v_2(d) \geq \max\{2j + v_2(j) + e_j - 1 \mid 0 \leq j \leq m\} .$$

Since

$$2j + v_2(j) + e_j - 1 = \begin{cases} 2j+v_2(j) \geq 2j+1, & \text{if } j \text{ is even,} \\ 2j+1 & \text{, if } j \text{ is odd,} \end{cases}$$

the equivalence (ii)' $\Leftrightarrow$ (iii) is established and (6.4) is proved.

Theorem 6.5. *The* $\mathfrak{J}$*-order of the canonical line bundle over* $\mathbb{H}P^m$, $m \geq 1$, *is equal to the integer* c_m *given by*

$$v_2(c_m) = \max\left\{2j + v_2(j),\ 2m+1 \mid 0 \leq j \leq m\right\}$$

$$v_p(c_m) = \max\left\{s + v_p(s) \mid 0 \leq s \leq \left[\frac{2m}{p-1}\right]\right\},\ p \quad \textit{odd.}$$

In particular, $v_p(c_m) = 0$ *if* $p > 2m+1$ *and* $v_p(c_m) \geq 1$ *if* $p \leq 2m+1$.

Proof. Since $\mathfrak{J}(z)$ and $\mathfrak{J}'(z)$ have the same order, the Theorem follows from (6.2), (6.3) and (6.4).

Remark. The first three values of c_m are: $c_1 = 24$, $c_2 = 1440$ and $c_3 = 362\,880$.

We are now in a position to give a solution of the cross-section problem for the quaternionic Stiefel fibrations.

Theorem 6.6. *The quaternionic Stiefel fibration*

$$\mathbb{H}_{n,k} \longrightarrow S^{4n-1}, \quad n \geq k \geq 2,$$

has a cross-section if, and only if, n *is a multiple of the integer* c_{k-1}, *called the Quaternionic James number, defined in* (6.5).

Proof. If $\mathbb{H}_{n,k} \to S^{4n-1}$ has a cross-section, then by (2.5) and (6.5) the integer n is a multiple of c_{k-1}.

Conversely, let n be a multiple of c_{k-1}. Then,

$$v_2(n) \geq v_2(c_{k-1}) \geq 2k-1 \quad \text{and} \quad v_3(n) \geq v_3(c_{k-1}) \geq 1 .$$

Hence, for all $k \geq 2$, one has $n \geq 2^{2k-1} \cdot 3 > 2k$ and with (2.6) we conclude that $\mathbb{H}_{n,k} \to S^{4n-1}$ has a cross-section.

Remarks. (1) In view of (2.2), Theorem 6.6 gives the following result:

"*The sphere* S^{4n-1} *admits* m *linearly independent quaternionic vector fields if, and only if,* n *is a multiple of* c_m *.*"

(2) As for the complex case, no explicit formulas are known giving the quaternionic vector fields over S^{4n-1} (even for the case $m = 1$).

We conclude this section investigating some relations between the complex and the quaternionic James numbers b_{2k-1} and c_{k-1} .

Proposition 6.7. *The quaternionic James number* c_{k-1} *is either equal to* b_{2k-1} *or to* $(1/2)b_{2k-1}$.

Proof. The Proposition follows at once from (5.8) and (6.5). However, we give a second proof which does not depend on the explicit knowledge of c_{k-1} and b_{2k-1} .

The homomorphism $\tilde{J}(g) : \tilde{J}(\mathbb{H}P^{k-1}) \to \tilde{J}(\mathbb{C}P^{2k-1})$ induced by the canonical map $g : \mathbb{C}P^{2k-1} \to \mathbb{H}P^{k-1}$ sends $\tilde{J}(z)$ onto $\tilde{J}(2y)$ (see (3.7)). Hence $(1/2)b_{2k-1}$ divides c_{k-1}. According to (3.10) the homomorphism $h^! = (g|\mathbb{C}P^{2k-2})^!$ identifies $\widetilde{KO}(\mathbb{H}P^{k-1})$ with the ψ - subring $A_{k-1}(y)$ of $\widetilde{KO}(\mathbb{C}P^{2k-2}) \cong \mathbb{Z}[y]/\langle y^k\rangle$; moreover, $h^!(z) = 2y$. Since b_{2k-2} is equal to the order of $\tilde{J}'(y)$, there is a polynomial $w \in \tilde{\mathbb{Z}}[y]/\langle y^k\rangle$ such that

$$\theta_p(b_{2k-2}y) = \frac{\psi^p(1+w)}{1 + w}$$

for all primes p , and hence,

$$\theta_p(b_{2k-2}2y) = \frac{\psi^p((1+w)^2)}{(1+w)^2} ,$$

for all primes p . The subring $A_{k-1}(y)$ contains $2w + w^2$ for any $w \in \tilde{\mathbb{Z}}[y]/\langle y^k\rangle$ and we conclude that c_{k-1} divides b_{2k-2} (compare with (6.2)). But b_{2k-2} divides b_{2k-1} (actually, $b_{2k-2} = b_{2k-1}$) and it follows that $c_{k-1} = (1/2)b_{2k-1}$ or $c_{k-1} = b_{2k-1}$.

Remark. Geometrically the equality $c_{k-1} = (1/2)b_{2k-1}$ means that if $2n \equiv 0$ (mod. b_{2k-1}), the complex vector fields on S^{4n-1} can be chosen "quaternionic", that is to say, S^{4n-1} admits $2k-1$ complex vector fields that stem from $k-1$ quaternionic vector fields. (The identification of the quaternions with the complex 2×2-matrices of the form

$$\begin{pmatrix} u & v \\ -\bar{v} & \bar{u} \end{pmatrix}$$

provides a canonical embedding of $\mathbb{H}_{n,k}$ into $\mathbb{C}_{2n,2k}$ compatible with the projections onto S^{4n-1}. A cross-section of $\mathbb{H}_{n,k} \to S^{4n-1}$ therefore gives rise to a cross-section of $\mathbb{C}_{2n,2k} \to S^{4n-1}$).

We now give a description of the integers k for which $c_{k-1} = (1/2)b_{2k-1}$.

Theorem 6.8. *Let* $k \geq 2$. *Then,* $c_{k-1} = (1/2)b_{2k-1}$ *if, and only if,* k *belongs to the set*

$$A = \{s + r\cdot 2^{2s-1} \mid s,r \in \mathbb{N}\}$$

(Here $\mathbb{N}$ is the set of integers ≥ 1).

Before giving the proof, let us draw some consequences which ilustrate the result.

(a) If k is odd, $c_{k-1} = (1/2)b_{2k-1}$. Indeed, all odd integers belong to A.

(b) The first even integer k for which $c_{k-1} = (1/2)b_{2k-1}$ is 10, that is, $c_9 = (1/2)b_{19}$.

Proof of (6.8). We first recall from (5.8) and (6.5) that

$$v_2(b_{2k-1}) = \max\{j + v_2(j) \mid 0 \leq j \leq 2k-1\}$$

and

$$\begin{aligned} v_2(c_{k-1}) &= \max\{2t + v_2(t),\ 2k-1 \mid 0 \leq t \leq k-1\} \\ &= \max\{j + v_2(j) - 1, 2k-1 \mid 0 \leq j \leq 2k-1\} \end{aligned}$$

We now show that for any integer $k \geq 2$ the following four properties are equivalent.

(i) $c_{k-1} = (1/2)b_{2k-1}$.

(ii) $v_2(b_{2k-1}) > 2k - 1$.

(iii) There exists an integer s such that $0 < s < k$ and $k \equiv s \pmod{2^{2s-1}}$.

(iv) $k \in A$.

The equivalence (i) $\Leftrightarrow$ (ii) follows immediately from the comparison of $v_2(c_{k-1})$ and $v_2(b_{2k-1})$. The equivalence (iii) $\Leftrightarrow$ (iv) is trivial.

(ii) $\Rightarrow$ (iii) : If $v_2(b_{2k-1}) > 2k - 1$ there exists an even integer j such that $1 \leq j < 2k - 1$ and $j + v_2(j) > 2k - 1$. Setting $j = 2k - 2s$ we obtain a number s such that $0 < s < k$ and $2k - 2s + v_2(2k-2s) > 2k - 1$. This latter condition is equivalent to $k - s \equiv 0 \pmod{2^{2s-1}}$.

(iii) $\Rightarrow$ (ii) : If $j = 2k - 2s$ then

$$v_2(b_{2k-1}) \geq (2k-2s) + v_2(2k-2s) \geq (2k-2s) + 2s > 2k - 1 .$$

The theorem is now proved.

Finally, we turn to an evaluation of the density of the set A in $\mathbb{N}$. Any integer of A belongs to exactly one arithmetic progression $A_q = \{q + r\cdot 2^{2q-1}\}$ with $q \notin A$. Moreover the set A is the disjoint union

$$A = \bigcup_{q \notin A} A_q$$

and its density α is equal to $\sum_{m \notin A} 2^{-(2m-1)}$. We then write

$$\alpha = \sum_{m \notin A} 2^{-(2m-1)} = \sum_{m \in N} 2^{-(2m-1)} - \sum_{m \in A} 2^{-(2m-1)}$$

$$= \frac{2}{3} - \sum_{n \notin A} \sum_{m \in A_n} 2^{-(2m-1)}$$

$$= \frac{2}{3} - \sum_{n \notin A} \sum_{m=1}^{\infty} 2^{-(2n-1)-m\cdot 2^{2n}}$$

$$= \frac{2}{3} - \sum_{n \notin A} 2^{-(2n-1)}\left(2^{(2^{2n})} - 1\right)^{-1}$$

$$= \frac{2}{3} - \left\{ \frac{1}{2(2^4-1)} + \frac{1}{2^3(2^{16}-1)} + \frac{1}{2^7(2^{256}-1)} \cdots \right\} .$$

The series converges in an extraordinarily rapid fashion.

Using the fact that A contains sequences of consecutive integers of arbitrary length, one deduces easily from Liouville's criterion that α is a transcendental number. The approximate value of α is 0.63. Hence, in about 63 % of all cases, the James number c_{k-1} is equal to $(1/2)b_{2k-1}$.

CHAPTER 5

SPAN OF SPHERICAL FORMS

1. Introduction and Generalities about Spherical Forms

In this chapter we will study the vector field problem for the quotient manifolds associated to free actions of a finite group G on the sphere S^n. Among the manifolds S^n/G to be considered, we find an important class of riemannian manifolds with constant curvature, namely the *Spherical Space Forms* which, for reasons of simplicity, we shall just call *Spherical Forms*. The present chapter aims at determining the span of the spherical forms, that is to say, the maximum number of linearly independent vector fields on these manifolds.

Except for the dimension 7, the span of an n-dimensional spherical form depends only on the dyadic valuation of $n+1$ and on the order $|G|$ of its defining group G. Complete results have been given by J.C.Becker [40] under the assumption that whenever the set of the 2-Sylow subgroups of G contains a generalized quaternion group Q_m then G is a metabelian group of rank 2. These results, stated in §§ 5 and 6, are independent of the choice of the free action of G on S^n. This is not the case in dimension 7, where the action must be taken into account; we shall see this in the following chapter, in which we discuss the problem of the parallelizability of the spherical forms.

The methods used to compute the span of spherical forms are just extensions towards the G-equivariant side of methods developped by J.F.Adams in [2]. Basically the ideas we present here are very close to those explained in the previous chapter. Thus, after a brief preliminary on spherical forms we shall recall some of the properties concerning the notion of G-fibre homotopy type (§ 3) ; then, we shall imitate the work of J.F.Adams to deal with G-(co)reducibility (§ 4). In all this, we adopt and follow Becker's point of view.

The reader who is interested in the various problems related to vector fields on manifolds is referred to the initial articles of H.Hopf [120] and E.Stiefel [233] (written, respectively, in 1927 and 1936), then to the papers of E.Thomas [240] and M.F.Atiyah [18] which state general results; next, he or she should go to [53] for π - manifolds, to [239] for low-dimensional manifolds and finally, to [222] for quotient manifolds. The notion 'Spherical Space Form' came up in connection with the Clifford-Klein Problem, which consisted in giving a description of all connected, complete riemannian manifolds of constant curvature. Indeed, the work of Killing [150] and Hopf [119] show that a riemannian manifold M of dimension $n \geq 2$ is a connected, complete manifold of constant curvature $k > 0$ if, and only if, M is isometric to S^n/G, where G is a finite subgroup of $O(n+1)$ which acts freely on the sphere S^n. It is such a manifold M that we call a *Spherical Form;* by abuse of language, any representative of the isometry class of S^n/G will also be called a Spherical Form. Furthermore S^1, which is homeomorphic to S^1/G, will also be considered as a Spherical Form. The classical properties of group actions give the following result.

Proposition 1.1. *Every Spherical Form* $M^n = S^n/G$ *is a connected, compact, orientable manifold without boundary, with a* CW-*structure and such that* $\pi_1(M) = G$ *and* $\pi_i(M) \cong \pi_i(S^n)$ *if* $i \neq 1$.

The classification of Spherical Forms can be reduced to the determination of all finite groups having fixed point free real orthogonal representations and therefore, of all those groups admitting fixed point free unitary representations. (Our reference for all the facts concerning representation theory of finite groups is [67]). This work, done first in dimension 3 by Seifert and Threlfall [211], was continued by J.A.Wolf [259] after the decisive contribution of G.Vincer [247]. The results obtained may be stated as follows.

<u>Proposition 1.2.</u> *Let* G *be a non-trivial finite group acting linearly and freely on* S^N .

1. *If* $N = 2n$, G *is isomorphic to* $\mathbb{Z}_2$ *and the only even-dimensional spherical forms are the real projective spaces* $\mathbb{R}P^{2n}$.
2. *If* $N = 2n + 1$ *the following hold:*

    (i) *if* G *is abelian, then* G *is cyclic;*

    (ii) *if* G *is non-abelian, it has the following equivalent properties:*

    (a) *all the abelian subgroups of* G *are cyclic;*

    (b) *the* p - *Sylow subgroups of* G *are of one of the following two types:*

    (I) *all of them are cyclic;*

    (II) *they are cyclic for* $p \neq 2$ *and generalized quaternionic groups for* $p = 2$.

The spherical forms corresponding to 2(i) are called *Lens Spaces*; if G is a generalized quaternionic group Q_m the associated spherical form is called *Quaternionic Spherical Form* or, more simply, Q_m - *spherical form.*
Let us mention that the complete classification of the groups satisfying 2(ii)(a) results from the papers of Zassenhaus [270] and Suzuki [235] : there are six classes of groups among which two are constituted by non-solvable groups (see [260; pages 179 and 195]).

Let M be a spherical form viewed as S^n/G where G acts freely on S^n by way of an orthogonal fixed point free representation ρ . Let G_p be a p - Sylow subgroup of G ; the canonical inclusion $i_p : G_p \to G$ induces a ring homomorphism $i_p^* : RO(G) \to RO(G_p)$ and defines a fixed-point free orthogonal representation $\rho_p = i_p^*(\rho)$ of G_p, by way of which G_p acts freely on S^n. Let $M_p = S^n/G_p$ be the quotient manifold thus obtained. If G_p' is another p - Sylow subgroup of G , the spherical form M_p' associated to G_p' is isometric to M_p ; indeed, since G_p and G_p' are conjugated in G , the representations ρ_p and ρ_p' are equivalent. This justifies us to call M_p the *Spherical* p - *form associated to* M.

Notice that because of Proposition 1.2, the spherical p-forms associated to a given spherical form are real projective spaces, lens spaces of order p^m or Q_m-spherical forms. We finally note that $\pi_p : S^n/G_p = M_p \to S^n/G = M$ is a covering fibration; this is one of the interesting facts about associated spherical p-forms.

Lemma 1.3. *Let* M *be an* n-*dimensional spherical form with* $\pi_1(M) = G$. *Then the integral (ordinary) cohomology groups of* M *are:*

$$H^i(M;\mathbb{Z}) = \begin{cases} \mathbb{Z}\ , & \textit{if}\ \ i = 0\ \ \textit{or}\ \ i = n\ ; \\ H^i(G;\mathbb{Z})\ , & \textit{if}\ \ 0 < i < n\ ; \\ 0\ , & \textit{otherwise}. \end{cases}$$

The lemma follows easily from the fact that S^n is a universal G-covering space of M and from (1.1).

Let now h^* be a generalized reduced cohomology theory. Suppose that the spherical form is $(2n+1)$-dimensional and let M_o be the $2n$-skeleton of M. We shall denote the order of $\pi_1(M) = G$ by q. Then the Atiyah-Hirzebruch spectral sequence for M or M_o is trivial in each of the following two cases:

(i) $h^* = \widetilde{K}O^*$ and q odd;

(ii) $h^* = \widetilde{K}U^*$

(for more general conditions see [162]). Under the preceeding assumptions we also have:

Proposition 1.4. (i) $h^{2s+1}(M) = H^{2n+1}(M;h^{2(s-n)}(S^o))$,

$$h^{odd}(M_o) = 0\ ;$$

(ii) *if* $\mathcal{G}$ *denotes the graded group associated to a convenient filtration,*

$$\mathcal{G}(h^{2s}(M)) = \mathcal{G}(h^{2s}(M_o)) \oplus H^{2n+1}(M;h^{2(s-n)-1}(S^o))\ ,$$

$$\mathcal{G}(h^{2s}(M_o)) = \bigoplus_{i=1}^{n} H^{2i}(G;h^{2(s-i)}(S^o))\ .$$

Using the Atiyah-Hirzebruch spectral sequence, we shall see how the K-theory of a spherical form is related to that one of its associated spherical p-forms.

<u>Lemma 1.5.</u> *Let* N *be a normal subgroup of a finite group* G *such that* G/N *is isomorphic to a* p - *Sylow subgroup* G_p *of* G . *Then, for every* $i > 0$, *the inclusion* $G_p \to G$ *induces an isomorphism*

$$H^i(G;\mathbf{Z})_{(p)} \longrightarrow H^i(G_p;\mathbf{Z}) \quad ,$$

where $H^i(G;Z)_{(p)}$ *is the* p - *primary component of* $H^i(G,Z)$.

<u>Proof.</u> Let r (resp. s) be the canonical inclusion of N (resp. G_p) into G . Let us consider the Hochschild-Serre spectral sequence

$$E_2^{i,j} = \hat{H}^i(G/N \; ; \; \hat{H}^j(N,\mathbf{Z})) \Rightarrow \hat{H}^{i+j}(G;\mathbf{Z})$$

which relates the Artin-Tate cohomology of a group to those of one of its normal subgroups. Since $(|N|,p) = 1$, we are led to a split exact sequence [127; page 127]

$$0 \longrightarrow \hat{H}^i(G_p;\mathbf{Z}) \xrightarrow{t_i} \hat{H}^i(G;\mathbf{Z}) \xrightarrow{r_i} \hat{H}^i(N;\mathbf{Z})^G \longrightarrow 0$$

for any $i \in \mathbf{Z}$. In the preceeding sequence, t_i and r_i are respectively, the natural transfer and restriction homomorphisms (the latter induced by r), $\hat{H}^i(N;\mathbf{Z})^G$ is the subgroup of $\hat{H}^i(N;\mathbf{Z})$ defined by the elements which are invariant under the action of G . Let u_i (resp. s_i) be the section of t_i (resp. the homomorphism induced by s); the definition of the transfer homomorphism t_i (cf. [62;page 255]) shows that $\mu_i = t_i s_i$ is multiplication by $[G : G_p] = |N|$ in $\hat{H}^i(G;\mathbf{Z})$. It follows that μ_i is onto; moreover, by killing the $|N|$ - primary component of $\hat{H}^i(G;\mathbf{Z})$ we obtain an automorphism of $\hat{H}^i(G;\mathbf{Z})_{(p)}$. Since

$$s_i = (u_i t_i) s_i = u_i \mu_i \quad ,$$

it is clear that s_i is onto and that $s_i | \hat{H}^i(G;\mathbf{Z})_{(p)}$ is an isomorphism.

<u>Theorem 1.6.</u> [169] *Let* $M = S^{2n+1}/G$ *be a spherical form with* G *a semi-direct product of a* p - *group* G_p *and of a group* N *whose orders are relatively prime. Let* M_p *be the covering space* S^{2n+1}/G_p *of* M *induced by the inclusion of*

G_p *in* G. *Then,* $\widetilde{K\mathbb{F}}(M_p)$ *is a direct summand of* $K\mathbb{F}(M)$ *(where* $\mathbb{F} = \mathbb{R}$ *or* $\mathbb{C}$*). More precisely,*

(i) $\widetilde{KU}(M)_{(p)}$ *is isomorphic to* $\widetilde{KU}(M_p)$;

(ii) $\widetilde{KO}(M)_{(2)}$ *is isomorphic to* $\widetilde{KO}(M_2)$;

(iii) *if* $p \neq 2$,

$$\widetilde{KO}(M)_{(p)} = \begin{cases} \widetilde{KO}(M_p) & \text{if } n \not\equiv 0 \pmod{4}, \\ \widetilde{KO}((M_p)_o) & \text{if } n \equiv 0 \pmod{4}. \end{cases}$$

Proof. Let π be the covering map $M_p \to M$. The homomorphisms $\pi^i : \tilde{H}^i(M;\mathbf{Z}) \to \tilde{H}^i(M_p;\mathbf{Z})$ induced by π are onto for $i \neq 2n+1$ this follows from 1.3 and 1.5, and for $i = 2n+1$, because π is orientation preserving. Since the functor $\otimes_{\mathbf{Z}}$ (resp. $\mathrm{Tor}_1^{\mathbf{Z}}$) is right exact (resp. additive), it follows that for every integer i the homomorphisms induced by

$$\tilde{\pi}_2^i : E_2^{i,-i}(M) = \tilde{H}^i(M;\widetilde{K\mathbb{F}}^{-i}(S^o)) \longrightarrow E_2^{i,-i}(M_p) = \tilde{H}^i(M_p;\widetilde{K\mathbb{F}}^{-i}(S^o))$$

in the Atiyah-Hirzebruch spectral sequence are onto. More precisely,except for the case in which $\mathbb{F} = \mathbb{R}$, $p \neq 2$, $i = 2n+1$ and $n \equiv 0 \pmod{4}$, the restriction of $\tilde{\pi}_2^i$ to the p-primary component $(E_2^{i,-i}(M))_{(p)}$ of $E_2^{i,-i}(M)$ is an isomorphism onto $E_2^{i,-i}(M_p)$. The additivity of the functor H^* from the abelian category of complexes of abelian groups into the category of abelian groups, shows that the same is still true for $\tilde{\pi}_r^i : E_r^{i,-i}(M) \to E_r^{i,-i}(M_p)$ for any $r \geq 2$; hence the result. When $\mathbb{F} = \mathbb{R}$, $p \neq 2$ and $n \equiv 0 \pmod{4}$, the term $E_2^{2n+1,-2n-1}(M_p)$ is isomorphic to $\mathbf{Z}_2$ and so, the preceeding argument is still valid if M_p is replaced by its 2n-skeleton $(M_p)_o$.

Remark. The proof of the previous Theorem does not require any particular assumption on the Atiyah-Hirzebruch spectral sequences. Besides, it does not depend on the (free) action of G on S^{2n+1} ; indeed, it still holds for any manifold having the same homotopy type as a spherical form.

Applying the Theorem to the lens spaces $S^{2n+1}/\mathbb{Z}_a$ $(a > 1)$, we obtain immediately (compare with [166]),

Corollary 1.7. *Let* $a = \prod_{i=1}^{s} p_i^{r_i}$ *, with* $2 \le p_1 < \dots < p_s$ *, be the prime decomposition of* a *. Then,*

1. $\widetilde{KU}(S^{2n+1}/\mathbb{Z}_a) \cong \bigoplus_{i=1}^{s} \widetilde{KU}(S^{2n+1}/\mathbb{Z}_{p_i^{r_i}})$;

2. (i) *if* $n \not\equiv 0 \pmod{4}$,

$$\widetilde{KO}(S^{2n+1}/\mathbb{Z}_a) \cong \bigoplus_{i=1}^{s} \widetilde{KO}(S^{2n+1}/\mathbb{Z}_{p_i^{r_i}}) \ ;$$

(ii) *if* $n \equiv 0 \pmod{4}$,

$$\widetilde{KO}(S^{2n+1}/\mathbb{Z}_a) \cong \begin{cases} \mathbb{Z}_2 \oplus \left[\bigoplus_{i=1}^{s} \widetilde{KO}\Big((S^{2n+1}/\mathbb{Z}_{p_i^{r_i}})_o \Big) \right] & \\ \textit{if } p_1 > 2 , & \\ \widetilde{KO}(S^{2n+1}/\mathbb{Z}_{2^{r_i}}) \oplus \left[\bigoplus_{i=2}^{s} \widetilde{KO}\Big((S^{2n+1}/\mathbb{Z}_{p_i^{r_i}})_o \Big) \right] & \\ \textit{if } p_1 = 2 . & \end{cases}$$

Theorem 1.6 often enables us to determine some of the primary components of the Grothendieck groups of other spherical forms. Let us mention, for instance, the case of the Dihedral Spherical Forms S^{4n+3}/D_a^* for which $K\mathbb{F}(S^{4n+3}/D_a^*)_{(2)}$ can be computed (an example of the computation of the odd primary components is given for Tetrahedral Spherical Forms in [168]). For completeness, we recall the definition of the group D_a^*.

Definition 1.8. *The generalized binary dihedral group* D_a^* *is the group generated by* x *and* y *, with presentation*

$$x^{2a} = 1,\ x^a = y^2,\ yxy^{-1} = x^{-1} \ .$$

For $m \ge 2$ *the group* $D^*_{2^{m-1}}$ *is the generalized quaternionic group* Q_m .

Suppose that $a = 2^{m-1}u$, with $m \ge 1$ and u odd;

then (1.6) gives the following.

Corollary 1.9.

$$\tilde{K}\mathbb{F}(S^{4n+3}/D_a^*)_{(2)} = \begin{cases} \tilde{K}\mathbb{F}(S^{4n+3}/\mathbb{Z}_4) & \text{if } m = 1 ; \\ \tilde{K}\mathbb{F}(S^{4n+3}/Q_m) & \text{if } m \geq 2 \end{cases}$$

Proof. Since $[D_a^*, D_a^*] = \{1, yxy^{-1}x^{-1} = x^{-2}\}$, we see that, for a odd, the quotient group $D_a^*/[D_a^*, D_a^*]$ is isomorphic to the 2-Sylow subgroup $\mathbb{Z}_4$ generated by y. When $m \geq 2$, the subgroup H_u of D_a^* generated by x^{2^m} is a normal subgroup of D_a^* and it is isomorphic to $\mathbb{Z}_u$. It is now easy to show that D_a^*/H_u is isomorphic to Q_m.

2. Vector Fields on Spherical Forms

In the previous chapter we have seen how to construct real vector fields on spheres by means of the Clifford algebras C_k (see 4.4). We may proceed in the same way with the $\mathbb{F}$-algebras $C_k \otimes_{\mathbb{R}} \mathbb{F}$, where $\mathbb{F} = \mathbb{C}$ or $\mathbb{H}$; whenever $\mathbb{R}^{d\cdot n} \cong \mathbb{F}^n$ ($d = \dim_{\mathbb{R}}(\mathbb{F})$) is a $C_k \otimes_{\mathbb{R}} \mathbb{F}$ module, we obtain k vector fields on $S^{d\cdot n-1} \subset \mathbb{R}^{d\cdot n}$, which are linearly independent *over* $\mathbb{R}$ (not $\mathbb{F}$!). Moreover, these fields are *compatible with the action of* S^{d-1} *on* $S^{d\cdot n-1}$ *and* $R^{d\cdot n}$, respectively.

Let m be a positive integer. The *Hurwitz-Radon number* $\rho(m,\mathbb{F})$(*) *associated to* m *and* $\mathbb{F}$ is the greatest integer q such that $\mathbb{R}^m$ admits a $C_{q-1} \otimes_{\mathbb{R}} \mathbb{F}$-module structure. It is given as follows (see [26] and [134]): Write $v_2(m) = 4c + e$ with $0 \leq e \leq 3$, then

		e=0	e=1	e=2	e=3
(2.0)	$\rho(m,\mathbb{R})$	8c+1	8c+2	8c+4	8c+8
	$\rho(m,\mathbb{C})$	8c	8c+2	8c+4	8c+6
	$\rho(m,\mathbb{H})$	8c−2 (c > 0)	8c	8c+4	8c+5

(*) See Chapter 4, § 4

The previous table will enable us to obtain some interesting upper and lower bounds of the span of some spherical forms. We begin by observing that for any N-dimensional spherical form M, we have

(2.1) $$\operatorname{span}(M) \leq \operatorname{span}_{\mathbb{R}}(S^N) = \rho(N+1,\mathbb{R})-1 .$$

Next, notice that the methods used to determine $\operatorname{span}_{\mathbb{R}}(S^n)$ prove that a system of $\rho(N+1,\mathbb{R})-1$ vector fields on S^N remains invariant under the antipodal action. It follows that

(2.2) $$\operatorname{span}(\mathbb{R}P^N) = \operatorname{span}_{\mathbb{R}}(S^N) ,$$

for any N.

Let us now suppose that N is odd, that is $N = 2n+1$, and let us consider the rotation γ on the sphere $S^{2n+1} = \left\{(z_o,\dots,z_n) \in \mathbb{C}^{n+1} \mid \sum_{k=o}^{n} |z_k|^2 = 1\right\}$, defined by

$$\gamma(z_o,z_1,\dots,z_n) = \left(\zeta^{a_o}\cdot z_o,\ \zeta^{a_1}\cdot z_1,\dots,\ \zeta^{a_n}\cdot z_n\right) ,$$

where $\zeta = \exp(2\pi i/a)$, with $a,a_o,a_1,\dots,a_n \in \mathbb{N}^*$ and such that $a \neq 1$ and for every $k \in \{0,1,\dots,n\}$, $(a_k,a) = 1$.

Under these conditions, the group $\Gamma = \{\gamma^i\}_{1<i<a}$ is cyclic of order a, isomorphic to $\mathbb{Z}_a$, acting freely and differentiably on S^{2n+1}.

<u>Definition 2.3.</u> *The spherical form* S^{2n+1}/Γ *is called Generalized Lens Space of order* a (or, mod.a); *we denote it by* $L^n(a;a_o,a_1,\dots,a_n)$. *When* $a_o = a_1 = \dots = a_n = 1$, $L^n(a;1,1,\dots,1)$ *is denoted simply by* $L^n(a)$ *and called Ordinary Lens Space of order* a.

Thus $L^n(a)$ is obtained as an orbit space when $\mathbb{Z}_a$ acts on S^{2n+1} by way of the standard action $\gamma(z) = \zeta z$. Equivalently, one can consider the standard unitary representation of degree $n+1$, $\sigma : \mathbb{Z}_a \to U(n+1)$ defined by

$$\sigma(\overline{1}) = \zeta \cdot 1_{\mathbb{C}^{n+1}}$$

where $\overline{1}$ represents the coset $1 + a\mathbb{Z}$; if we take $\mathbb{Z}_a$ as a subgroup of the multiplicative group S^1 of the complex numbers with norm 1, σ can be identified with the restriction to $\mathbb{Z}_a$ of the standard representation of S^1 on $\mathbb{C}^{n+1} \cong \mathbb{R}^{2n+2}$. Notice that the complex vector fields (referred to in 2.0) remain invariant by such action; hence we conclude that

$$(2.4) \qquad \operatorname{span}(L^n(a)) \geq \rho(2n+2, \mathbb{C}) - 1 \quad .$$

<u>Remark.</u> Yoshida [265] established the inequality (2.4) noting that

$$\rho(2n+2,\mathbb{C}) - 1 = 2t + 1$$

with $t = v_2(n+1)$, and working directly on the $(2t+1)$ - linearly independent vector fields exhibited by Eckmann [85; page 365]. Indeed, he proved the existence of $2t + 1$ unitary matrices $A_1, A_2, \ldots, A_{2t+1}$ in $U(n+1)$ such that

$$A_k^2 = -1_{n+1} \; , \; A_k A_\ell + A_\ell A_k = 0 \quad ,$$

for any $1 \leq k, \ell \leq 2t + 1$ and $\ell \neq k$. Under these conditions, we have for every $z \in S^{2n+1}$,

$$(z|A_k(z)) = \left(A_k(z)|A_k^2(z)\right) = (A_k(z)|-z) = -\overline{(z|A_k(z))} .$$

In other words, for any $z \in S^{2n+1}$, the inner product $(z|A_k(z))$ has a trivial real part and so, A_k is a vector field on S^{2n+1}. This field remains invariant by the standard action of $\mathbb{Z}_a$ because $A_k(\zeta \cdot z) = \zeta \cdot A_k(z)$, and hence, it defines a vector field on $L^n(a)$. Now, the $2t + 1$ vector fields thus built on $L^n(a)$ are linearly independent since, for any $z \in S^{2n+1}$, we have for $1 \leq k \neq \ell \leq 2t + 1$

$$(A_k(z)|A_\ell(z)) = (A_\ell A_k(z)|-z) = (-A_k A_\ell(z)|-z) = \overline{(z|A_k A_\ell(z))}$$

$$(A_k(z)|A_\ell(z)) = -(A_k(z)|A_\ell(z)) \quad ,$$

that is to say, $\operatorname{Re}(A_k(z)|A_\ell(z)) = 0$.

Finally, if we compare (2.1) and (2.4) we see that table (2.0) gives the span of the ordinary lens spaces mod. a (a beeing not necessarily prime) when $v_2(2n+2) \equiv 1$ or 2 (mod.4). We shall see later that $\operatorname{span}(L^n(a))$ is always equal to $\operatorname{span}_R(S^{2n+1})$ provided that $n \neq 3$ and a odd (for $a = 2$, see (2.2)) ; this result, conjectured by Sjerve [222; page 104] under a general form, has been proved by Yoshida [266]. (Using the techniques of obstruction theory developed in [221], a partial result has been given before in [135]).

We are now going to study the results which can be deduced from our knowledge of $\rho(n,\mathbb{H})$. As for the complex case, we will get a lower bound for the minimal number of linearly independent vector fields on certain spherical forms, namely, on some Q_m-spherical forms.

We take this opportunity to describe the ring $RU(Q_m)$ of the (non-equivalent) unitary representations of the generalized quaternion group Q_m (cf. Def. 1.8).

The order of Q_m is 2^{m+1} $(m \geq 2)$ and, writing its elements as $x^k y^\ell$ with $0 \leq k \leq 2^m - 1$, $\ell = 0,1$, the conjugacy classes of Q_m are

$$C_o = \{x^{2k}y \mid 0 \leq k < 2^{m-1}\} \ , \ C_1 = \{x^{2k+1}y \mid 0 \leq k < 2^{m-1}\}$$

$$C_{k+2} = \{x^k, x^{-k}\} \text{ with } 0 \leq k \leq 2^{m-1} \ .$$

Since the commutator group $[Q_m,Q_m]$ is isomorphic to $\{1,x^2\}$, it follows that Q_m has 4 irreducible unitary representations of degree 1 : ξ_o,ξ_1,ξ_2 and ξ_3 defined by

$$\xi_o(x) = 1 \ ; \ \xi_1(x) = 1 \ ; \ \xi_2(x) = -1 \ ; \ \xi_3(x) = -1 \ ;$$

$$\xi_o(y) = 1 \ ; \ \xi_1(y) = -1 \ ; \ \xi_2(y) = 1 \ ; \ \xi_3(y) = -1 \ .$$

The number of conjugacy classes of Q_m shows that Q_m has $2^{m-1} - 1$ other irreducible unitary representations (all of degree 2), given by

$$\zeta_{r+3}(x) = \begin{pmatrix} \zeta^r & 0 \\ 0 & \zeta^{-r} \end{pmatrix} \ , \quad \zeta_{r+3}(y) = \begin{pmatrix} 0 & (-1)^r \\ 1 & 0 \end{pmatrix}$$

where $\zeta = \exp(i\pi/2^{m-1})$ and $1 \leq r \leq 2^{m-1}$. The construction of their character table shows that the representations quoted before are not equivalent; furthermore, it enables us to obtain the multiplicative structure of $RU(Q_m)$ by means of the tensor product of characters.

<u>Proposition 2.5.</u> (cf [196]) *As a group,* $RU(Q_m)$ *is the free abelian group generated by* $(\xi_r)_{0 \leq r \leq 2^{m-1}+2}$. *Its multiplicative structure is given by the following relations:*

$$\xi_0 = 1,\ \xi_1^2 = \xi_2^2 = 1,\ \xi_1\xi_2 = \xi_3,\ \xi_1\xi_4 = \xi_4,\ \xi_2\xi_4 = \xi_{2^{m-1}+2},$$

$$\xi_4^2 = \begin{cases} 1 + \xi_1 + \xi_2 + \xi_3 & \textit{if} \quad m = 2, \\ 1 + \xi_1 + \xi_5 & \textit{if} \quad m > 3, \end{cases}$$

$$\xi_r\xi_s = \xi_s\xi_r, \quad \textit{if} \quad 0 \leq r, s \leq 2^{m-1} + 2,$$

$$\xi_4\xi_r = \xi_{r+1} + \xi_{r-1}, \quad \textit{if} \quad 5 \leq r \leq 2^{m-1} + 1 \quad (\textit{when} \quad m \geq 3).$$

It is clear that $\xi_0, \xi_1, \xi_2, \xi_3$ are not fixed point free representations of Q_m; indeed, the only (inequivalent) irreducible fixed point free unitary representations of Q_m are the ξ_{r+3} where r *is odd* [260; pages 171-172]. It follows that the only spheres S^N on which Q_m acts freely are the spheres S^{4n+3} $(n \geq 0)$, the action being given by way of a virtual representation ρ of the type

$$\rho = \sum_{i=1}^{s} n_i\rho_i$$

with $\sum_{i=1}^{s} n_i = n + 1$ and $\rho_i \in \left\{\xi_{r+3} \mid 1 \leq r\text{=odd} \leq 2^{m-1}-1\right\}$, $1 \leq i \leq s$.

<u>Definition 2.6.</u> *Let* $r_0, \ldots, r_n$ *be* $n + 1$ *integers (not necessarily distinct) of* $\{1,3,5,\ldots,2^{m-1}-1\}$. *Then,* $N^n(m;r_0,r_1,\ldots,r_n)$ *is the Generalized* Q_m *- Spherical Form* S^{4n+3}/Q_m *where* Q_m *acts by way of* $\rho = \sum_{i=0}^{n} \xi_{r_i+3}$. *When all the* r_i's *are equal*

to a same number r *, we shall write* $N^n(m;r)$ *for* $N^n(m;r,r,\dots,r)$ *; finally, if* $r = 1$*, the form* $N^n(m;1)$ *will be denoted by* $N^n(m)$ *and called Ordinary* Q_m *- Spherical Form.*

If we consider the unit sphere S^3 as the (multiplicative) group of quaternions with norm 1 , that is to say, $S^3 = Sp(1)$, one can identify Q_m with the subgroup of S^3 generated by $\zeta = \exp(i\pi/2^{m-1})$ and j , by the correspondence $x \to \zeta$, $y \to j$. With these assumptions, the ordinary Q_m - spherical form $N^n(m)$ is no other than the form which corresponds to the restriction to Q_m of the standard representation of S^3 in $\mathbb{R}^{4n+4} \cong \mathbb{H}^{n+1}$. When $N \equiv 3$ (mod.4), the quaternionic vector fields on S^N (referred to in 2.0) are invariant under such an action; we conclude that

(2.7) $\quad \mathrm{span}(N^n(m)) \geq \rho(4n+4,\mathbb{H}) - 1$.

The reader will notice that the span of $N^n(m)$ coincides with $\mathrm{span}_{\mathbb{R}}(S^{4n+3})$ when $v_2(n+1) \equiv 0$ (mod.4) (use table (2.0)). For convenience, we summarize (2.2) and the inequalities (2.1), (2.4) and (2.7) in the following Proposition.

<u>Proposition 2.8.</u> (i) *For any* N *- dimensional spherical form* M *,* $\mathrm{span}(M) \leq \mathrm{span}_{\mathbb{R}}(S^N)$*. In particular,* $\mathrm{span}(\mathbb{R}P^N) = \mathrm{span}_{\mathbb{R}}(S^N)$.

(ii) *Moreover,*

$$\mathrm{span}(L^n(a)) \geq \rho(2n+2,\mathbb{C}) - 1 \quad ,$$

$$\mathrm{span}(N^n(m)) \geq \rho(4n+4,\mathbb{H}) - 1 \quad .$$

The above proposition and the arguments which preceeded it show how the determination of the span of an arbitrary spherical form depends on the kind of its associated spherical 2 - forms. For this reason, we shall classify the spherical forms according to the following terminology.

<u>Definition 2.9.</u> *A Spherical Form is said to be of cyclic type (resp. of quaternionic type) if its associated spherical 2 - form has a cyclic group* $\mathbb{Z}_{2m}$ $(m > 0)$ *(resp. a generalized quaternionic group* Q_m $(m \geq 2)$*) as its defining group.*

<u>Remark 2.10.</u> The reader who is familiar with Clifford Algebras (cf. [26] or [134]) will easily deduce the preceeding results

from the fact that if $\mathbb{R}^{N+1}$ is a $C_{k-1} \otimes \mathbb{F}$ - module then S^N admits $k - 1$ linearly independent vector fields which are S^{d-1} - equivariant under the standard action of S^{d-1} on $\mathbb{F}^{N+1/d} \cong \mathbb{R}^{N+1}$ (see beginning of § 2).

In this context, denoting by $a_k(\mathbb{F})$ the smallest integer n such that $\mathbb{R}^{dn}$ is a $C_{k-1} \otimes \mathbb{F}$ - module, the integer $\rho(m,\mathbb{F})$ appears as the greatest integer k such that

$$m \equiv 0 \pmod{d \cdot a_k(\mathbb{F})} .$$

The reader can find more observations on Clifford Algebras in the previous chapter.

3. G - Fibre Homotopy J - Equivalence.

Two vector bundles $\xi = (E,p,X)$ and $\xi' = (E',p',X)$ over a finite connected CW - complex X are said to be J - equivalent, if the associated sphere bundles $S(\xi) = (S(E),p_S,X)$ and $S(\xi') = (S(E'),p_S',X)$ are of the same stable fibre homotopy type (see chapter 3, § 3). Let us recall the following result due to Dold [74]. Suppose that we can find a fibrewise map $f : S(E) \to S(E')$ such that, for each $x \in X$ the map $f_x : S(E_x) \to S(E_x')$ induced on the fibres is a homotopy equivalence; then, f is a fibre homotopy equivalence.

If the sphere bundles $S(\xi)$ and $S(\xi')$ have the same fibre homotopy type, then the Thom spaces $T(\xi)$ and $T(\xi')$ are homotopy equivalent; in the stable range this assertion admits a partial converse (see [12], pages 298-299).

Let G be a finite group (with discrete topology) and let $\xi = (E,p,X)$ be a G - vector bundle such that X is a finite, path-connected CW - complex and G acts trivially on X. We also suppose that ξ has a riemannian metric such that the G - action restricted to every fibre is orthogonal; with these conditions G will act (always by restriction) orthogonally on each fibre of the associated sphere bundle $S(\xi) = (S(E),p,X)$. In this context, if $\dim_{\mathbb{F}} \xi = n$ and G acts freely on its fibres $\mathbb{F}^n$, we can associate to ξ the spherical form S^{dn-1}/G, where $d = 1,2,$ or 4, according to

$\mathbb{F} = \mathbb{R}, \mathbb{C}$ or $\mathbb{H}$, respectively. These considerations motivate the following terminology.

Definition 3.1. *Two G - vector bundles* $\xi = (E,p,X)$ *and* $\xi' = (E',p',X)$ *are G - fibre homotopy J - equivalent if, and only if, there is a G - fibre homotopy equivalence between the associated sphere bundles* $S(\xi)$ *and* $S(\xi')$.

By this we mean the following:

(i) there exist maps $S(E) \underset{\psi'}{\overset{\psi}{\rightleftarrows}} S(E')$ which are fibre-preserving (that is to say, such that $p'\psi = p$, $p\psi' = p'$) and G - equivariant (that is, for every $g \in G$, $x \in X$, $y \in S(E_x)$ (resp. $y' \in S(E'_x)$), $\psi(gy) = g\psi(y)$ (resp. $\psi'(gy') = g\psi'(y')$)),

(ii) there exist homotopies $h_t : S(E) \to S(E')$ and $h'_t : S(E') \to S(E)$ which are fibre-preserving, G - equivariant and such that

$$h_o = \psi'\psi, \; h_1 = 1_{S(E)}, \; h'_o = \psi\psi' \text{ and } h'_1 = 1_{S(E')} .$$

Concerning this notion of G - fibre homotopy J - equivalence, we shall utilize the following criterion which makes use of the Bott classes θ_k and the Adams operations ψ^k, whose main properties were described in Chapter 0.

Proposition 3.2. (Compare with [48; Th. II]) - *Let* ξ *and* ξ' *be two complex (resp. real,* 8n - *dimensional; G - action factors through* Spin(8n)) *G - vector bundles over the trivial G - space* X . *If* ξ *and* ξ' *are G - fibre homotopy J - equivalent, then there is a unit* $u \in KU_G(X)$ *(resp.* $u \in KO_G(X)$) *such that*

$$\theta_k(\xi) \cdot \psi^k(u) = u \cdot \theta_k(\xi') .$$

Proof. Let ψ be a G - fibre homotopy equivalence between $S(\xi)$ and $S(\xi')$. By extending radially ψ to the total space $D(E)$ of the disk bundle $D(\xi)$ associated to ξ, we obtain a map $\overline{\psi}$ between the pairs $(D(E),S(E))$ and $(D(E'),S(E'))$ inducing the homomorphism

$$\overline{\psi}^* : K_G(D(E'), S(E')) \longrightarrow K_G(D(E), S(E)).$$

Using the Thom isomorphisms $K_G(X) \cong K_G(D(E),S(E))$ and $K_G(X) \cong K_G(D(E'),S(E'))$, we see that there exists an element $u \in K_G(X)$ such that

$$(3.3) \qquad \overline{\psi}^*(\mu(\xi')) = u \cdot \mu(\xi) \quad ,$$

where $\mu(\xi)$ is the Thom class associated with ξ. The definition of the Bott classes now shows that

$$\theta_k(\xi) \cdot \psi^k(u) = u \cdot \theta_k(\xi') \; .$$

Let ψ' be an inverse G - fibre homotopy equivalence of ψ; then, as before we have that

$$(3.4) \qquad \overline{\psi}'^*(\mu(\xi)) = u' \cdot \mu(\xi')$$

for some element $u' \in K_G(X)$. Since $\overline{\psi}'^* \circ \overline{\psi}^*$ and $\overline{\psi}^* \circ \overline{\psi}'^*$ are the identities, (3.3) and (3.4) show that $uu' = 1 = u'u$ in $K_G(X)$.

After this technical result we are going to consider the relations between the G - fibre homotopy and G_p - fibre homotopy J - equivalences, where G_p is a p - Sylow subgroup of G. Since the problems we shall examine turn around the possibibity of extending these homotopy J - eqúivalences, the machinery used to measure the possible obstructions to such extensions will be obviously based on the manipulation of the homotopy groups of appropriated spaces. We shall begin by defining some objects which will prove to be very useful.

Let X and Y be two G - spaces and let $M_G(X,Y)$ be the space of all G - equivariant maps from X to Y. The inclusion $G_p \subset G$ defines clearly a map from $M_G(X,Y)$ into $M_{G_p}(X,Y)$ which, for a fixed $f \in M_G(X,Y)$, induces the homomorphisms

$$\iota_r : \pi_r(M_G(X,Y),f) \longrightarrow \pi_r(M_{G_p}(X,Y),f) \; .$$

Let $(\iota_r)_p$ be the restriction of ι_r to the p - primary component $\pi_r(M_G(,Y),f)_{(p)}$ of $\pi_r(M_G(X,Y),f)$. Moreover, let A be the subspace of X which remains invariant under the action of G; denote by $M_G(X,A;Y;f)$ the subspace $\{h \in M_G(X,Y) | h|A = f\}$.

<u>Lemma 3.5.</u> *Let* X *and* Y *be* G - *spaces such that:*

(i) X *is a finite* CW - *complex on which* G *acts freely and cellularly;*

(ii) Y *is simply connected;*

(iii) *the action of* G *on* X *(resp.* Y*) is such that for every* $g \in G$, *the automorphism of* X *(resp.* Y*) defined by* $x \to gx$ *(resp.* $g \to gy$*) is homotopic to* 1_X *(resp.* 1_Y*)* ;

(iv) *the inclusion* $G_p \subset G$ *induces an isomorphism from* $\hat{H}^*(G;\mathbf{Z})_{(p)}$ *onto* $\hat{H}^*(G_p;\mathbf{Z})$.

Then, for a fixed G - *equivariant map* $f : X \to Y$ *and for every* $r > 0$, *the homomorphism*

$$(\iota_r)_p : \pi_r(M_G(X,Y),f)_{(p)} \longrightarrow \pi_r(M_{G_p}(X,Y),f)$$

is an isomorphism.

<u>Proof.</u> We shall only sketch it. Let f_{s-1} be the restriction of f to the (s - 1) - skeleton X_{s-1} of X ; the homotopy exact sequence of the fibration

$$M_G(X_s,X_{s-1};Y;f_{s-1}) \xrightarrow{i^s} M_G(X_s,Y) \xrightarrow{j^s} M_G(X_{s-1},Y) \qquad (s > 1)$$

gives rise to an exact couple [171] in which

$$E_1^{r,s} = \pi_r(M_G(X_s,X_{s-1};Y;f_{s-1})) .$$

By the exponential law, there is a bijection between $E_1^{r,s}$ and a certain subset of the set of the homotopy classes of maps in $M_G(X_s,M(S_r,Y))$. Then, by (ii) and (iii), we can exhibit an isomorphism between $E_1^{r,s}$ and $H^s(X_s/G,X_{s-1}/G;\pi_{r+s}(Y))$, so that the E_2 term of the spectral sequence is

$$E_2^{r,s} \cong H^s(X/G;\pi_{r+s}(Y)) . \qquad (3.6)$$

Considering also the isomorphism (3.6) for G_p , the reader will see that τ_r may be viewed as the homomorphism

$$\iota_r^* : H^s(X/G;\pi_{r+s}(Y)) \longrightarrow H^s(X/G_p;\pi_{r+s}(Y))$$

induced by the canonical projection $X/G_p \to X/G$. As condition (iii) for X implies that G acts trivially on $\pi_{r+s}(Y)$, the Universal Coefficient Theorem and (iv) show that the restriction of ι_r^* to $H^s(X/G;\pi_{r+s}(Y))_{(p)}$ is an isomorphism; then the result holds for ι_r.

We shall suppose, *from now to the end of this section*, that the groups G and G_p satisfy condition (iv) of lemma 3.5.

Let us consider two n-dimensional G-vector bundles ξ and ξ' with base X, such as described at the beginning of the section. We also suppose that G acts freely on their fibres (which are isomorphic to $\mathbb{F}^n$) by equivalent actions. Hence, we can consider the fibres as n-dimensional G-modules, which will be denoted simply by L. Finally, if s is the cellular dimension of X, let e^s be an s-cell of X and set $X^s = X - \text{int } e^s$.

<u>Lemma 3.7.</u> *Let ϕ be a G-fibre homotopy J-equivalence between $\xi|X^s$ and $\xi'|X^s$ $(s \geq 1)$. If ϕ is G_p-fibre homotopic to $\psi|X^s$ where ψ is a G_p-fibre homotopy J-equivalence between ξ and ξ', then we can find an integer a prime to p and a G-fibre homotopy J-equivalence $\overline{\psi}$ between $a\xi$ and $a\xi'$, such that $\overline{\psi}$ is an extension of $\phi^{(a)}$.*
Furthermore,

<u>Lemma 3.8.</u> *Let ϕ be a G_p-fibre homotopy J-equivalence between ξ and ξ'. If $\phi|X^s$ $(s \geq 1)$ is G_p-fibre homotopic to $\psi|X^s$ where ψ is a G-fibre homotopy J-equivalence between ξ and ξ', then we can find an integer a prime to p and a G-fibre homotopy J-equivalence $\overline{\psi}$ between $a\xi$ and $a\xi'$ such that $\overline{\psi}$ is G_p-fibre homotopic to $\phi^{(a)}$.*

These two Lemmas, as well as Lemma 3.5 from which they derive, are due to J.C.Becker [40; pages 1018-1024]. Since the proofs of (3.7) and (3.8) are obtained following the same kind of considerations, we shall only establish the proof for the first of these two results.

<u>Proof of (3.7).</u> Clearly there are equivalences of G-bundles $f : (\partial e^s) \times S(L) \to S(E|\partial e^s)$ and $f' : S(E'|\partial e^s) \to (\partial e^s) \times S(L)$; the G-fibre homotopy equivalence

$$\phi|\partial e^s : S(E|\partial e^s) \longrightarrow S(E'|\partial e^s)$$

and let $\tilde{\phi}$ be the composition

$$(\partial e^s) \times S(L) \xrightarrow{f} S(E|\partial e^s) \xrightarrow{\phi|\partial e^s} S(E''|\partial e^s) \xrightarrow{f'} (\partial e^s) \times S(L) \to S(L)$$

where the last map is the canonical projection. Now, since $\partial e^s \cong S^{s-1}$, we associate to $\tilde{\phi}$ an element $h(\phi)$ in $\pi_{s-1}(M_G(S(L),S(L))$ (with base point $1_{S(L)}$), via the exponential law. Obviously, ϕ can be extended to a G - fibre homotopy J - equivalence between ξ and ξ' if, and only if, $h(\phi) = 0$. But, by assumption, ϕ being G_p - fibre homotopic to $\psi|X^s$, the homomorphism

$$\iota_{s-1} : \pi_{s-1}(M_G(S(L),S(L))) \longrightarrow \pi_{s-1}(M_{G_p}(S(L),S(L)))$$

yelds $\iota_{s-1}(h(\phi)) = 0$. Thus, for $s > 1$, the injectivity of $(\iota_{s-1})_p$ (Lemma 3.5) shows that there is an integer a prime to p such that $ah(\phi) = 0$, that is to say, $\phi^{(a)}$ has an extension $\overline{\psi}$ from $a\xi$ to $a\xi'$. If $s = 1$, the result is trivial, for ι_o is itself injective.

<u>Proposition 3.9.</u> *Let* ϕ *be a* G_p *- fibre homotopy* J *- equivalence between* ξ *and* ξ'*. We assume that* ϕ *, up to* G *- homotopy equivalence, induces maps of degree one on* S(L) *. Then, there are an integer* a *prime to* p *, and a* G *- fibre homotopy* J *- equivalence* $\overline{\psi}$ *between* $a\xi$ *and* $a\xi'$ *such that* $\overline{\psi}$ *is* G_p *- fibre homotopic to* $\phi^{(a)}$.

In this case, the proof is obtained by induction on the cellular dimension s of X using Lemmas 3.7 and 3.8. We need only to see that the assertion is true when $s = 0$: this is guaranted by the assumption made on the degree of ϕ . Indeed, it means that we can find G - homotopy equivalences $S(L) \underset{f'}{\overset{f}{\rightleftarrows}} S(L)$ such that $f'^{-1}\phi f$ has degree one. Hence ϕ is G_p - homotopic to $f'f^{-1}$. Actually, the reader can see that this assumption is not necessary (!) as soon as $p \geq 3$.

Our objective is now to give a characterization of the G - fibre homotopy equivalence of Corollary 3.16 which will be useful later on . We shall first explain the hypothesis of that result and introduce some notation.

Let $\xi = (E,X,p,F)$ and α be G-bundles with base space X, $\xi' = (E',X,p|E',F)$ a G-sub-bundle of ξ, and $\phi' : \xi' \to \alpha$ a G-bundle morphism; we shall denote the set of homotopy classes of G-bundle morphism $\phi : \xi \to \alpha$ such that $\phi|\xi' = \phi'$ by $[\xi,\xi';\alpha;\phi']_G$. If α' is another G-bundle and $f : \alpha \to \alpha'$ is a G-bundle morphism, let

$$f_! : [\xi,\xi';\alpha;\phi']_G \longrightarrow [\xi,\xi';\alpha;f\cdot\phi']_G \tag{3.10}$$

be the map induced by f. Furthermore, we shall assume that X is a finite CW-complex on which G acts cellularly and that A is an acyclic CW-complex (i.e., $H^i(A;\mathbb{Z}) = 0$ for every $i > 0$) on which G acts freely.

Definition 3.11. *A pair of topological spaces* (Y,Z) *is called a* G-*pair if there is an action of* G *on* Y *leaving* Z *invariant. If* G *acts freely on* Y *(resp.* $Y-Z$*) then* (Y,Z) *is said to be a free (resp. relatively free)* G-*pair.*

Definition 3.12. *A* G-*pair* (Y,Z) *is said to be* n-*coconnected for* G *if, and only if,*

$$H^i(Y\times A/G,\ Z\times A/G;\Pi) = 0$$

for every $i \geq n$ *and every* G-*group* Π.

Now, let S^n be an n-sphere on which G is supposed to act; denote by $\xi * S^n$ the bundle obtained by taking the join of each fibre of ξ with S^n. Note that $\xi * S^n$ is a G-bundle for which the action of G is given by

$$g((1-t)e+tz) = (1-t)ge + tgz$$

for every $g \in G$, $e \in E$, $z \in S^n$ and $t \in [0,1]$.

Let

$$\sigma^n_* : [\xi,\xi';\alpha;\phi']_G \longrightarrow \left[\xi*S^n,\xi'*S^n;\alpha*S^n;\alpha*S^n;\phi'*1_{S^n}\right]_G \tag{3.13}$$

be the map given by $\sigma^n_*(\phi) = \phi * 1_{S^n}$, for every $\phi \in [\xi,\xi';\alpha;\phi']_G$.

Proposition 3.14. *Let* ξ,α *be given* G-*bundles and let* ξ' *be a* G-*sub-bundle of* ξ; *we assume that* (E,E') *is a relatively free* G-*pair, that the* G-*actions are cellular and that the fibre type of* α *is* $(r-1)$-*connected. Then,*

(i) *if* (E,E') *is* (2r-1) - *coconnected for* G, σ_*^n *in injective;*
(ii) *if* (E,E') *is* 2r - *coconnected for* G, σ_*^n *is surjective.*

Proof. For any given $x \in X$, let A_x be the fibre of α over x and let β_x be the space of the maps $b : D^{n+1} \to A_x * S^n$ such that $b|\partial D^{n+1}$ is the natural inclusion of S^n into the join $A_x * S^n$; then, $\beta = (B = \bigcup_{x\in X} \beta_x, q, X)$ with $q(b) = x$, is a G - bundle. In fact, we radially extend the G - action on $S^n \cong \partial D^{n+1}$ (in other words, for every $g \in G$, every $z \in S^n$ and $0 \leq \rho \leq 1$, we set $g \cdot \rho z = \rho \cdot gz$); then G acts on B by $(g,b) \to gbg^{-1}$. Let us consider the G - bundle morphism $f : \alpha \to \beta$ defined by

$$f(a)(\rho z) = (1-\rho)a + \rho z \ ,$$

for every $a \in E_\alpha$, $z \in S^n$ and $0 \leq \rho \leq 1$ (E_α is the total space of α). Then we construct the isomorphism

$$\varepsilon : [\xi,\xi';\beta;f\cdot\phi']_G \longrightarrow \left[\xi * S^n, \xi' * S^n; \alpha * S^n; \phi' * 1_{S^n}\right]_G$$

by setting, for every $e \in E$, $z \in S^n$ and $t \in [0,1]$,

$$\varepsilon(\phi)((1-t)e + tz) = \phi(e)(tz) \ .$$

Because of (3.10) and (3.13) we have that $\varepsilon \circ f_! = \sigma_*^n$. The reader should notice that the conclusion of the Proposition can be now obtained from the following Lemma.

Lemma 3.15. *Let* ξ, α *and* β *be given* G - *bundles,* ξ' *a* G - *sub-bundle of* ξ *and* $f : \alpha \to \beta$ *a* G - *bundle morphism. We assume that* (E,E') *is a relatively free* G - *pair, that the* G - *actions are cellular, and the restriction of* f *to every fibre is an* (r-1) - *equivalence. Then,*

$$f_! : [\xi,\xi';\alpha;\phi']_G \longrightarrow [\xi,\xi';\beta;f\cdot\phi']_G$$

is injective (resp. surjective) if (E,E') *is* r - *coconnected for* G *(resp.* (r+1) - *coconnected for* G) .

This result is based on an comparison theorem due to I.M.James [142; page 374] .

Observe that Proposition 3.14 clearly implies the following.

Corollary 3.16. *Let* ξ *and* η *be* G-*bundles with base* X. *Suppose that* G *acts freely and cellularly on their total spaces and on* S^n, *that their fibres are* $(r-1)$-*coconnected, and that the cellular dimension of* X *is* $< r$. *If* $\xi * S^n$ *and* $\eta * S^n$ *are* G-*fibre homotopically equivalent, then* ξ *and* η *are* G-*fibre homotopically equivalent.*

4. G - (Co)Reducibility.

We bring the attention of the reader to the fact that the definitions, notations and results contained in this section consist roughly in a transposition into G-equivariant terms of analogous ideas developped by J.F.Adams in his famous and classical paper [2]. This explains why we shall not enter into the discussion of certain details, except whenever a new situation will require some explanation. For the reader's benefit we recall the following.

Definition 4.0. (see [12]) - *A space* X *with base point* x_o *is said to be reducible (resp. coreducible) if there is a based map*

$$\alpha : (S^n, a) \longrightarrow (X, x_o)$$

(resp. $\beta : (X, x_o) \longrightarrow (S^n, a)$*) such that the induced homomorphisms*

$$\tilde{H}_p(S^n) \longrightarrow \tilde{H}_p(X)$$

(resp. $\tilde{H}^p(S^n) \longrightarrow \tilde{H}^p(X)$*) are isomorphisms for every* $p \geq n$ *(resp.* $p \leq n$*). In that case, we say that* α *(resp.* β*) is a reduction (resp. a coreduction).*

We also note that if X is a finite CW-complex of dimension n posessing a unique n-cell and if

$$\pi : X \longrightarrow X/X_{n-1} = S^n$$

is the canonial projection (X_{n-1} is the $(n-1)$-skeleton of X) then, to say that $\alpha : S^n \to X$ is a reduction is the same as saying that the composition

$$S^n \xrightarrow{\ \alpha\ } X \xrightarrow{\ \pi\ } S^n$$

has degree 1. On the other hand, if X is a finite CW-complex such that $X_n = S^n$, $\beta : X \to S^n$ is a coreduction if, and only if the composition

$$S^n \xrightarrow{i} X \xrightarrow{\beta} S^n$$

is a map of degree 1.

The results we are aiming at in this section are those expressed in Propositions 4.7, 4.11 and 4.12; they correspond respectively to Theorems 2.1, 2.2 and 2.3 of Becker's work [40] .

As for the notation, given a G - module L^n on which G is supposed to act freely, let us denote by $S(L^n)$, $P(L^n)$ and $V_k(L^n)$, the sphere, the projective space, and the Stiefel manifold of k - frames associated to L^n (we work in the context of section 3). Naturally, these objects have G - structures induced by the G - action on L^n. If $m \geq k$, let

$$P_k(L^n \oplus \mathbb{R}^m) = P(L^n \oplus \mathbb{R}^m)/P(L^n \oplus \mathbb{R}^{m-k})$$

be the "truncated projective space"; this quotient space is a G - space and moreover, $(P_k(L^n \oplus \mathbb{R}^m), P_k(\mathbb{R}^m))$ is a relatively free G - pair.

We shall write

$$(4.1) \qquad \lambda_k : \left(P_k(L^n \oplus \mathbb{R}^m), P_k(\mathbb{R}^m)\right) \longrightarrow \left(P_1(L^n \oplus \mathbb{R}^m), P_1(\mathbb{R}^m)\right)$$

for the usual collapsing map.

We now deal with the Stiefel manifolds. In what follows we shall denote by

$$p_k : V_k(L^n) \longrightarrow S(L^n)$$

the projection on the last factor, i.e., $p_k(v_1,\dots,v_k) = v_k$.Next, we define the *Intrinsic Map*

$$(4.2) \qquad h_k : V_k(L^n) * V_k(L^m) \longrightarrow V_k(L^n \oplus L^m) .$$

(Here $V_k(L^n) * V_k(L^m)$ is the *join* of $V_k(L^n)$ and $V_k(L^m)$, already defined in Chapter 2, § 2.) Given $(u,v,t) \in V_k(L^n) * V_k(L^m)$, $t \in [0,1]$, we define $h_k(u,v,t) = w = (w_1,\dots,w_k)$, where $w_i = (u_i \cos\alpha, v_i \sin\alpha)$, with $\alpha = \frac{1}{2}\pi t$ (the reader is referred to [138; pages 513 and seq.], for a study of the properties of the intrinsic map). One can verify that the map

$$\theta_k : P_k(L^n \oplus R^m) \longrightarrow V_k(L^n \oplus R^m)$$

given by $\theta_k[x] = (v_1, v_2, \ldots, v_k)$ with $x \in S(L^n \oplus R^m)$ and

$$v_i = e_{n+m-k+i} - 2(e_{n+m-k+i}|x) \cdot x$$

$1 \leq i \leq k$, defines a morphism of G - space pairs (denoted in the same way)

(4.3) $$\theta_k : \left(P_k(L^n \oplus \mathbb{R}^m), P_k(\mathbb{R}^m)\right) \longrightarrow \left(V_k(L^n \oplus \mathbb{R}^m), V_k(\mathbb{R}^m)\right).$$

It can be shown that the homomorphisms $\pi_i(P_k(L^n \oplus \mathbb{R}^m)) \to \pi_i(V_k(L^n \oplus \mathbb{R}^m))$ induced by θ_k are isomorphisms for $i \leq 2(n+m-k-1)$ (see [213]).

Definition 4.4. *The pair* $(P_k(L^n \oplus \mathbb{R}^m), P_k(\mathbb{R}^m))$ *is* G - *reducible (resp.* G - *coreducible) if there is a* G - *morphism*

$$\alpha_k : (S(L^n \oplus \mathbb{R}^m), S(\mathbb{R}^m)) \longrightarrow (P_k(L^n \oplus \mathbb{R}^m), P_k(\mathbb{R}^m))$$

$$\left(\textit{resp. } \beta_k : (P_k(L^n \oplus \mathbb{R}^m), P_k(\mathbb{R}^m)) \to (S(L^n \oplus \mathbb{R}^{m-k+1}), S(\mathbb{R}^{m-k+1}))\right)$$

such that α_k *and* $\alpha'_k = \alpha_k | S(\mathbb{R}^m)$ *(resp.* β_k *and* $\beta'_k = \beta_k | P_k(\mathbb{R}^m)$) *are reductions (resp. coreductions).*

Under certain conditions on the G - action, and on n and m, the reducibility of α'_k is sufficient to induce the reducibility of α_k. More precisely,

Lemma 4.5. *Let* $n \equiv 0 \pmod{a_k(\mathbb{R})}$ (*), $m \equiv 0 \pmod{a_k(\mathbb{R})}$ *with* $m > k$, *and let us suppose that* G *acts simply on the homotopy groups* $\pi_i\left(P_k(L^n \oplus \mathbb{R}^m)\right)$ (+). *Then, if there is a* G-*map*

(*) See Remark 2.10 for the definition of $a_k(\mathbb{R})$.

(+) This means that $\pi_i(P_k(L^n \oplus \mathbb{R}^m))$ as a G - module is simple.

$$\alpha_k : (S(L^n \oplus \mathbb{R}^m), S(\mathbb{R}^m)) \longrightarrow (P_k(L^n \oplus \mathbb{R}^m), P_k(\mathbb{R}^m))$$

such that α'_k *is a reduction, the pair* $(P_k(L^n \oplus \mathbb{R}^m), P_k(\mathbb{R}^m))$ *is* G - *reducible.*

Proof. Let f be the map $(S(L^n \oplus \mathbb{R}^m), S(\mathbb{R}^m))$ $(S(L^n \oplus \mathbb{R}^m), S(\mathbb{R}^m)) = (V_1(L^n \oplus \mathbb{R}^m), V_1(R^m))$ defined by $f = \theta_1 \cdot \lambda_k \cdot \alpha_k$. By (3.10), this composition leads to the map

$$\left(1_{S(L^n)} * f'\right)_! : \left[S(L^n \oplus \mathbb{R}^m), S(\mathbb{R}^m); S(L^n \oplus \mathbb{R}^m) ; 1_{S(\mathbb{R}^m)}\right]_G \rightarrow$$

$$\left[S(L^n \oplus \mathbb{R}^m), S(\mathbb{R}^m); S(L^n \oplus \mathbb{R}^m) ; f'\right]_G$$

which, according to (3.15), is an isomorphism. Besides (see (3.13)), the map

$$\sigma_*^{m-1} : [S(L^n); S(L^n)]_G \longrightarrow \left[S(L^n \oplus \mathbb{R}^m), S(\mathbb{R}^m); S(L^n \oplus \mathbb{R}^m); 1_{S(\mathbb{R}^m}\right]_G$$

being also bijective, it follows that $\left(1_{S(L^n)} * f'\right)_! \circ \sigma_*^{m-1}$ is an isomorphism.
Since, for every $[\phi] \in [S(L^n); S(L^n)]_G$, we have that

$$\left(1_{S(L^n)} * f'\right)_! \cdot \sigma_*^{m-1}[\phi] = \left(1_{S(L^n)} * f'\right)_! \left[\phi * 1_{S(R^m)}\right] = [\phi * f'],$$

we see that it is possible to find a G - map ϕ of $S(L^n)$ into itself (unique up to G - homotopy) such that f is G - homotopic to $\phi * f'$.

Now, since α'_k is a reduction, we obtain that $\deg(\phi * f') = \deg(\phi) \cdot \deg(f') = \pm \deg(\phi)$, and consequently, $\deg(f) = \pm \deg(\phi) = a \cdot |G| \pm 1$, with $a \in \mathbf{Z}$. If $a = 0$, it follows easily that α_k is a reduction. Otherwise, by modifying α_k slightly, we can find a G - morphism

$$\tilde{\alpha}_k : (S(L^n \oplus \mathbb{R}^m), S(\mathbb{R}^m)) \longrightarrow (P_k(L^n \oplus \mathbb{R}^m), P_k(\mathbb{R}^m))$$

which is a reduction (take $\tilde{\alpha}'_k = \alpha'_k$). Since $n \equiv 0 \pmod{a_k(\mathbb{R})}$ and $m \equiv 0 \pmod{a_k(\mathbb{R})}$, there is an element $u \in \pi_{n+m-1}(P_k(L^n \oplus \mathbb{R}^m))$ such that $(\theta_1 \cdot \lambda_k)_!(u) = \left[1_{S(L^n \oplus \mathbb{R}^m)}\right]$.
We need only to choose $\tilde{\alpha}_k$ so that

(i) the restrictions of α_k and $\tilde{\alpha}_k$ to the $(n+m-2)$-skeleton are equal,

(ii) by setting $\tilde{f} = \theta_1 \cdot \lambda_k \cdot \tilde{\alpha}_k$, the difference cohomology class $[\tilde{f}] - [f]$ in $H^{n+m-1}(S(L^n)/G * S(\mathbb{R}^m), S(\mathbb{R}^m); \pi_{n+m-1}(P_k(L^n \oplus \mathbb{R}^m))$ is precisely $-au$.

Indeed, these conditions guarantee that $\deg(\tilde{f}) = \deg(f) - a|G| = \pm 1$; the argument is standard.

As far as the relation existing between the span of the spherical form $S(L^n)/G$ and the G-reducibility of a pair $(P_k(L^n \oplus \mathbb{R}^m), P_k(\mathbb{R}^m))$ is concerned, an answer is given by the following lemma.

<u>Lemma 4.6.</u> *Let* $m > k$ *and* $m \equiv 0 \pmod{a_k(\mathbb{R})}$. *If* $\operatorname{span}(S(L^n)/G) \geq k-1$, *then* $(P_k(L^n \oplus \mathbb{R}^m), P_k(\mathbb{R}^m))$ *is* G-*reducible. If* $n \geq 2k$, *the converse is true.*

<u>Proof.</u> As $m \equiv 0 \pmod{a_k(\mathbb{R})}$, the results obtained for the spheres imply immediately the existence of a reduction $\tilde{\alpha}'_k : S(\mathbb{R}^m) \to P_k(\mathbb{R}^m)$. To establish the G-reducibility of $(P_k(L^n \oplus \mathbb{R}^m), P_k(\mathbb{R}^m))$, we must extend $\tilde{\alpha}'_k$ to a G-morphism $\alpha_k : S(L^n \oplus \mathbb{R}^m) \to S(\mathbb{R}^m)$ so that α_k is also a reduction. Let us notice that the map

$$(\theta'_k)_! : \left[S(L^n \oplus \mathbb{R}^m), S(\mathbb{R}^m); P_k(L^n \oplus \mathbb{R}^m); \tilde{\alpha}'_k\right]_G \longrightarrow$$
$$\left[S(L^n \oplus \mathbb{R}^m), S(\mathbb{R}^m); V_k(L^n \oplus \mathbb{R}^m); \theta'_k \cdot \tilde{\alpha}'_k\right]_G$$

induced by θ'_k (4.3), is an isomorphism (3.15). So, we only need to extend $\theta'_k \circ \tilde{\alpha}'_k$ into a G-morphism between $S(L^n \oplus \mathbb{R}^m)$ and $V_k(L^n \oplus \mathbb{R}^m)$. Since $\operatorname{span}(S(L^n)/G) \geq k-1$, it is obvious that the projection $p_k : V_k(L^n) \to S(L^n)$ has a G-equivariant cross-section $s_k : S(L^n) \to V_k(L^n)$. Now, the extension wanted is no other than the composition given by the following diagram

$$\begin{array}{ccc} S(L^n \oplus \mathbb{R}^m) \cong S(L^n) * S(\mathbb{R}^m) & \longrightarrow & V_k(L^n \oplus \mathbb{R}^m) \\ s_k * (\theta'_k \circ \tilde{\alpha}'_k) \searrow & & \nearrow h_k \\ & V_k(L^n) * V_k(\mathbb{R}^m) & \end{array} .$$

To prove the converse, that is to say, that the G - reducibility of $(P_k(L^n \oplus \mathbb{R}^m), P_k(\mathbb{R}^m))$ implies the existence of a G - equivariant cross-section s_k, the reader needs only to exhibit a G - map $\tilde{s}_k : S(L^n) \to V_k(L^n)$ such that $p_k \circ \tilde{s}_k$ is G - homotopic to the identity. The assumption that $n \geq 2k$ becomes then necessary to obtain $\tilde{s}_k$ (use (3.14)(ii) and (3.15)).

Proposition 4.7. *Let* L_1^n *and* L_2^n *be two* n - *dimensional* G - *modules on which* G *acts freely. Then, if* $n \geq 2k$, $\operatorname{span}(S(L_1^n)/G) \geq k - 1$ *is equivalent to* $\operatorname{span}(S(L_2^n)/G) \geq k - 1$.

Proof. Suppose that $\operatorname{span}(S(L_1^n)/G) \geq k - 1$; to get the same lower bound for the span of the other spherical form $S(L_2^n)/G$, the two Lemmas just given show that we need only to find a G - map

$$\alpha_2 : \left(S(L_2^n \oplus \mathbb{R}^m), S(\mathbb{R}^m)\right) \longrightarrow \left(P_k(L_2^n \oplus \mathbb{R}^m), P_k(\mathbb{R}^m)\right)$$

such that α_k' is a reduction. By assumption, we know that there is a G - reduction

$$\alpha_1 : \left(S(L_1^n \oplus \mathbb{R}^m), S(\mathbb{R}^m)\right) \longrightarrow \left(P_k(L_1^n \oplus \mathbb{R}^m), P_k(\mathbb{R}^m)\right) .$$

We are now going to build G - maps γ and δ so that $\delta' \cdot \alpha_1' \cdot \gamma'$ is a reduction; then the following diagram will define α_2 :

$$\begin{array}{ccc}
\left(S(L_2^n \oplus \mathbb{R}^m), S(\mathbb{R}^m)\right) & \xrightarrow{\ \gamma\ } & \left(S(L_1^n \oplus \mathbb{R}^m), S(\mathbb{R}^m)\right) \\
\Big\downarrow \alpha_2 & & \Big\downarrow \alpha_1 \\
\left(P_k(L_2^n \oplus \mathbb{R}^m), P_k(\mathbb{R}^m)\right) & \xleftarrow{\ \delta\ } & \left(P_k(L_1^n \oplus \mathbb{R}^m), P_k(\mathbb{R}^m)\right)
\end{array}$$

with γ defined as the join of a G - equivariant map $S(L_2^n) \to S(L_1^n)$ with $1_{S(\mathbb{R}^m)}$. Likewise, to obtain δ, we choose a G - equivariant map $S(L_1^n) \to S(L_2^n)$, which commutes with the antipodal maps (such a map always exists), and take its join with $1_{S(\mathbb{R}^m)}$ before passing to the quotient; we shall have

$\delta' = 1_{P_k(\mathbb{R}^m)}$.

Remark 4.8. Invoking the table given at the beginning of section 2 , the reader will notice that

(i) if $n \neq 1,2,4,8,16$, then $n \geq 2 \cdot \rho(n, \mathbb{R})$,

(ii) if $n \neq 1,2,4,8$, then $n \geq 2 \cdot \rho(n, \mathbb{F})$, with $\mathbb{F} = \mathbb{C}$ or $\mathbb{H}$.

Let us examine now the relations between G - coreducibility, G - reducibility and G - fibre homotopy type. Let us recall that we have already indicated that the G - fibre homotopy J - equivalence between two G - vector bundles ξ and ξ' induces a G - homotopy equivalence between their Thom spaces $T(\xi)$ and $T(\xi')$. In this context, if we denote by η_k the canonical line bundle over the (k-1) - dimensional projective space, we have:

Lemma 4.9. *Suppose that* $n > k$ *and* $m \geq k$, *with* $m \equiv k$ $(\mathrm{mod}.\, a_k(\mathbb{R}))$. *Then* $\eta_k \otimes L^n$ *and* L^n *are* G - *fibre homotopically* J - *equivalent if, and only if, the pair* $(P_k(L^n \oplus \mathbb{R}^m),$ $P_k(\mathbb{R}^m))$ *is* G - *coreducible.*

Proof. The asumption $m \equiv k$ $(\mathrm{mod}.\, a_k(\mathbb{R}))$ means that there is a coreduction

$$\tilde{\beta}'_k : P_k(\mathbb{R}^m) \longrightarrow S(\mathbb{R}^{m-k+1}) .$$

Then, we must characterize β_k so that $[\beta_k] \in \left[P_k(L^n \oplus \mathbb{R}^m), P_k(\mathbb{R}^m); S(L^n \oplus \mathbb{R}^{m-k+1}), \tilde{\beta}'_k\right]_G$. Because of the well-known identification

$$\left(P_k(L^n \oplus \mathbb{R}^m), P_k(\mathbb{R}^m)\right) = \left(T(\eta_k \otimes (L^n \oplus \mathbb{R}^{m-k})), T(\eta_k \otimes \mathbb{R}^{m-k})\right)$$

[12; page 304] the assertion will follow quickly by using on one hand, the construction of the Thom space, and on the other hand, (3.13) and (3.16).

The duality between reducibility and coreducibility can be expressed in the following way

Lemma 4.10. *Let* k, ℓ *and* m *be integers such that* $\ell \geq 2k$, $m \geq k$, $\ell \equiv 0$ $(\mathrm{mod}.\, a_k(\mathbb{R}))$ *and* $m \equiv k$ $(\mathrm{mod}.\, a_k(\mathbb{R}))$. *Then, the pair* $(P_k(L^n \oplus \mathbb{R}^\ell), P_k(\mathbb{R}^\ell))$ *is* G - *reducible if, and only if,* $(P_k(L^n \oplus \mathbb{R}^m), P_k(\mathbb{R}^m))$ *is* G - *coreducible*

As a consequence we have the following.

Proposition 4.11. *If* $\mathrm{span}(S(L^n)/G) = k - 1$ *then* $\eta_k \otimes L^n$ *are* G - *fibre homotopically* J - *equivalent; on the other hand, if* $\eta_k \otimes L^n$ *and* L^n *are* G - *fibre homotopically* J - *equivalent and* $n \geq 2k$, *then* $\mathrm{span}(S(L^n)/G \geq k - 1$.

This proposition, given by Becker as a generalization of Atiyah's work [12; § 6], is an immediate consequence of Lemmas 4.6, 4.9 and 4.10.

Finally, we wish to notice that, while Proposition 4.7 shows that the spans of two spherical forms with the same dimension and same defining group have the same lower bound (which is independent of the free representation selected), the next Proposition proves that a similar result holds for some spherical forms and their associated spherical 2 - forms. More precisely:

Proposition 4.12. *Let* G_2 *be a* 2 - *Sylow subgroup of* G *and let* $n \geq 2k$; *suppose that the canonical inclusion* $G_2 \to G$ *induces an isomorphism between the groups* $H^i(G;\mathbf{Z})_{(2)}$ *and* $H^i(G_2;\mathbf{Z})$, *for every* $i \geq 0$. *Then,* $\mathrm{span}(S(L^n)/G) \geq k - 1$ *if, and only if,* $\mathrm{span}(S(L^n)/G_2) \geq k - 1$.

Proof. The only implication that we have to consider is, of course, the one for which we assume that $\mathrm{span}(S(L^n)/G_2) \geq k - 1$. Since $n \geq 2k$, according to Proposition 4.11, $\eta_k \otimes L^n$ is G_2 - fibre homotopy J - equivalent to L^n ; then, because of (3.9), there is an odd integer a such that $a\eta_k \otimes L^n$ is G - fibre homotopy J - equivalent to aL^n.

Let $\hat{KO}_G(\mathbb{R}P^{k-1})$ be the Grothendieck group associated to the class of all G - vector bundles over $\mathbb{R}P^{k-1}$ (the G - actions being free); then, by analogy to the standard construction of the J - theory, let us denote by $\hat{J}_G(\mathbb{R}P^{k-1})$ the quotient of $\hat{KO}_G(\mathbb{R}P^{k-1})$ by the subgroup generated by the elements of type $[\xi] - [\xi']$ where ξ and ξ' are G - fibre homotopically J - equivalent, and let us denote by

$$\hat{J}_G : \hat{K}O_G(\mathbb{R}P^{k-1}) \longrightarrow \hat{J}_G(\mathbb{R}P^{k-1})$$

the corresponding "J - homomorphism". With this terminology, the G - fibre homotopy J - equivalence between $a\eta_k \otimes L^n$ and aL^n is written

$$\hat{J}_G\left(\left[a\eta_k \otimes L^n\right] - [aL^n]\right) = 0.$$

Since $\hat{K}O_G(\mathbb{R}P^{k-1})$ has only 2 - torsion, it follows that

$$\hat{J}_G\left(\left[\eta_k \otimes L^n\right] - [L^n]\right) = 0 .$$

In other words, there is a G - module M (the G - action being fixed point free) such that $(\eta_k \otimes L^n) + M$ and $L^n + M$ are G - fibre homotopically J - equivalent. Indeed, by (3.16), this J - equivalence still holds for $\eta_k \otimes L^n$ and L^n ; Proposition 4.11 now shows that $\mathrm{span}(S(L^n)/G) \geq k - 1$.

5. Span of Spherical Forms of Cyclic Type.

In this section, as well as in the following one, we shall eventually assume that the defining group G of the spherical form S^n/G is not trivial. Notice that $\mathrm{span}(S^1/G) = \mathrm{span}_R(S^1) = 1$ and that, because any 3 - dimensional orientable manifold is parallelizable [233], $\mathrm{span}(S^3/G) = \mathrm{span}_{\mathbb{R}}(S^3) = 3$.

We must examine the spherical forms whose defining group has a 2-Sylow subgroup G_2 isomorphic to a cyclic group $\mathbb{Z}_{2^m}$ $(m \geq 0)$ (see (2.9)). When $m = 0$ or 1, we always have that $\mathrm{span}(S^{2n+1}/G_2) = \mathrm{span}_{\mathbb{R}}(S^{2n+1})$. From Remark 4.8 we conclude that if $n \neq 0, 1, 3$ or 7 ,

$$n + 1 \geq \rho(2n+2, \mathbb{R}) ;$$

then, Propositions 4.7 and 4.12 show that

$$\mathrm{span}(S^{2n+1}/G) \geq \mathrm{span}_{\mathbb{R}}(S^{2n+1}) .$$

Furthermore, using (2.8) we conclude the following.

Lemma 5.1. *If* $n \neq 3$ *or* 7 *and* $v_2(|G|) = 0$ *or* 1 , $\mathrm{span}(S^{2n+1}/G) = \rho(2n+2, \mathbb{R}) - 1$.

Let us now suppose that $m \geq 2$ and $n \neq 0, 1$ or 3; in this case we also have that $n + 1 \geq \rho(2n+2, \mathbb{C})$. Because

$\text{span}(L^n(2^m)) \geq \rho(2n+2,\mathbb{C}) - 1$ (see Proposition 2.8), if $n \neq 3$, (4.7) and (4.12) prove that

$$\text{span}(S^{2n+1}/G) \geq \rho(2n+2,\mathbb{C}) - 1$$

If we consider the canonical inclusion $\mathbb{Z}_4 \to \mathbb{Z}_{2^m}$ (here $m \geq 2$) which induces the surjective map $L^n(4) \to L^n(2^m)$, we see that $\text{span}(L^n(2^m)) \leq \text{span}(L^n(4))$. Thus, by using (4.12) (and eventually (4.7)), we also have (for $n \neq 3$) that

$$\text{span}(S^{2n+1}/G) \leq \text{span}(L^n(4)) .$$

This shows that we must study the span of $L^n(4)$ for $n \neq 0,1$ and 3 . Let ξ_1 be the unitary representation of degree 1 of $\mathbb{Z}_4$ defined by $\xi_1(\bar{1}) = \zeta \cdot 1_{\mathbb{C}}$, with $\zeta = \exp(i\pi/2)$; we know that $L^n(4)$ is obtained from the standard complex representation $\sigma = (n+1)\xi_1$ (cf. (2.3)). We would like to establish the fact that

$$\text{span}(L^n(4)) = \rho(2n+2,\mathbb{C}) - 1 .$$

Since we already know that

$$\text{span}(L^n(4)) \geq \rho(2n+2,\mathbb{C}) - 1 ,$$

let us suppose that $\text{span}(L^n(4)) \geq k - 1$, with $k > \rho(2n+2,\mathbb{C})$. Then (4.1) shows that $\eta_k \otimes \sigma$ and σ are $\mathbb{Z}_4$-fibre homotopy J-equivalent. But then, this means that we can find a unit u of $KU_{\mathbb{Z}_4}(\mathbb{R}P^{k-1})$ such that

$$\theta_j(\eta_k \otimes \sigma) \cdot \psi^j(u) = u \cdot \theta_j(\sigma)$$

(see Proposition 3.2). In order to go on, we must use the ψ-structure of $KU_{\mathbb{Z}_4}(\mathbb{R}P^{k-1})$; the following lemma gives the necessary information.

<u>Lemma 5.2.</u> (i) $\widetilde{KU}(\mathbb{R}P^{k-1})$ *is isomorphic to* $\mathbb{Z}_{a_k(\mathbb{C})}$ *with generator* $x_k = [c(\eta_k)] - 1 = [\eta_k \otimes \mathbb{C}] - 1$ *; moreover* $x_k^2 + 2x_k = 0$.

(ii) *For every* $x \in KU_{\mathbb{Z}_4}(\mathbb{R}P^{k-1})$ *we have:* $\psi^{j+4}(x) = \psi^j(x)$.

The first assertion was already given in the previous chapter. The other assertion, which means that ψ^s is but the identity on $KU_{\mathbb{Z}_4}(\mathbb{R}P^{k-1})$, results from chapter 4, § 3 and from the following.

<u>Proposition 5.3.</u> [36; page 274] - *Let* G *be a finite group; then the Adams operations in* RU(G) *are periodic; in particular,* $\psi^{|G|+j} = \psi^j$.

All this implies that

$$\theta_5[\eta_k \otimes (n+1)\xi_1] = \theta_5[(n+1)\xi_1] \quad .$$

On the other hand, $RU(\mathbb{Z}_4)$ is additively generated by $\xi_o = 1, \xi_1, \xi_2 = \xi_1^2$ and $\xi_3 = \xi_1^3$; hence, the properties of the Bott operations (see Chapter O) show that

$$\theta_5(\xi_1) = 1 + \xi_1 + \xi_1^2 + \xi_1^3 + \xi_1^4 = 1 + \xi_o + \xi_1 + \xi_2 + \xi_3 \ ,$$

$$\theta_5(\eta_k \otimes \xi_1) = 1 + [\eta_k \otimes \mathbb{C}]\cdot\xi_1 + \xi_2 + [\eta_k \otimes \mathbb{C}]\cdot\xi_3 + \xi_o,$$

$$= 1 + \xi_o + \xi_1 + \xi_2 + \xi_3 + x_k(\xi_1 + \xi_3) \ ,$$

because $[\eta_k \otimes \mathbb{C}]^2 = 1$ (5.2)(i). Writing

$$\xi_o + \xi_1 + \xi_2 + \xi_3 = \alpha \quad \text{and} \quad \xi_1 + \xi_3 = \beta \ ,$$

we obtain the following relation

$$(1+\alpha+x_k\cdot\beta)^{n+1} = (1+\alpha)^{n+1}$$

that is to say

$$\sum_{i=1}^{n+1} \binom{n+1}{i} x_k^i \, \beta^i \, (1+\alpha)^{n+1-i} = 0 \ ,$$

and since $x_k^i = (-2)^{i-1} x_k$,

$$\text{(5.4)} \qquad \sum_{i=1}^{n+1} (-1)^{i-1} \binom{n+1}{i} 2^{i-1} \, x_k \, \beta^i (1+\alpha)^{n+1-i} = 0 \ .$$

Using the conjugate complex representation $\bar{\alpha} = t\alpha$ of α , the reader can prove that (5.4) may be written simply as

$$\frac{1}{2}\left[(1+\alpha)^{n+1} - (1+\bar{\alpha})^{n+1}\right] x_k = 0.$$

Since $(1+\alpha)^{n+1} - 1 = (1/4)(5^{n+1}-1)\alpha$ (because, for $i > 1$, $\alpha^i = 2^{2i-2}\alpha$) we obtain that

$$(1/4)(5^{n+1} - 1)\alpha x_k = 0.$$

Using (5.2)(i), the last relation implies that

$$(5.5) \qquad (1/4)(5^{n+1} - 1) \equiv 0 \qquad (\text{mod.}\, a_k(\mathbb{C})).$$

From this congruence we conclude that

$$n + 1 \equiv 0 \qquad (\text{mod.}\, a_k(\mathbb{C})).$$

But the integer $\rho(2n+2,\mathbb{C})$ is such that

$$\rho(2n+2,\mathbb{C}) = \sup\{k \mid n+1 \equiv 0 \quad (\text{mod.}\, a_k(\mathbb{C}))\}$$

(cf. Remark 2.10); hence, there is a contradiction with the hypothesis $k > \rho(2n+2,\mathbb{C})$. Thus

$$\text{span}(L^n(4)) = \rho(2n+2,\mathbb{C}) - 1 .$$

More generally, we can state the following.

<u>Proposition 5.6.</u> *If* $n \neq 3$ *and* $G_2 \cong \mathbb{Z}_{2^m}$ *with* $m \geq 2$,

$$\text{span}(S^{2n+1}/G) = \rho(2n+2,\mathbb{C}) - 1 .$$

To conclude this study of the span of the spherical forms of cyclic type, we determine the 15 - spherical forms for which G_2 is trivial or isomorphic to $\mathbb{Z}_2$. Indeed, these are the cases in which $\rho(16,\mathbb{R}) = 9 \geq n + 1 = 8$. Notice that for these forms we still have $\text{span}(S^{15}/G) = \rho(16,\mathbb{R}) - 1 = 8$. Although this result has been established by methods different from those we developed previously, we give here its proof for the benefit of the reader.

Let us consider the following commutative diagram

$$\begin{array}{ccc} T_8(S^{15}/G_2) & \xrightarrow{T_8(\pi)} & T_8(S^{15}/G) \\ p_2 \downarrow & & \downarrow p \\ S^{15}/G_2 & \xrightarrow{\pi} & S^{15}/G \end{array}$$

where $(T_8(S^{15}/G), p, S^{15}/G)$ $\left(\text{resp. } (T_8(S^{15}/G_2), p_2, S^{15}/G_2)\right)$ is the tangent 8-frame bundle of S^{15}/G (resp. S^{15}/G_2) and π, $T_8(\pi)$ are the maps induced by the natural inclusion $G_2 \to G$. It is obvious that the fibre of these tangent bundles is the Stiefel manifold $V_8(R^{16})$. To show that $\operatorname{span}(S^{15}/G) = 8$, we are reduced to prove the existence of a cross-section of p; this may be done by an argument of obstruction theory. (The reader is referred to [225] for the relevant definitions and development of obstruction theory). If p has a cross-section over the $(k-1)$-skeleton of S^{15}/G $(k \leq 15)$, the obstruction to its extension to the k-skeleton is measured by a certain subset E_k of the cohomology group $H^k(S^{15}/G; \pi_{k-1}(V_8(R^{16})))$. In fact this obstruction vanishes if E_k contains the zero element of this group. Here, the homomorphisms π_k^* induced by π,

$$\pi_k^* : H^k\left(S^{15}/G; \pi_{k-1}(V_8(R^{16}))\right) \to H^k\left(S^{15}/G_2; \pi_{k-1}(V_8(R^{16}))\right),$$

are isomorphisms for $2 \leq k \leq 15$, because $\pi_i(V_8(R^{16})) = 0$ for $1 < i < 7$ and $\pi_8(V_8(R^{16})) \cong \mathbb{Z}$ (these facts are derived easily from the homotopy exact sequence of the fibration $SO(16) \to V_8(R^{16})$), whereas the groups $\pi_i(V_8(R^{16}))$ contain only 2-torsion if $9 \leq i \leq 14$ (cf. [222; page 98]. Now, since $\operatorname{span}(S^{15}/G_2) = \operatorname{span}_{\mathbb{R}}(S^{15}) = 8$, the fibration p_2 has a cross-section and hence it is clear that $\pi_k^*(E_k)$ contains the zero element; it follows that E_k also contains 0 and ultimately, p has a cross-section over S^{15}/G. This proof is due to Denis Sjerve for $G_2 = 0$ [222; page 104] and to Becker for $G_2 \cong \mathbb{Z}_2$ [40; page 998].

All the informations obtained in this section (notably (5.1) and (5.6)) can be summarized as follows.

Theorem 5.7. *Let* n *be a positive integer* $\neq 7$ *and let* S^n/G *be a spherical form of cyclic type. Then*

(i) *if* $v_2(|G|) = 0$ *or* 1, $\operatorname{span}(S^n/G) = \rho(n+1, \mathbb{R}) - 1$;

(ii) *if* $v_2(|G|) \geq 2$, $\operatorname{span}(S^n/G) = \rho(n+1, \mathbb{C}) - 1$.

The assumption that $n \neq 7$ is necessary; we shall

see in the next chapter that in dimension 7 there are forms which are parallelizable and others which are not.

6. Span of Spherical Forms of Quaternionic Type

In this section we shall leave out the discussion of the spherical forms of quaternionic type of dimension ≤ 3, for which the results are trivial; besides, the spherical forms of quaternionic type we are going to speak about here are only those for which the inclusion of G_2 into G (G_2 is isomorphic to the generalized quaternionic group Q_m (see (2.9)) induces an isomorphism $H^i(G;\mathbb{Z})_{(2)} \cong H^i(G_2;\mathbb{Z})$. In this respect, the reader will notice that this condition is always fulfilled when the group G contains a normal subgroup whose order is precisely the odd part of the order of G (see Lemma 1.5); this algebraic property characterizes the groups G which are metabelian of rank 2 [247; page 132].

In this context, for $n \neq 1$, the arsenal constituted by (2.8), (4.8) and (4.12) gives immediately that

$$\operatorname{span}(S^{4n+3}/G) \geq \rho(4n+4,\mathbb{H}) - 1 .$$

Observe that if n is even, say $n = 2\ell$,

$$\rho(8\ell+4,\mathbb{H}) - 1 = 3 = \operatorname{span}_{\mathbb{R}}(S^{8\ell+3})$$

(see (2.0)). Hence, the inequality $\operatorname{span}(S^{8\ell+3}/G) \leq \operatorname{span}_{\mathbb{R}}(S^{8\ell+3})$ implies that

(6.1) $$\operatorname{span}(S^{8\ell+3}/G) = \rho(8\ell+4,\mathbb{H}) - 1 .$$

We shall suppose that from now on n is odd, $n = 2\ell+1$. If we proceed as in section 5, the computation of $\operatorname{span}(S^{8\ell+7}/G)$ is reduced to consider $\operatorname{span}(S^{8\ell+7}/Q_2)$, where $S^{8\ell+7}/Q_2$ is the Q_2-spherical form $N^{2\ell+1}(2)$ obtained by the action of Q_2 on $S^{8\ell+7}$ defined by the complex representation $\rho = (n+1)\xi_4$ (cf. (2.6)). Consider the additive structure of $RU(Q_2)$ given by (2.5) and put $\alpha = \xi_o + \xi_1 + \xi_2 + \xi_3$, so that we have

(6.2) $$\begin{cases} \alpha^2 = 4\alpha = \xi_4^2 , & \alpha \cdot \xi_4 = 4\xi_4 , \\ \xi_4 \cdot \xi_i = \xi_4 , & \xi_i\xi_j = \xi_{i+j} , \end{cases}$$

where $0 \leq i, j \leq 3$ and $i + j$ possibly reduced mod. 4. Since we want to show that $\mathrm{span}(N^{2\ell+1}(2)) = \rho(8\ell+8,\mathbb{H}) - 1$ and we already know that this span is $\geq \rho(8\ell+8,\mathbb{H}) - 1$, let us assume that $\mathrm{span}(N^{2\ell+1}(2)) \geq k - 1$ with $k > \rho(8\ell+8,\mathbb{H})$. Then, because of (4.11), we conclude that $\eta_k \otimes \rho$ and ρ are Q_2-fibre homotopically J-equivalent. Thus, there is a unit $u \in KO_{Q_2}(\mathbb{R}P^{k-1})$ such that

$$\theta_j(\eta_k \otimes \rho) \cdot \psi^j(u) = u\theta_j(\rho) .$$

In particular, it follows that

$$(6.3) \qquad \theta_5[\eta_k \otimes (n+1)\xi_4] = \theta_5[(n+1)\xi_4] .$$

We obtain (6.3) by proceeding just as in section 5 ; here the assertion (ii) of Lemma 5.2 is translated as follows: ψ^5 being the identity on $RU(Q_2)$ (see (5.3)), it will still be such in $RO(Q_2)$ since the complexification homomorphism $c : RO(Q_2) \to RU(Q_2)$ is injective. Hence, we have:

Lemma 6.4. (i) $KO_{Q_2}(\mathbb{R}P^{k-1})$ *is generated by elements of the type* ζ *and* $\eta_k \otimes \zeta$, $\zeta \subset RO(Q_2)$. *In* $KO_{Q_2}(\mathbb{R}P^{k-1})$, *the element* $y_k \cdot \xi_4$, *with* $y_k = [\eta_k] - 1$, *is of order* $a_k(\mathbb{H})$;
(ii) ψ^5 *is the identity on* $KO_{Q_2}(\mathbb{R}P^{k-1})$.

The relation (6.3) allows us to compute $\theta_5(2\xi_4)$ and $\theta_5[\eta_k \otimes (2\xi_4)]$ since $n + 1$ is even. By (0.13)

$$\theta_5(2\xi_4) = \prod_{i=1}^{4} \lambda_{-\zeta_i}(\xi_4), \quad \theta_5[\eta_k \otimes (2\xi_4)] = \prod_{i=1}^{4} \lambda_{-\zeta_i}(\eta_k \otimes \xi_4) ,$$

with $\zeta_1 = \exp(2\pi i/5)$ and $\zeta_2 = \zeta_1^2$, $\zeta_3 = \zeta_1^3$, $\zeta_4 = \zeta_1^4$. The reader is invited to verify the validity of these equalities (existence of a reduction through $RSpin(Q_2)$).
Now, if $x = \eta_k \otimes \xi_4$ or ξ_4, we have that

$$\lambda_a(x) = 1 + xa + (2+\alpha)a^2 + xa^3 + a^4$$

and consequently,

$$\prod_{i=1}^{4} \lambda_{-\zeta_i}(x) = (1+3\alpha+3x)^2 .$$

Equality (6.3) thus becomes

$$(1+3\alpha+3\xi_4)^{n+1} = (1+3\alpha+3\xi_4+3y_k\xi_4)^{n+1} .$$

In other words,

$$\sum_{i=1}^{n+1} \binom{n+1}{i} (3y_k\xi_4)^i (1+3\alpha+3\xi_4)^{n+1-i} = 0 .$$

Since $y_k^i = (-2)^{i-1} y_k$ for $i \geq 1$, this can be written as

$$\frac{1}{2}\left\{ \sum_{i=1}^{n+1} (-1)^{i-1} \binom{n+1}{i} (6\xi_4)^i (1+3\alpha+3\xi_4)^{n+1-i} \right\} y_k = 0 ,$$

that is to say,

(6.5) $$\frac{1}{2}\left\{ (1+3\alpha+3\xi_4)^{n+1} - (1+3\alpha-3\xi_4)^{n+1} \right\} y_k = 0 .$$

From (6.2) we conclude that $\alpha\xi_4 = 4\xi_4$ and $\alpha^i = 2^{2i-2} \cdot \alpha$ $(i \geq 1)$ and so,

(6.6) $$(1+3\alpha)^i \cdot \xi_4 = \sum_{j=o}^{i} \binom{i}{j} 3^j \alpha^j \xi_4 = \sum_{j=o}^{i} \binom{i}{j} 12^j \xi_4 = 13\xi_4 .$$

With the aid of (6.6) we can now easily verify that (6.5) can be written simply as

(6.7) $$(1/8)(25^{n+1} - 1) y_k \xi_4 = 0 .$$

Since the order of $y_k\xi_4$ is $a_k(\mathbb{H})$ (Lemma 6.4 (i)), relation (6.7) implies that $n + 1 \equiv 0 \pmod{a_k(\mathbb{H})}$. But the definition of $\rho(8\ell+8,\mathbb{H})$ (cf. Remark 2.10) shows that the inequality $k > \rho(8\ell+8,\mathbb{H})$ is not true. Thus,

$$\mathrm{span}(N^{2\ell+1}(2)) = \rho(8\ell+8,\mathbb{H}) - 1$$

and consequently,

(6.8) $$\mathrm{span}(S^{8\ell+7}/G) = \rho(8\ell+8,\mathbb{H}) - 1$$

(with $\ell \neq 0$).

We now state the main result of this section by summarizing (6.1) and (6.8).

Theorem 6.9. *Let* S^n/G *be a spherical form of quaternionic type with* $n \neq 7$. *We assume that the natural inclusion of* $G_2 \cong Q_m$ *into* G *induces an isomorphism* $H^i(G;\mathbb{Z})_{(2)} \cong H^i(G_2;\mathbb{Z})$. *Then,*

$$\mathrm{span}(S^n/G) = \rho(n+1,\mathbb{H}) - 1 .$$

The case $n = 7$ has to be excluded in our theorem; in fact, we shall see in the next chapter that no Q_m - spherical form S^7/Q_m is parallelizable.

CHAPTER 6

IMMERSIONS AND EMBEDDINGS OF MANIFOLDS

Throughout this chapter, except when otherwise specified, all manifolds and morphisms are of class C^∞. The dimension of a manifold M will often appear as a superscript of M.

1. Background.

Let $f : M \to N$ be a morphism between two manifolds and let x be a point of M. By definition, the *rank* of f at x is the rank of the tangent linear transformation $T_x f : T_x M \to T_{f(x)} N$. Clearly,

$$\mathrm{rank}_x f = \dim(\mathrm{Im}\, T_x f) \leq \inf(\dim_x M, \dim_{f(x)} N) .$$

It seems then natural to study the morphisms f whose rank coincides with the local dimension of one of the manifolds considered. We say that f is a *local immersion* at x (resp. a *local submersion* at x) whenever $\dim_x M \leq \dim_{f(x)} N$ and $\mathrm{rank}_x f = \dim_x M$ (resp. $\dim_{f(x)} N \leq \dim_x M$ and $\mathrm{rank}_x f = \dim_{f(x)} N$); in other words, f is a local immersion at x (resp. a local submersion at x) when $T_x f$ is injective (resp. surjective). If f is a local immersion (resp. a local submersion) at every point $x \in M$, we say that f is an *immersion* (resp. a *submersion*) of M into N. It follows that an immersion is always locally injective (and thus, a topological immersion); however, a morphism can be globally injective without being an immersion (for example, this is the case of the function $f : \mathbb{R} \to \mathbb{R}$ given by $f(x) = x^3$). From now on we shall be dealing only with immersions; some of the facts that we shall recall can be transposed "mutatis mutandis" to submersions. Furthermore, we shall consider only *connected* manifolds.

The reader should observe that the composition of two immersions is an immersion; moreover, if $f : M \to N$ and $f' : M' \to N'$ are immersions, $f \times f' : M \times M' \to N \times N'$ is also an immersion. Finally, given a morphism $f : M \to N$, its graph $\Gamma_f : M \to M \times N$ defined by $\Gamma_f(x) = (x, f(x))$, for every $x \in M$, is an immersion. From this last property it follows that the diagonal morphism,

$$\Delta_M : M \longrightarrow M \times M \ , \ \Delta_M(x) = (x,x) \ ,$$

is an immersion. The notion of immersion can be characterized geometrically as follows: a morphism $f : M \to N$ is an immersion if, and only if, there exists an open covering $\{U_i\}_{i \in I}$ of M such that, for every $i \in I$, $f|U_i$ is an isomorphism of U_i onto a sub-manifold of N. In particular, it follows that if M is a sub-manifold of N, the canonical injection $i : M \to N$ is an immersion.

If $f : M \to N$ induces a homeomorphism of M onto $f(M)$, then $f(M)$ is a sub-manifold of N. In this situation, f is an isomorphism (of manifolds) from M onto $f(M)$; we say that f is an *embedding* of M into N. In order to emphasize the topological difference between the notions of immersion and embedding, we recall the example of the curve C on the torus $T = S^1 \times S^1 \subseteq \mathbb{R}^3$ obtained from a line of irrational slope: C is a 1-dimensional manifold immersed in T; however, this immersion is not an embedding because the open sets of C are not necessarily the intersections of open sets of T with C. However, applying the well-known "inverse function theorem" one can easily see that, for every $x \in M$, the restriction of an immersion of M to a certain neighborhood of x is an embedding.

One of the most important results on immersions and embeddings is the famous theorem of H.Whitney: "every n-dimensional manifold can be embedded in R^{2n} [255] and, if $n \geq 2$, such a manifold can be immersed in R^{2n-1} [256]." We shall call this theorem the Embedding (resp. Immersion) Theorem. At this point we want to observe that in the preceeding remarks we have not made any statements in terms of local coordinates; however, their proofs are local indeed and based on the classical "Rank Theorem" (see, for instance, [72]).

We now take the point of view of fibre bundles. We begin by observing that if $f : M \to N$ is an immersion, then there is a monomorphism of vector bundles over M, namely $\tau(M) \rightarrowtail f^*\tau(N)$, where $\tau(M)$ (resp. $\tau(N)$) is the tangent bundle of M (resp. N). Because this monomorphism has constant rank, we can form the quotient bundle $\nu(f)$ of $f^*\tau(N)$ by the image of $\tau(M)$; the bundle $\nu(f)$ is called the *normal bundle associated to the immersion* f. Hence, attached to each immersion $f : M \to N$, there is an exact sequence of bundles

(1.1) $$0 \longrightarrow \tau(M) \longrightarrow f^*\tau(N) \longrightarrow \nu(f) \longrightarrow 0 \quad .$$

It is well-known that such a sequence is split whenever M is paracompact:

(1.2) $$\tau(M) \oplus \nu(f) \simeq f^*\tau(N) \ .$$

In particular, if M is an n-dimensional paracompact manifold immersed into the euclidean space $\mathbb{R}^{n+k}$ via a morphism f, then

(1.3) $$\tau(M) \oplus \nu(f) \simeq \varepsilon^{n+k} \ .$$

Now let $f : M \to \mathbb{R}^{n+k}$ be an embedding; the total space of its associated normal bundle can be viewed as the set of pairs $(a,b) \in \mathbb{R}^{n+k} \times \mathbb{R}^{n+k}$ such that: (i) there is $x \in M$ such that $f(x) = a$; (ii) for every $z \in \operatorname{Im} T_x f$, the scalar product $(b|z) = 0$. Hence, M can be identified with the image of the trivial cross-section s_o of $\nu(f)$. From this standpoint, whenever M is compact we can show that there exists an embedding $f : M \to \mathbb{R}^{n+k}$ (k sufficiently large) and a homeomorphism $\phi : D(\nu(f)) \to V$ - where $D(\nu(f))$ is the total space of the disc bundle associated to $\nu(f)$ and V is a neighborhood of $f(M)$ in $\mathbb{R}^{n+k}$ - such that $\phi s_o = f$; V is a *tubular neighborhood* of $f(M)$ in $\mathbb{R}^{n+k}$ (see, for instance, [161]).

The relevance of (1.3) in K-theory is clear: it just means that the stable classes of $\tau(M)$ and $\nu(f)$ are inverse to each other in $\widetilde{KO}(M)$. This remark shows that the stable class of $\nu(f)$ is independent of the immersion f selected; it depends only on the manifold M. Indeed, this result is even more interesting since it has an inverse: we are referring to the well-known theorem of M.W.Hirsch [111; § 5].

Theorem 1.4. *An* n-*dimensional manifold* M *can be immersed in* $\mathbb{R}^{n+k}$ *if, and only if, there is a real* k-*vector bundle* ξ *such that* $\tau(M) \oplus \xi$ *is trivial* $(k \geq 1)$.

This characterization enables us to obtain several non-immersion and non-embedding criteria of a manifold into an euclidean space. They consist in taking the appropriate characteristic classes relative to the (co)-homology theory being utilized and then, observing the nullity of these classes for the tangent bundle.

In this chapter we shall state the criteria of the type mentioned before obtained by M.F.Atiyah, via K-theory and Grothendieck operations [13]. Hence, we shall apply them to certain spherical forms introduced in the previous chapter. We should mention that the method we just referred to presupposes our ability to determine the tangent bundle of a given manifold (or, at least, its stable class). This determination is not always easy; besides the case of homogeneous spaces, whose tangent bundles have been studied by A.Borel and F.Hirzebruch [44], we shall work based in two general situations. First, if M is a manifold with boundary, there is a relation between the tangent bundles of M and ∂M, namely,

$$\tau(M) \,|\, \partial M \simeq \tau(\partial M) \oplus \varepsilon^1 \;;$$

this relation can be established by observing that there is an embedding $f : \partial M \times I \to M$ such that $f(x,0) = x$. The second general situation is that of the quotient manifolds. Let $\xi = (E,p,M)$ be a (differentiable) G-bundle over a manifold M, with fibre F and E compact. Let $\tau_F(\xi) = (T_F(\xi),\pi,M)$ be the vector bundle defined by all vectors of $\tau(E)$ which are tangent to the fibres. Finally, let $f : E \to \mathbb{R}^n$ be a G-equivariant embedding relative to an orthogonal representation $\rho : G \to O(n)$ (such an embedding always exists: see [185] or [192]). Giving to E the riemannian metric induced by f, we observe that the action of G on the bundles $\nu(f) = (N_f M, q, M)$ and $\tau_F(\xi)$ is such that

$$(N_f M/G, q_G, M/G) = \nu(f)/G \quad \text{and} \quad (T_F(\xi), \pi_G, M/G) = \tau_F(\xi)/G$$

are principal G-bundles. Let ξ_ρ be the real vector n-bundle associated to the representation ρ : then,

<u>Proposition 1.5.</u> (see [237]) - $\tau(M) \oplus \tau_F(\xi)/G \oplus \nu(f)/G \simeq \xi_\rho$.

This relation is extremely usefull to the determination of the tangent bundle of a spherical form. In particular, if η is the canonical real vector line bundle over $\mathbb{R}P^n$, if ξ is the unitary representation of rank 1 of $\mathbb{Z}_a$ defined by $\xi(\overline{1}) = \xi \cdot 1_{\mathbb{C}}$ where $\xi = \exp(2\pi i/a)$ and, if ξ_{r_i+3} $(0 \leq i \leq n)$ are the representations defined in § 2 of the last chapter, we obtain the following.

<u>Corollary 1.6.</u>

(i) $\tau(\mathbb{R}P^n) \oplus \varepsilon^1 \simeq (n+1)\eta$;

(ii) $\tau\left(L^n(a;a_o,a_1,\ldots,a_n)\right) \oplus \varepsilon^1 \simeq \bigoplus_{i=o}^{n} r(\xi^{a_i})$;

(iii) $\tau\left(N^n(m;r_o,r_1,\ldots,r_n)\right) \oplus \varepsilon^1 \simeq \bigoplus_{i=o}^{n} (2\xi_{r_i+3})$.

With regard to statement (iii) we have written $2\xi_{r_i+3}$ instead of the realification of ξ_{r_i+3} because the unitary representations without fixed point ξ_{r_i+3} of Q_m are self-conjugate. We wish to remind the reader that the irreducible unitary representations ρ of a finite group G are divided into three classes:

(C_1) ρ is equivalent to a real representation;

(C_2) ρ is equivalent to its conjugate $\overline{\rho}$ but is not equivalent to a real representation;

(C_3) ρ is not equivalent to $\overline{\rho}$.

If σ is an orthogonal representation of G , let $\sigma_{\mathbb{C}}$ be its complexification ; the following classification theorem holds:

Proposition 1.7. *The irreducible orthogonal representations of* G *are divided into the following classes:*

1. $\sigma_{\mathbb{C}}$ *is equivalent to a unitary representation of type* (C_1);
2. $\sigma_{\mathbb{C}}$ *is equivalent to* $\rho \oplus \bar{\rho}$, *with* ρ *of type* (C_2);
3. $\sigma_{\mathbb{C}}$ *is equivalent to* $\rho \oplus \bar{\rho}$, *with* ρ *of type* (C_3).

2. A brief Historical Survey

The fundamental problem of the theory of immersions (resp. embeddings) is the following: given two manifolds M and X, determine the equivalence classes defined by regular homotopy (*) (resp. regular isotopy) of all immersions (resp. embeddings) of M into X. It is clear that this problem contains the problem of the existence of an immersion (resp. embedding) of M into X.

H.Whitney gave general answers to this problem for the case in which X is an euclidean space. His methods were based on arguments of "general position". That approach allowed Whitney, on the one hand, to study carefully the singularities of a differentiable map and on the other hand, to approximate a given map by a more "regular" differentiable map. For the latter question, the Weierstrass Approximation Theorem reveals itself to be very useful. Since 1935 Whitney observed a certain number of interesting facts. For example, if $k \geq 2n+2$, any two homotopic immersions $f_o, f_1 : M^n \to X^k$ are regularly homotopic; if $k \geq 2n$, there exists always an immersion of M into X; finally, every n-dimensional manifold (of class C^r, $r \geq 1$) can be embedded in $\mathbb{R}^{2n+1}$ [254]. In that paper and in [255] one can find the definitions necessary for the proof of the Embedding Theorem of M^n into $\mathbb{R}^{2n}$ (see § 1); anyway the argument goes as follows. A map $f : M^n \to \mathbb{R}^{2n}$ is said to be *regular* if its Jacobian matrix has rank n everywhere; if for this regular map f there are two points $x_1, x_2 \in M$ with the same image $y = f(x_1) = f(x_2)$ and if the tangent n-planes $T_f(x_1)$ and $T_f(x_2)$ have only y in common, then y is called

(*) For the definition of regular homotopy (isotopy) the reader is also referred to [186].

a *regular self-intersection*. Finally, if a regular map f has no triple points and has only regular self-intersections, f i said to be *completely regular*. The existence of completely regular maps is guaranteed by Theorem 3 of [254]. Now, if M is orientable, we can distinguish these self-intersections according to their sign; the *intersection number* I_f of f is then defined as the algebraic number of self-intersections.(*) If M is not orientable or if its dimension n is odd, I_f is the mod. 2 - number of self-intersections. The procedure followed by Whitney in [255] to obtain the proof of the Embedding Theorem consists in showing that every completely regular proper map $M \to \mathbb{R}^{2n}$ can be modified into a completely regular proper map(**) without self-intersections. In order to get rid of the self-intersections, he observed the following crucial facts: (i) for every compact manifold M^n $(n \geq 1)$, it is always possible to find a completely regular proper map with intersection number equal to an integer given "a priori"; (ii) if M^n is a closed manifold $(n \geq 3)$, every proper regular map $f : M \to \mathbb{R}^{2n}$ with finite I_f, is regularly homotopic to a completely regular proper map $g : M \to \mathbb{R}^{2n}$. If the number of self-intersections of g is strictly larger than $|I_f|$ (resp., 0), whenever M is orientable with n even (resp. M is not orientable or n is odd), then the number of self-intersections can be decreased by 2. Finally, let us point out that in order to alter at will the number of self-intersections, Whitney exhibited the following example of a regular map $\phi : \mathbb{R}^n \to \mathbb{R}^{2n}$ with only one self-intersection (which, incidentally, is regular):

$$\phi(x_1,\dots,x_n) = (y_1,\dots,y_{2n}) ,$$

where:

$$y_1 = x_1 - 2x_1/u ,$$

$$y_i = x_i , \text{ if } 2 \leq i \leq n ,$$

(*) The reader interested in studying the relationship between this "intersection number" and that defined in Chapter 1 should read Theorem 1 [255].

(**) See [254] and [255] for the definition of *proper map*.

$$y_{n+1} = 1/u \ ,$$
$$y_{n+i} = x_1 x_i/u \ , \quad 2 \le i \le n \ ,$$

with $u = \prod_{i=1}^{n} \left(1+x_i^2\right)$.

To prove the Immersion Theorem, Whitney in his paper [256], lets the (completely) *semi-regular maps* play the same role as that played before by the (completely) regular maps; he defines a proper C^∞ - map $f : M^n \to \mathbb{R}^{2n-1}$ to be *semi-regular* if for every singular point x_o, there exists a system of local coordinates $(x_1,\ldots,x_n)$ on a convenient neighborhood of x_o such that $(\partial f/\partial x_1)_{x=x_o} = 0$ and the $2n-1$ vectors

$$\left(\partial^2 f/\partial x_1^2\right)_{x=x_o} , \ (\partial f/\partial x_i)_{x=x_o} \quad \text{for} \quad 2 \le i \le n \ ,$$
$$\left(\partial^2 f/\partial x_1 \partial x_i\right)_{x=x_o} \quad \text{for} \quad 2 \le i \le n$$

are linearly independent. Whenever a map f is semi-regular, every singular point x_o has a special characteristic: on the neighborhoods of x_o and $f(x_o)$ we can find local coordinates $(u_1,\ldots,u_n)$ and $(v_1,\ldots,v_{2n-1})$ such that $v_1 = u_1^2$, $v_i = u_i$ and $v_{n+i-1} = u_1 u_i$, if $2 \le i \le n$. On the other hand, every continuous function $f : M^n \to \mathbb{R}^{2n-1}$ can be replaced by a completely semi-regular map. Next, in order to obtain the desired immersion, it is necessary to enumerate the algebraic number $\mathcal{L}_f(M)$ of the singular points. For example, if f is semi-regular and M is closed of dimension $n =$ odd (resp. n even), we have $\mathcal{L}_f(M) = 0$ (resp. $\mathcal{L}_f(M) \equiv 0 \pmod{2}$). Actually, the method followed by Whitney is very delicate since one can proceed only by modifications which allow us to put ∂M into the desired position. Nevertheless, the results obtained were a good omen for the developments in the theory of singularities and differential topology, yet to come.

The next important developments in the theory of immersions were due to S.Smale and M.W.Hirsch. As a consequence of their work, the fundamental problem of the classification of immersions is now completely transposed into the realm of

Homotopy Theory. In [224] (see also [223]), Smale obtained the classification for the immersions of S^n into $\mathbb{R}^k$. For a given base point $x_0 \in M^n$, let $f,g : M^n \to X^k$ be two base-preserving immersions such that their first order derivatives at x_0 are equal and given. Let $Im(M,X,x_0)$ be the function space defined by such immersions. Notice that if M is the unit n-ball $\mathbb{D}^n$, $Im(\mathbb{D}^n,X,x_0)$ is contractible; this property allows us to associate to each pair of immersions $f,g : S^n \to \mathbb{R}^k$ as before, a map of S^n into the Stiefel manifold $V_{k,n}$ or, in other words, an element $\omega(f,g) \in \pi_n(V_{k,n})$. The statement of Smale's result is then the following: (i) f and g are (base preserving) regularly homotopic if, and only if, $\omega(f,g) = 0$; (ii) given $f \in Im(S^n,\mathbb{R}^k,x_0)$ and $\alpha \in \pi_n(V_{k,n})$, there exists an element $g \in Im(S^n,\mathbb{R}^k,x_0)$ such that $\omega(f,g) = \alpha$. As a consequence of this theorem one shows that the immersions of S^n into $\mathbb{R}^{2n}$ $(n > 1)$ are classified by the number of self-intersections; furthermore, any two immersions of S^2 into $\mathbb{R}^3$ are regularly homotopic, because $\pi_2(V_{3,2}) = 0$.

Hirsch obtained the *Fundamental Theorem* of the theory of immersions [111] trying to generalize Smale's result:

Theorem 2.1. *If* $k > n$, *then there is a bijection between the regular homotopy classes of immersions of* M^n *into* X^k *and the homotopy classes of monomorphisms from* $\tau(M)$ *into* $\tau(X)$.

The method employed by Hirsch consists in constructing a model for an immersion of a triangulated manifold M^n into X^k by working thru the successive skeleta; the obstructions take values in the homotopy groups of Stiefel manifolds. One of the difficulties of this method is that the skeleta of a manifold are not necessarily manifolds; hence, one is forced to replace the immersions by a class of functions which are immersions on a neighborhood of the skeleton considered.

Theorem 2.1 has several interpretations; in particular, it can be seen as reducing the classification problem of the immersions from M into X to the study of the "homotopy classes of cross-sections of the bundle associated to the bundle of n-frames of M, having for fibres the n-frames of X " (this was a conjecture due to C.Ehresmann [89]). This theorem, besides showing again the result of Whitney about immersions, has many consequences; for example, Theorem 1.4 is one of its Corollaries. It also shows that every parallelisable manifold M^n can be immersed in $\mathbb{R}^{n+1}$.

We now turn to the work done on the question of reducing the classification of the embeddings $M^n \to X^k$ into a homotopy problem. After the work of Whitney, we should quote first of all, the papers of A.Shapiro [217] and Wu Wen Tsün [263]. The latter author showed that two arbitrary embeddings of M^n (connected) into $\mathbb{R}^{2n+1}$ are necessarily regularly homotopic. However, the decisive step was taken by A.Haefliger, for $k \geq (3/2)\cdot(n+1)$ (see [99], [100] and [101]). In order to simplify the ideas, let us restrict ourselves to the case in which $X = \mathbb{R}^k$; then, one of Haefliger's essential results can be explained as follows. A map $\phi : M \times M - \Delta_M \to S^{k-1}$ (where Δ_M is the diagonal of $M \times M$) is said to be *equivariant* if $\phi(x_1,x_2) = -\phi(x_2,x_1)$, for every $(x_1,x_2) \in M \times M - \Delta_M$. Let E_M (resp. B_M) be the quotient space obtained by identification of the points (x_1,x_2,z) and $(x_2,x_1,-z)$ (resp. (x_1,x_2) and (x_2,x_1)) of $(M \times M - \Delta_M) \times S^{k-1}$ (resp. $M \times M - \Delta_M$) ; we then obain a fibration $\xi_M = (E_M,\pi_M,B_M)$ with fibre S^{k-1}. It is easy to see that the equivariant maps $\phi : M \times M - \Delta_M \to S^{k-1}$ correspond canonically to the cross-sections of ξ_M. Haefliger's theorem reads now as follows.

Theorem 2.2. *If* $k > (3/2)\cdot(n+1)$ *(resp.* $k \geq (3/2)\cdot(n+1)$*) there is a bijective (resp. surjective) correspondence between the set of isotopy classes of embeddings from* M^n *into* $\mathbb{R}^k$ *and the set of homotopy classes of cross-sections of* ξ_M.

Applying Haefliger's method, E.Rees [201] has shown that every closed manifold M^n such that $\tilde{H}_i(M) \otimes \mathbb{Z}_2 = 0$ for $i < n$, can be embedded in $\mathbb{R}^k$, provided that $k \geq (3/2)\cdot(n+1)$.

Once the classification problems are reduced to questions of Homotopy Theory, the standard method of approach for their resolution is to use Obstruction Theory, either in its classical form, or under the more sophisticated format of Postnikov Towers; as an example, we refer the reader to the papers of M.Mahowald [170] and J.C.Becker [39].

For the case of embeddings, another approach was conceived: surgery. In this context we quote, in particular, the important papers of S.P.Novikov [191] and W.Browder [58]. Using this kind of technique C.T.C.Wall [248] proved that every 3-dimensional manifold can be embedded in $\mathbb{R}^5$; this is the answer to an old conjecture of Whitney [256; § 1].

We want to observe that from Hirsch's work (theorem 1.4) it is possible to obtain several criteria of non-immersion and non-embedding of a manifold into an euclidean space. These results can be formulated in terms of the appropriate characteristic classes: Stiefel-Whitney, Chern or Pontrjagin for cohomology, Grothendieck operations for K-theory, Anderson-Brown-Peterson classes for cobordism, etc.. The central idea is that the existence of an immersion induces the annihilation of the characteristic classes from a certain dimension on. For example, if there is an immersion (resp. embedding) $f : M^n \to \mathbb{R}^{n+k}$, the Stiefel-Whitney dual classes $\bar{w}_i(M)$ are trivial for $i > k$ (resp. $i \geq k$). Expanding this kind of considerations M.F.Atiyah and F.Hirzebruch formulated a beautiful non-embedding criterion for connected, compact, oriented manifolds of even dimension: the main result is that if M^{2n} is embedded in $\mathbb{R}^{2n+2k}$, the number $2^{n+k-1} H(\frac{1}{2})$, where $H(t)$ is the Hilbert polynomial of M, is an integer [27]. These results have been extended to immersions by B.J.Sanderson and R.Schwarzenberger [203]; analogous results were obtained by K.H.Mayer [173] as applications of the Atiyah-Singer Index Theorem.

To conclude this review - from which we deliberately excluded every reference to the many results obtainded for particular manifolds - let us bring up the following question of Hirsch. Let M^n be an orientable stably parallelizable manifold (this means that the tangent bundle of M is stably trivial, see chapter 3, § 2):

If $k = n + \left[\frac{n+1}{2}\right]$, can M be embedded into $\mathbb{R}^k$?

Meanwhile, one of the objectives being presently pursued is the study of the following "conjecture" (see § 7):

Conjecture 2.3. *Every* n-*dimensinal manifold* $(n \geq 2)$ *can be immersed (resp. embedded) into* $\mathbb{R}^{2n-\alpha(n)}$ *(resp.* $\mathbb{R}^{2n-\alpha(n)+1}$*) where* $\alpha(n)$ *is the number of non-trivial terms in the dyadic development of* n .

3. Atiyah's Criterion

As an application of Grothendieck's operations γ^i in real K-theory, M.F.Atiyah [13] obtained the following result.

Theorem 3.1. *Let* M *be an* n-*dimensional compact manifold and let* $\tau_o \in \widetilde{KO}(M)$ *be the stable class of the tangent bundle* $\tau(M)$. *Then,*

(i) *if* M *is immersible in* $\mathbb{R}^{n+k}$, $\gamma^i(-\tau_o) = 0$, *for every* $i > k$;

(ii) *if* M *is embeddable in* $\mathbb{R}^{n+k}$, $\gamma^i(-\tau_o) = 0$, *for every* $i \geq k$.

The proof of this theorem consists essentially in studying the geometric dimension of $-\tau_o = n - [\tau(M)]$; we shall see it in a little while.

More generally, let X be a (finite) connected CW-complex and let $\mathcal{J} : \mathrm{Vect}_{\mathbb{R}}(X) \to KO(X)$ be the canonical morphism which associates to the isomorphism class of a (real) vector bundle over X its class in KO(X); then, we say that an element $x \in KO(X)$ is *positive* $(x \geq 0)$ provided $x \in \mathrm{Im}\,\mathcal{J}$. Now, let $x \in \widetilde{KO}(X)$ be given; recalling that $KO(X) \simeq \widetilde{KO}(X) \oplus \mathbb{Z}$, we define the *geometric dimension* of x (notation: gdim x) to be the smallest integer a such that $x + a \geq 0$. (Recall, that each element of KO(X) is of the form $[\xi] - n$ (see (0.1)). Hence, for any $x \in \widetilde{KO}(X)$ there are integers n such that $x+n \geq 0$; this implies that the geometric dimension exists for any $x \in \widetilde{KO}(X$ Moreover, for all $x \in \widetilde{KO}(X)$: gdim $x \leq \dim X$; see (0.4).)

Proposition 3.2. *If* $x \in \widetilde{KO}(X)$, *then* $\gamma^i(x) = 0$ *for every* $i >$ gdim x.

Proof. Suppose that gdim x = a; then

$$\gamma_t(x) = \lambda_{t/(1-t)}(x) = \lambda_{t/(1-t)}(-a)\lambda_{t/(1-t)}(x+a)$$

and hence,

(3.3) $$\gamma_t(x) = \sum_{i\geq o} \lambda^i(x+a)t^i(1-t)^{a-i} .$$

Since $x + a \geq 0$ there is a real vector bundle ξ such that $[\xi] = x + a$; but $\text{rank}[\xi] = a$, so $\lambda^i(\xi) = 0$ for every $i > a$. Relation (3.3) shows that $\gamma_t(x)$ is a polynomial of degree $\leq a$, which is precisely our statement.

Corollary 3.4. *Let* $\xi = (E,p,X)$ *be a real vector bundle of dimension* k *and let* $S(\xi) = (S(E),\pi,X)$ *be its associated sphere bundle. If* $x = [\xi] - k \in \widetilde{KO}(X)$, $\gamma^i(\pi^! x) = 0$ *for every* $i \geq k$.

Proof. It is easy to see that the vector bundle $\pi^*(\xi)$ has an everywhere non-zero cross-section; hence, $\pi^*(\xi) \cong \eta \oplus 1$ for some $(k-1)$ - dimensional vector bundle over $S(E)$. Then $\pi^!(x) = [\eta] + 1 - k$ and therefore,

$$\text{gdim } \pi^!(x) = \text{gdim } ([\eta] + 1-k) \leq k - 1$$

since $[\eta] \geq 0$. By (3.2) $\gamma^i(\pi^!(x)) = 0$ for every $i > \text{gdim } \pi^!(x)$.

Proof of Theorem 3.1. (i) If f is an immersion of M^n into $\mathbb{R}^{n+k}$, (1.3) shows that $\tau(M) \oplus \nu(f) \simeq \varepsilon^{n+k}$. Hence, $[\nu(f)] = k - \tau_o$ in $KO(M)$ and therefore, $\text{gdim}(-\tau_o) \leq k$. Now (3.2) shows that $\gamma^i(-\tau_o) = 0$ for every $i > k$.

(ii) Let us assume now that f is an embedding of M^n into $\mathbb{R}^{n+k}$. Let $S(\nu(f)) = (S(N_f M),\pi,M)$ be the sphere bundle associated to the normal bundle $\nu(f)$; because of (3.4) we obtain, for every $i \geq k$,

(3.5) $$\gamma^i(\pi^!(-\tau_o)) = 0$$

But $S(N_f M)$ can be viewed as the boundary of a tubular neighborhood N of M and so, from considerations developed in § 1 , we conclude that the trivial cross-section s is actually a homotopy equivalence between M and N , so that the diagram below is homotopy-commutative:

$$\begin{array}{ccc} \partial N & \hookrightarrow & N \\ & \overset{\pi}{\searrow} & \uparrow s \\ & & M \ . \end{array}$$

This shows that the exact sequence of the pair $(N,\partial N)$ in $\widetilde{KO}$-theory can be written as

$$(3.6) \qquad \cdots \longrightarrow \widetilde{KO}(N,\partial N) \xrightarrow{\alpha^!} \widetilde{KO}(M) \xrightarrow{\pi^!} \widetilde{KO}(\partial N) \longrightarrow \cdots$$

where $\alpha^! = s^! j^!$, with $j^! : \widetilde{KO}(N,\partial N) \to \widetilde{KO}(N)$. Now let us map $(N,\partial N)$ into $(D^{n+k}, \overline{D^{n+k}-N})$ where D^{n+k} is a sufficiently large ball of $\mathbb{R}^{n+k}$ it is then clear that

$$\begin{array}{ccc} \widetilde{KO}(D^{n+k}, \overline{D^{n+k}-N}) & \longrightarrow & \widetilde{KO}(D^{n+k}) \\ \rho^! \Big\downarrow \simeq & & \Big\downarrow \\ \widetilde{KO}(N,\partial N) & \xrightarrow{\ \alpha^!\ } & \widetilde{KO}(N) \end{array}$$

commutes, where $\rho^!$ is the excision isomorphism. It follows that $\alpha^! = 0$ because $\widetilde{KO}(D^{n+k}) = 0$ and hence, the morphism $\pi^!$ of (3.6) is injective. The naturality of the γ^i operations and (3.5) now show that $\gamma^i(-\tau_o) = 0$, for every $i \geq k$.

Of course, it is very tempting to compare the criterion just obtained to the classical ones. For example, if we consider rational Pontrjagin classes, we know that if M^n is immersed (resp. embedded) in $\mathbb{R}^{n+k}$, then

$$(3.7) \qquad p_i(-\tau_o) = 0 \ \text{ if } \ 2i > k \quad (\text{resp. } \ 2i \geq k).$$

This relation shows that the Grothendieck operations have the advantage of giving more informations for $k > [n/2]$.

More generally, let x be an element of $\widetilde{KO}(X)$ and let let $\underline{ch} : KO(X) \to H^*(X;\mathbb{Q})$ be the composition of the complexification homomorphism with the Chern character. We claim that

$$(3.8) \qquad \underline{ch}\ \gamma^{2j}(x) = (-1)^j\, p_j(x) + \text{higher terms.}$$

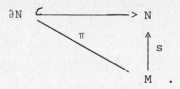

This relation shows that (3.7) is a consequence of Atiyah's theorem.

To prove (3.8) let η be an m-dimensional vector bundle such that

$$x = \eta - m \ ,$$

and let us suppose for the present, that the complexification $\eta_{\mathbb{C}}$ is a sum of complex line bundles,

$$\eta_{\mathbb{C}} \cong \eta_1 \oplus \dots \oplus \eta_m \ , \qquad \dim \eta_i = 1 \ .$$

We denote the first Chern class of η_i by a_i ; for the total Chern class of $\eta_{\mathbb{C}}$ we then have

$$c(\eta_{\mathbb{C}}) = \prod_{i=1}^{m} (1+a_i) \ ,$$

and the j-th Pontrjagin class of x is therefore given by

$$(3.9) \qquad p_j(x) = (-1)^j c_{2j}(\eta_{\mathbb{C}}) = (-1)^j s_{2j}(a_1,\dots,a_m) \ ,$$

where s_{2j} denotes the $2j$-th elementary symmetric function. For the complexification $\gamma_t(x)_{\mathbb{C}}$ of $\gamma_t(x)$ we compute

$$\gamma_t(x)_{\mathbb{C}} = \gamma_t^{\mathbb{C}}(\eta_{\mathbb{C}} - m) = \prod_{i=1}^{m} \gamma_t^{\mathbb{C}}(\eta_i - 1) = \prod_{i=1}^{m} (1+(\eta_i-1)t) .$$

Hence,

$$(3.10) \qquad \begin{aligned} \underline{\mathrm{ch}}\ \gamma_t(x) &= \prod_{i=1}^{m} \{1+(\exp(a_i)-1)t\} \\ &= \sum_k (s_k(a_1,\dots,a_m) + \text{higher terms})\ t^k . \end{aligned}$$

Relations (3.9) and (3.10) imply our claim (3.8), in case $\eta_{\mathbb{C}}$ decomposes into a sum of line bundles. The general case follows invoking the splitting principle.

Finally, whenever we restrict ourselves to *integral* Pontrjagin classes, the previous result can be stated as follows.

<u>Proposition 3.11.</u> *Suppose that* X *is torsion-free (i.e.,* $H^*(X;\mathbb{Z})$ *is free). Let* ξ_s *be the stable class of a real vec-*

tor bundle ξ *over* X. *Then, the assertion*

$$\gamma^{2i}(\xi_s) = 0 \quad \textit{for every} \quad i > k \quad ,$$

implies that

$$p_i(\xi_s) = 0 \quad \textit{for every} \quad i > k \quad .$$

We shall see, later on, based on the examples mentioned in [155], that for spaces with torsion (3.11) is not always valid; other comments about 3.11 will also be made.

We give next a first application of Theorem 3.1. Let us recall that a *Dold manifold* D(m,n) *of dimension* m + 2n is obtained from $S^m \times \mathbb{C}P^n$ by identification of every $(x,z) \in S^m \times \mathbb{C}P^n$ with $(-x,\bar{z})$. Clearly D(m,0) (resp. D(0,n)) is just the real (resp. complex) projective space $\mathbb{R}P^m$ (resp. $\mathbb{C}P^n$). The cellular structure and the mod.2 cohomology ring of D(m,n) were described by Albrecht Dold in [75] : let (a^i,b^j), $0 \leq i \leq m$, $0 \leq j \leq n$, be the $(i+2j)^{th}$ - integral cohomology class of D(m,n) which corresponds to the (i+2j) - cell $\left(e_0^i,e_1^j\right)$ of D(m,n); then, using the same notation for such a class or for its reduction mod.2, we obtain:

<u>Proposition 3.12.</u> *The cohomology ring* $H^*(D(m,n);\mathbb{Z}_2)$ *coincides with the truncated polynomial ring* $\mathbb{Z}_2[a,b]/\langle a_1^{m+1},b_1^{n+1}\rangle$, *where* $a_1 = (a^1,b^0)$ *and* $b_1 = (a^0,b^1)$.

Let $\pi : D(m,n) \to \mathbb{R}P^m$ be the canonical projection. The reader can check that $(D(m,n),\pi,\mathbb{R}P^m)$ is a fibre bundle with fibre $\mathbb{C}P^{n-1}$ and group $\mathbb{Z}_2$; we shall denote by i the inclusion of the fibre $\mathbb{C}P^{n-1}$ into D(m,n). Now let η (resp. ξ) be the canonical line bundle over $\mathbb{R}P^m$ (resp. $\mathbb{C}P^{n-1}$). M.Fujii [92; Theorem 2.2] and J.Ucci [244; Proposition 1.4] proved independently that there is a 2 - dimensional real vector bundle μ over D(m,n) such that $i^*\mu \simeq r\xi$ (if n = 0, $\mu \simeq \varepsilon \oplus \pi^*\eta$). From the two papers quoted before it follows, firstly , that the tangent bundle $\tau = \tau(D(m,n))$ can be expressed by the relation

$$\tau \oplus \varepsilon \oplus \pi^*\eta \simeq (\pi^*\tau(\mathbb{R}P^m)) \oplus (n+1)\mu \quad ,$$

or, in other words (using 1.5),

(3.13) $\tau \oplus \varepsilon^2 \oplus \pi^*\eta \simeq ((m+1)\pi^*\eta) \oplus (n+1)\mu$;

and secondly, assuming that $m,n \geq 2$, the total Stiefel-Whitney classes of $\pi^*\eta$ and μ are given by

(3.14) $w(\pi^*\eta) = 1 + a_1$, $w(\mu) = 1 + a_1 + b_1$.

This last result implies that

$$c((\pi^*\eta)_{\mathbb{C}}) = 1 + a_2 \ , \ c(\mu_{\mathbb{C}}) = 1 + a_2 - b_2 \ ,$$

where $a_2 = (a^2,b^o)$ and $b_2 = (a^o,b^2)$. Hence,

(3.15) $c(\tau_{\mathbb{C}}) = (1+a_2)^m(1+a_2-b_2)^{n+1}$.

Let us now compute $\gamma^i(-\tau_o)$, where $\tau_o = [\tau] - (m+2n) \in \widetilde{KO}(D(m,n))$. We observe first that the KU - theory of the Dold manifolds was studied by Ucci [244] and Fujii [92], [93]; as for their KO - theory, very complete results were obtained by M.Fujii and T.Yasui in [94]. For our purposes it will be sufficient to know that $\widetilde{KO}(D(m,n))$ contains a direct factor which is isomorphic to

$$\mathbb{Z}^{[n/2]} \oplus \mathbb{Z}_{2^{f(m)}}$$

generated by

$$x = [\mu] - [\pi^*\eta] - 1, x^2,\dots,x^{[n/2]}, \ \sigma = [\pi^*\eta] - 1$$

with the relations

(3.16) $\sigma^{f(m)+1} = 0, \ \sigma^2 = -2\sigma, \ \sigma x = 0, \ x^{[n/2]+1+h} = 0$,

where $h = 0$ or 1 according to $n \not\equiv 1$ or $n \equiv 1$ (mod.4) and also where $f(m)$ is the Radon-Hurwitz number defined by

$$f(m) = \text{cardinality } \{0 < s \leq m \mid s \equiv 0,1,2 \text{ or } 4 \pmod{8}\}$$

(see Chapter 4, Proposition 3.12).
From (3.13) we conclude that

(3.17)
$$\gamma_t(-\tau_o) = \gamma_t(-(m+n+1)\sigma - (n+1)x)$$
$$= (\gamma_t(\sigma))^{-(m+n+1)}(\gamma_t(x))^{-(n+1)} \ .$$

Because $\text{gdim}\ \sigma = 1$ and $\text{gdim}\ x = 2$, it follows that

$$\gamma_t(\sigma) = 1 + \sigma t, \ \gamma_t(x) = 1 + xt + \gamma^2(x)t^2 .$$

We now compute $\gamma^2(x)$. To this end, we take (0.12) to obtain that

$$\gamma^2(x) = \gamma^2([\mu] - 2) + \gamma^1([\mu] - 2)\gamma^1(-\sigma) + \gamma^2(-\sigma) .$$

But

$$\gamma_t(-\sigma) = (1+\sigma t)^{-1} = 1 - \sigma t + \sigma^2 t^2 - \dots$$
$$= 1 - \sigma t - 2\sigma t^2 + \dots$$

(utilize 3.16) and so,

$$\gamma^2(x) = \gamma^2([\mu] - 2) - \sigma \cdot \gamma^1([\mu] - 2) - 2\sigma .$$

On the other hand

$$\gamma_t([\mu]-2) = \gamma_{(t/1-t)}([\mu]-2) = (1+(t/1-t))^{-2}(1+[\mu](t/1-t) +$$
$$+ [\Lambda^2\mu](t/1-t)^2) = (1-t)^2 + [\mu]t(1-t) + [\Lambda^2\mu]t^2 ,$$

which shows that

$$\gamma^1([\mu]-2) = [\mu] - 2 \quad \text{and} \quad \gamma^2([\mu]-2) = 1 - [\mu] + [\Lambda^2\mu] .$$

It follows that

$$\gamma^2(x) = 1 - [\mu] + [\Lambda^2\mu] - \sigma([\mu]-2) - 2\sigma$$
$$= 1 - [\mu] + [\Lambda^2\mu]$$

since, as one can easily prove, $\sigma[\mu] = 0$; thus,

$$\gamma^2(x) = [\Lambda^2\mu] - \sigma - 1 - x .$$

Actually, the line bundle $\Lambda^2\mu$ is isomorphic to $\pi^*\eta$. To see this, we compare their first Stiefel-Whitney classes; since $w_1(\pi^*\eta) = a_1 = w_1(\mu)$ (see 3.14), it suffices to show that

(3.18) $$w_1(\mu) = w_1(\Lambda^2\mu) .$$

But (3.18) holds for every real vector bundle of dimension 2 ; we leave the proof of this result to the reader. Putting together these facts, we conclude that $\gamma^2(x) = -x$. Going back to (3.17) we now see that

$$\gamma_t(-\tau_o) = (1+\sigma t)^{-(m+n+1)}(1+xt-xt^2)^{-(n+1)} ;$$

finally, because $\sigma x = 0$ and $\sigma^{i+1} = (-1)^i 2^i \sigma$, for every $i \geq 0$, it is not difficult to conclude that

$$(3.19) \qquad \gamma^i(-\tau_o) = (-1)^{i-1} 2^{i-1} \binom{m+n+i}{i} \sigma + \sum_{k=\left[\frac{i+1}{2}\right]}^{i} A_i^k x^k$$

where

$$A_i^k = (-1)^{i-k} \binom{k}{i-k} \binom{n+k+1}{k} .$$

Consider the integer

$$\bar{A}(m,n) = \sup\left\{ i \mid 2^{i-1} \binom{m+n+i}{i} \not\equiv 0 \pmod{2^{f(m)}} \right\}$$

and define

$$(3.20) \qquad A(m,n) = \begin{cases} \max(\bar{A}(m,n),\ 2[n/2]), & \text{if } m > 0 \\ 2[n/2] , & \text{if } m = 0 ; \end{cases}$$

then, because the order of σ is $2^{f(m)}$ and the term

$$A_{[n/2]}^{[n/2]} x^{[n/2]} = \binom{n+1+ n/2}{[n/2]} x^{[n/2]} \not\equiv 0$$

(this term is part of the coefficient of $t^{2[n/2]}$ in the development of $(1+xt-xt^2)^{-(n+1)}$) , it follows that $\gamma^{A(m,n)}(-\tau_o) \not\equiv 0$. This result together with (3.1) shows that

<u>Proposition 3.21.</u> -[244] - (i) $D(m,n)$ *cannot be immersed in* $\mathbb{R}^{m+2n+A(m,n)-1}$;

(ii) $D(m,n)$ *cannot be embedded in* $\mathbb{R}^{m+2n+A(m,n)}$.

Setting $A(m) = A(m,0) = \bar{A}(m,0)$, we obtain

<u>Corollary 3.22.</u> - [13] - (i) $\mathbb{R}P^m$ *cannot be immersed in* $\mathbb{R}^{m+A(m)-1}$;

(ii) $\mathbb{R}P^m$ *cannot be embedded in* $\mathbb{R}^{m+A(m)}$.

<u>Remarks.</u> For the benefit of the reader interested in comparing the results obtained above for $\mathbb{R}P^m$ and those obtained by different methods (the literature is full of examples) we trans-

cribe next some of the best possible known values; in the following tables, $m + k$ is the dimension of the smallest euclidean space in which $\mathbb{R}P^m$ can be immersed.

m	1	2	3	4	5	6	7	8	9	10	11	12	13	14	15	16	17	18	19	20
m+k	2	3	4	7	7	7	8	15	15	16	16	18	22	22	22	31	31	32	32	34

$m = 2^s-1$, $s \geq 4$	$s \equiv 0$ (mod.4)	$s \equiv 1$ (mod.4)	$s \equiv 2$ (mod.4)	$s \equiv 3$ (mod.4)
m+k	$2m - 2s$	$2m - 2s + 1$	$2m - 2s + 1$	$2m - 2s - 1$

m	2^s $s \geq 1$	2^s+1 $s \geq 2$	2^s+2 $s \geq 3$	2^s+3 $s \geq 2$	2^s+4 $s \geq 3$	2^s+5 $s \geq 3$	2^s+5 $s \geq 3$	2^s+7 $s \geq 3$
m+k	2m-1	2m-3	2m-4	2m-6	2m-6	2m-4	2m-6	2m-8

We want to observe that Atiyah had already indicated that in certain cases 3.22 could give better informations than the Stiefel-Whitney classes (particularly for $m = 2^s-1$), while in other cases (for example, $m = 2^s$) the results are less interesting. We leave to the reader the task of obtaining from 3.21, analogous criteria of non-immersion and non-embedding for the complex projective space $\mathbb{C}P^m$; at any rate, we shall go back to this problem at the end of the chapter.

We now revert back to the discussion of the result given in 3.11. Suppose that $X = \mathbb{R}P^m$ and let α be a real vector bundle over it. Since $\widetilde{KO}(\mathbb{R}P^m)$ is generated by $\sigma = [\eta] - 1$, the stable class α_s of α in $\widetilde{KO}(\mathbb{R}P^m)$ is an integral multiple of σ, say $\alpha_s = u\sigma$. Computing the total Chern class of $u\sigma_{\mathbb{C}}$ one obtains that the ith - integral Pontrjagin class of α_s is given by

$$p_i(\alpha_s) = (-1)^i c_{2i}(\alpha_{\mathbb{C}}) = \binom{u}{2i} y^{2i}$$

where y is the generator of $H^2(\mathbb{R}P^m;\mathbb{Z}) \simeq \mathbb{Z}_2$ (if $4i > m$ we have that $p_i(\alpha_s) = 0$). Moreover, (3.16) shows that

$$\gamma_t(\alpha_s) = (\gamma_t(\sigma))^u = \sum_{i \geq 0} \binom{u}{i} \sigma^i t^i$$

$$= 1 + \sum_{i \geq 1} (-1)^{i-1} 2^{i-1} \binom{u}{i} \sigma t^i$$

and hence,

$$\gamma^{2i}(\alpha_s) = -2^{2i-1}\binom{u}{2i}\sigma \quad .$$

Because $\widetilde{K}O(\mathbb{R}P^m) \cong \mathbb{Z}_{2^{f(m)}}$ (see Chapter 4), to say that $\gamma^{2i}(\alpha_s) = 0$ is equivalent to saying that $2^{2i-1}\binom{u}{2i} \equiv 0 \pmod{2^{f(m)}}$. But a scrutiny of the values assumed by $f(m)$ reveals that if $4i \leq m$, $2i - 1 < f(m)$; hence, $\gamma^{2i}(\alpha_s) = 0$ implies that $\binom{u}{2i} \equiv 0 \pmod{2}$ and consequently, $p_i(\alpha_s) = 0$. With this we observe that the conclusion of 3.11 holds for $X = \mathbb{R}P^m$, although such a space is not torsion-free. On the other hand, if $X = D(m,n)$ with $m = 2^s$, $n = 2^v$ $(v > s > 1)$ and $k = 2^{s-2} + 2^{v-1}$, (3.15) and (3.17) show that $\gamma^{2i}(-\tau_o) = 0$ for $i \geq k$, while $p_k(-\tau_o) \neq 0$. Finally, using $\mathbb{R}P^m$ we can give a counter-example to show that the converse of 3.11 is false, in general. It suffices, for instance, to take $m = 2^s - 1$, $n = 2^s$ and $i = 2^{s-3}$ $(s > 3)$; in fact, since the dyadic valuation of $\binom{2^s}{2^{s-2}}$ is

$$v_2\binom{2^s}{2^{s-2}} = 2 \qquad (*)$$

it follows that $p_{2^{s-3}}(\alpha_s) = 0$, while

$$v_2\left(2^{2^{s-2}-1}\binom{2s}{2^{s-2}}\right) = 2^{s-2} + 1 < f(2^s-1) = 2^{s-1} - 1 ,$$

i.e., $\gamma^{2^{s-2}}(\alpha_s) \neq 0$.

(*) see Lemma 4.5.

4. About Immersions and Embeddings of Lens Spaces

In order to apply Atiyah's Theorem (3.1) to lens spaces $L^n(a;a_o,a_1,\ldots,a_n)$ (see Definition 5,2.3) we must describe their γ-structure. We begin by recalling some basic facts about $KU(L^n(a;a_o,a_1,\ldots,a_n))$. Let λ and α_r, $0 \le r \le n$, be the rank 1 unitary representations of $\mathbb{Z}_a$ defined by

$$\lambda(\overline{1}) = \zeta \cdot 1_{\mathbb{C}} \ , \ \alpha_r(\overline{1}) = \zeta^{a_r} \cdot 1_C \ ,$$

where $\zeta = \exp(2\pi i/a)$. Hence, the $(n+1)$-unitary representation $\alpha = \bigoplus_{r=o}^{n} \alpha_r : \mathbb{Z}_a \to U(n+1)$ is precisely the representation which is canonically associated to the action of $\mathbb{Z}_a$ on S^{2n+1} defining $L^n(a;a_o,a_1,\ldots,a_n)$. Furthermore, let us consider the complex line bundle $\xi : S^{2n+1} \times_\lambda \mathbb{C} \to L^n(a;a_o,a_1,\ldots,a_n)$ associated by λ to the covering $S^{2n+1} \to L^n(a;a_o,a_1,\ldots,a_n)$; its class in $KU(L^n(a;a_o,a_1,\ldots,a_n))$ will be indicated by the usual notation $[\xi]$. If $a_o = a_1 = \ldots = a_n = 1$, ξ will be denoted by η; the reader will verify easily that $\eta \simeq \pi^*\mu$, where π is the canonical projection of $L^n(a)$ onto $\mathbb{C}P^n$ and μ is the canonical line bundle over $\mathbb{C}P^n$.

Since $RU(\mathbb{Z}_a) = \mathbb{Z}[\lambda]/\langle \lambda^a - 1\rangle$

and $\sum_{k=o}^{n} (-1)^k \Lambda^k \alpha = \prod_{r=o}^{n} (1-\alpha_r)$,

Atiyah's exact sequence

$$0 \longrightarrow KU^1\Big(L^n(a;a_o,a_1,\ldots,a_n)\Big) \longrightarrow RU(\mathbb{Z}_a) \xrightarrow{\Phi} RU(\mathbb{Z}_a) \longrightarrow KU^o\Big(L^n(a;a_o,a_1,\ldots,a_n)\Big) \longrightarrow 0$$

(chapter 0) gives the following description of $KU(L^n(a;a_o, a_1,\ldots,a_n))$.

Proposition 4.1. - [166] - (i) *The ring* $KU(L^n(a;a_o,a_1,\ldots,a_n))$ *is isomorphic to* $\mathbb{Z}[\xi]/\langle \prod_{r=o}^{n} (1-[\xi]^{a_r}), [\xi]^a - 1\rangle$ *where* $\langle \prod_{r=o}^{n} (1-[\xi]^{a_r}),\ \xi^a - 1\rangle$ *is the ideal generated by the polynomials*

$\prod_{r=0}^{n}\left(1-\xi^{a_r}\right)$ *and* $[\xi]^a - 1$ *in* $\mathbb{Z}[\xi]$;

(ii) *there is a canonical ring isomorphism*

$$\Psi : KU\left(L^n(a;a_0,a_1,\ldots,a_n)\right) \longrightarrow KU(L^n(a))$$

characterized by $\Psi([\xi]) = [\eta]$.

The second statement is a consequence of the invariance of the order of the groups $\widetilde{KU}(L^n(a;a_0,a_1,\ldots,a_n))$ when the action changes. More precisely, Proposition 5.1.4 shows the following.

Lemma 4.2. (i) *The order of the group* $\widetilde{KU}(L^n(a;a_0,a_1,\ldots,a_n))$ *is* a^n ;

(ii) $\widetilde{KU}^{-1}(L^n(a;a_0,a_1,\ldots,a_n)) \simeq \mathbb{Z}$ *and, denoting the* $2n$-*skeleton of* $L^n(a;a_0,a_1,\ldots,a_n)$ *by* $L^n_0(a;,a_0,a_1,\ldots,a_n)$, $\widetilde{KU}^{-1}(L^n_0(a;a_0,a_1,\ldots,a_n)) = 0$

(iii) *the canonical inclusion of* $L^n_0(a;a_0,a_1,\ldots,a_n)$ *into* $L^n(a;a_0,a_1,\ldots,a_n)$ *induces an isomorphism*

$$\widetilde{KU}(L^n(a;a_0,a_1,\ldots,a_n)) \xrightarrow{\simeq} \widetilde{KU}(L^n_0(a;a_0,a_1,\ldots,a_n)) .$$

Remarks. (1) F.Uchida has given a generalization of Proposition 4.1 to the"lens-like-spaces" $M(a_0,\ldots,a_m;b_0,\ldots,b_n)$ obtained as the orbit spaces of $S^{2m+1} \times S^{2n+1}$ under an action ϕ of S^1 (cf. [246]); if $a_0,\ldots,a_m,b_0,\ldots,b_n$ are positive integers such that each a_i is prime to every b_j , the action ϕ is defined by

$$\phi(z,(x_0,x_1,\ldots,x_m), (y_0,y_1,\ldots,y_n)) =$$

$$((z^{a_0}x_0,\ldots,z^{a_m}x_m) , (z^{b_0}y_0,\ldots,z^{b_n}y_n)) .$$

Uchida also shows that $KU(M(a_0,\ldots,a_m; b_0,\ldots,b_n))$ is canonically isomorphic to $KU(M(a_0,\ldots,a_m; 1,\ldots,1))$.

(2) On the other hand, M.Kamata [146], using 4.1 and the Conner-Floyd Isomorphism $\widetilde{KU}(X) \simeq \widetilde{U}^{even}(X) \otimes_{U^*}\mathbb{Z}$ which connects KU - theory to unitary cobordism, related the $\widetilde{KU}$ - theory of certain quotient manifolds $D_p(m,n) = S^{2m+1} \times S^n/D_p$ (where D_p is a dihedral group) to the $\widetilde{KU}$ - theory of lens spaces.

One of the interesting points of 4.1 is that it connects the KU - theory of the generalized lens spaces to that of the ordinary lens spaces $L^n(a)$. On the other hand, Corollary 5.1.7 shows that the KU - theory of $L^n(a)$ is related to that of the spaces $L^n(p^m)$ where p is a *prime number*. With this we legitimize our position to study the KU - theory of the spaces $L^n(p^m)$, to obtain informations about the KU - theory of the lens spaces.

Take $\sigma = [\eta] - 1 \in \widetilde{KU}(L^n(p^m))$; then, $\widetilde{KU}(L^n(p^m))$ is generated, as a group, by $\sigma, \sigma^2, \ldots, \sigma^s$ where $s = \inf(n, p^m - 1)$; furthermore, its ring structure is characterized by the relations

(4.3) $$\sigma^{n+1} = 0 \quad \text{and} \quad (\sigma + 1)^{p^m} - 1 = 0 .$$

Since $\widetilde{KU}(L^n(p^m))$ is an abelian group of order p^{mn}, we can write it as a direct sum $\bigoplus_{i=1}^{s} G_i$ (where $G_i = \mathbb{Z}_{p^{m_i}}$) whose generators are integral linear combinations of σ^i, $1 \leq i \leq s$. Generalizing the results obtained by Kambe [147] for $m = 1$, one of the authors was able to give precise values to the integers m_i (see [167]). For the present work, it will be sufficient to know that the first $N = \inf(n, p-1)$ cyclic groups G_i are respectively generated by $x_1 = \sigma, x_2 = \sigma^2, \ldots, x_N = \sigma^N$; moreover,

<u>Proposition 4.4.</u> *For* p *prime, the following hold in* $\widetilde{KU}(L^n(p^m))$:

(i) *if* $1 \leq i \leq n$, *the order of* σ^i *is* $p^{m+[(n-i)/(p-1)]}$

(ii) *if* $0 \leq i \leq n - p$,

$$p^{m+[i/(p-1)]}\sigma^{n-p+1-i} = -p^{m-1+[i/(p-1)]}\sigma^{n-i}.$$

This result is essentially obtained from (4.3) utilizing the following arithmetical property of p-adic valuations.

Lemma 4.5. *Let* p,m *and* j *be non-negative integers such that* p *is prime and* $1 \le j \le p^m$. *Then*

$$v_p\binom{p^m}{j} + v_p(j) = m .$$

For the benefit of the reader interested in a proof of this Lemma, we just observe that it can be obtained as an easy consequence of Theorem 4.2 [189].

With a few minor changes, some of the preceeding results can be immediately transposed into KO-theory. For example, the usual properties of the Bott homomorphisms and the classification of the orthogonal representations given in 1.7, allow us to see that the isomorphism

$$\Psi : KU(L^n(a;a_o,a_1,\ldots,a_n)) \simeq KU(L^n(a))$$

also induces an isomorphism in KO-theory, for every integer a. Because of Corollary 5.1.7 we can limit ourselves to the study of $\widetilde{KO}(L^n(p^m))$ (with p a prime number) or even, $\widetilde{KO}(L_o^n(p^m))$. One of the difficulties we encounter is the impossibility of using Proposition 6.1.4 whenever a is even; nevertheless, we can overcome the problem and obtain:

Proposition 4.6. (i) *If* a *is odd, the order of* $\widetilde{KO}(L_o^n(a;a_o,a_1,\ldots,a_n))$ *is* $a^{[n/2]}$;

(ii) *the order of* $\widetilde{KO}(L^n(2^m;a_o,a_1,\ldots,a_n))$ $\Big($*resp.* $\widetilde{KO}(L_o^n(2^m;a_o,a_1,\ldots,a_n))\Big)$ *is* $2^{f(2n+1)+(m-1)[n/2]}$ (*resp.* $2^{f(2n)+(m-1)[n/2]}$) .

The following results are very useful for a detailed description of the additive and multiplicative structures of $\widetilde{KO}(L^n(a))$.

Proposition 4.7. (i) *For every* a , *the element* $cr(\sigma) \in \widetilde{KU}(L^n(a))$ *is equal to* $\sigma^2/(1+\sigma)$;

(ii) *if* a *is odd, the Bott homomorphisms* $r : \widetilde{KU}(L_o^n(a)) \to$

$\widetilde{KO}(L_o^n(a))$, $c : \widetilde{KO}(L_o^n(a)) \to \widetilde{KU}(L_o^n(a))$ *are respectively, a surjection and an injection;*

(iii) *if* $c : \widetilde{KO}(L^n(a)) \to \widetilde{KU}(L^n(a))$ *is an injection then, for every* $k \geq 1$,

$$r([\eta]^k - 1) = \sum_{i=1}^{k} (k/i)\binom{k+i-1}{2i-1}(r\sigma)^i .$$

For an even a, the previous proposition leads us towards the question of the injectivity or not of c. Yasuo has recently proved [264] that

$$c : \widetilde{KO}(L^n(2^m)) \longrightarrow \widetilde{KU}(L^n(2^m))$$

is a monomorphism if $n \equiv 3 \pmod{4}$, thus confirming a conjecture stated in [169]. This result allows us to determine completely the order of $(r\sigma)^i$ in $\widetilde{KO}(L^n(2^m))$ whenever $m \geq 2$ (see [157]). Indeed, from Propositions 4.4 and 4.7 we can obtain the following.

<u>Proposition 4.8.</u> (i) *If* p *is an odd prime and* $1 \leq i \leq [n/2]$, *the order of* $(r\sigma)^i \in \widetilde{KO}(L^n(p^m))$ *is* $p^{m+[(n-2i)/(p-1)]}$ *and moreover,* $(r\sigma)^{[n/2]+1} = 0$;

(ii) *if* $1 \leq i \leq [n/2]$ *the order of* $(r\sigma)^i \in \widetilde{KO}(L^n(2^m))$ $(m \geq 2)$ *is:*

$$2^{m+n+1-2i}, \textit{ if } n \textit{ is even,}$$
$$2^{m+n-2i}, \textit{ if } n \textit{ is odd.}$$

Moreover, $(r\sigma)^{[n/2]+1} = 0$ *if* $n \not\equiv 1$ (*mod.*4); *otherwise* $(r\sigma)^{[n/2]+1}$ *has order* 2 *and* $(r\sigma)^{[n/2]+2} = 0$.

Finally, we should observe that the stable class κ_s of the non-trivial line bundle over $L^n(2^m)$ (it corresponds to a generator of $H^1(L^n(2^m);\mathbb{Z}_2) \simeq \mathbb{Z}_2$) also appears in the description of $\widetilde{KO}(L^n(2^m))$ and indeed,

$$c(\kappa_s) = (\sigma+1)^{2^{m-1}} - 1 . \tag{4.9}$$

The reader interested in obtaining more informations about the structure of $KF(L^n(a))$ is referred to the references given previously, as well as to [156] and [91] .

With the usual notation, it follows easily from (1.6)(ii) that

$$\tau_o = \left[\tau\left(L^n(a;a_o,a_1,\ldots,a_n)\right)\right] - (2n+1)$$

is such that

$$\gamma_t(-\tau_o) = \gamma_t\left(- \sum_{i=o}^{n} r([\xi]^{a_i} - 1)\right)$$

$$= \prod_{i=o}^{n} \left(\gamma_t\left(r([\xi]^{a_i} - 1)\right)\right)^{-1} .$$

Notice that, on one hand,

$$\gamma_t\left(r([\xi]^{a_i} - 1)\right) = \gamma_t(-2)\gamma_t\left(r([\xi]^{a_i})\right)$$

$$= (1-t)^2\gamma_t\left(r\left([\xi]^{a_i}\right)\right)$$

and on the other hand,

$$\gamma_t\left(r([\xi]^{a_i})\right) = \sum_{j\geq o} \Lambda^j\left(r([\xi]^{a_i})\right) (t/(1-t))^j$$

$$= 1 + r([\xi]^{a_i}) (t/(1-t)) + ((t/(1-t))^2$$

because $r(\xi^{a_i})$ is a real vector bundle of dimension 2 such that $\Lambda^2(r(\xi^{a_i})) = 1$; putting together this information we obtain:

$$(4.10)\qquad \gamma_t(-\tau_o) = \prod_{i=o}^{n} \left(1+r([\xi]^{a_i} - 1)\, t(1-t)\right)^{-1} .$$

Hence, whenever we limit ourselves to consider *ordinary lens spaces* $L^n(a)$,

$$(4.11)\qquad \gamma_t(-\tau_o) = (1+(r\sigma)t(1-t))^{-(n+1)}$$

$$= \sum_{i\geq o} (-1)^i \binom{n+1}{i} (r\sigma)^i (t-t^2)^i .$$

If, furthermore, $a = p^m$ with p = odd prime, we consider the integer

$$L(n,p,m) = \sup\left\{1 \leq i \leq [n/2] \,\middle|\, \binom{n+i}{i} \not\equiv 0 \ (\mathrm{mod.}\ p^{m+[(n-2i)/(p-1)]})\right\}$$

and observe that Atiyah's Criterion implies the following.

Proposition 4.12. - [167][(*)] - *For every odd prime* p ,

(i) $L^n(p^m)$ *cannot be immersed in* $\mathbb{R}^{2n+2L(n,p,m)}$;

(ii) $L^n(p^m)$ *cannot be embedded in* $\mathbb{R}^{2n+2L(n,p,m)+1}$.

We leave to the reader the formulation of an analogous criterion for an arbitrary positive odd integer a . Meanwhile, if p is a prime factor of a and $v_p(a) = r$, because the canonical inclusion of $\mathbb{Z}_{p^r}$ into $\mathbb{Z}_a$ induces a principal covering $(L^n(p^r), \pi, L^n(a))$ it is clear that if $L^n(a)$ could be immersed in $\mathbb{R}^{2n+k+1}$, the same would happen to $L^n(p^r)$. From this point of view, since (4.11) implies that

$$(4.13) \qquad \gamma^k(-\tau_o) = (-1)^k \sum_{i=[(k+1)/2]}^{k} \binom{n+i}{i}\binom{i}{i-k} (r\sigma)^i \quad ,$$

the relation $(r\sigma)^{[n/2]+1} = 0$ of 4.8 (i) allows us to state the following criterion:

Proposition 4.14. *Let* $a = \prod_{j=1}^{\ell} p_j^{r(j)}$ *be the prime decomposition of* a . *Then, if there is an integer* $i \in \{1,2,\ldots,\ell\}$ *such that* $p_i \neq 2$ *and*

$$\binom{n+[n/2]}{[n/2]} \not\equiv 0 \pmod{p_i^{r(i)}} \quad ,$$

$L^n(a)$ *cannot be immersed (resp. embedded) in* $\mathbb{R}^{2n+2[n/2]}$ (*resp.* $\mathbb{R}^{2n+2[n/2]+1}$) .

The previous proposition is considerably strengthened if we take into consideration the results of Sjerve ([221] and [222]). In fact, he has shown that for every odd prime p,

(i) every lens space $L^n(p;a_o,\ldots,a_n)$ can be immersed in $\mathbb{R}^{2n+2[n/2]+2}$;

(ii) if, moreover, $p \geq [n/2] + 3 \geq 4$, $L^n(p)$ can be immersed in $\mathbb{R}^{2n+2[n/2]+1}$ if, and only if,

(*) This result is due to Kambe [147] for m = 1 .

$\binom{n+[n/2]}{[n/2]}$ is a quadratic residue modulo p.

Thus, we can state the following.

Corollary 4.15. *Consider* $n \geq 2$ *and a prime number* p *such that* $p \geq [n/2] + 3$ *; also, suppose that* $\binom{n+[n/2]}{[n/2]} \not\equiv 0$ (mod. p) *but is a quadratic residue modulo* p. *Then* $L^n(p)$ *can be immersed in* $\mathbb{R}^{2n+2[n/2]+1}$ *but not in* $\mathbb{R}^{2n+2[n/2]}$.

Example: $L^3(5)$ can be immersed in $\mathbb{R}^9$ but not in $\mathbb{R}^8$; hence, it is not parallelizable!

However it may be that other arithmetic considerations would eventually allow us to improve on the criteria of non-immersion (or non-embedding) we discussed for ordinary lens spaces of odd order. The reader can find several other results concerning the spaces $L^n(p)$ with p an odd prime in the literature: let us quote those given in [188] for $p = 3$ and $n = 3^s$ or $3^s + 3^t$ $(s \geq t \geq 0,\ s \geq 1)$; those in [154] with $p = 5$ and $n = 3.5^{s+1} + 5^s$; those in [152], for $p \geq 5$ and $n = \alpha p^s + \beta p^t$ (with suitable conditions on α and β); finally, those given also by Kobayashi in [153], for $p \geq 3$. The best results concerning existence of immersions and embeddings are - to the best of our knowledge - those of Sjerve already mentioned (see also [187] and [245]).

Going through the literature one finds out, that until very recently hardly anything was known about immersions of lens spaces of even order, say $L^n(2^m)$ with $m > 2$. Actually, Proposition 4.8 (ii) and Atiyah's criterion imply the following detailed results. Let

$$L(n,m) = \sup\left\{1 \leq i \leq [n/2] \,\middle|\, \binom{n+i}{i} \not\equiv 0 \pmod{2^{m+n+\varepsilon-2i}}\right\}$$

where $\varepsilon = 1$ or 0 according to n being even or odd respectively. Then, similarly to (4.12) we have:

Proposition 4.16. - [157] - *If* $m \geq 2$, $L^n(2^m)$ *cannot be immersed (resp. imbedded) in* $R^{2n+2L(n,m)}$ *(resp.* $R^{2n+2L(n,m)+1}$*).*

For example, if we compute the dyadic valuation of

$\binom{n+[n/2]}{n}$ using the well-known arithmetic relation

(4.17) $$v_p(N!) = \sum_{i\geq 1} [N/p^i] \quad (p \text{ prime}) \quad (*)$$

we obtain the following.

Corollary 4.18. *Given* m *and* $s \geq 2$, *for every* $n = 2^s$ *(resp.* 2^s+1*)* $L^n(2^m)$ *cannot be immersed in* $\mathbb{R}^{3n}$ *(resp.* $\mathbb{R}^{3n-1}$*) and cannot be imbedded in* $\mathbb{R}^{3n+1}$ *(resp.* $\mathbb{R}^{3n}$*).*

Finally, the reader who wishes to state the analogous results for the *generalized lens spaces*, can do so by using (4.7)(iii) and (4.10). Furthermore, the expression for $r([\eta]^k-1)$ given in 4.7 (iii) can also be employed to show that, for every real vector bundle over $L^n(a)$, with a odd, the implication of 3.11 holds.

5. The Case of the Q_m - Spherical Forms

The results concerning the $K\mathbb{F}$ - theory of the Q_m-spherical forms which we are about to present follow essentially from the work of one of the authors (see [169]) and from D.Pitt's paper [196]. Another useful reference is [90]. We shall keep the notation introduced in 2.5 and 2.6 of Chapter 5, relative to the generalized Q_m - spherical forms $N^n(m;r_o,r_1,\dots,r_n)$ defined by the action $\rho = \bigoplus_{i=o}^{n} \xi_{r_i+3}$ of the generalized quaternionic group Q_m over S^{4n+3}. If we set

$$\alpha = \xi_1 - 1,\ \beta = \xi_2 - 1,\ \gamma = \xi_1 + \xi_2 + \xi_3 - 3 \quad \text{and}$$

$$\delta_r = \xi_{r+3} - 2 \quad \text{with} \quad 1 \leq r \leq 2^{m-1} - 1,$$

it follows that $RU(Q_m)$ is additively generated by α,β,γ and δ_r, while its multiplicative structure is given by the relations

(5.1)
$$\begin{gathered}
\alpha^2 = -2\alpha,\ \beta^2 = -2\beta,\ \gamma = \alpha\beta + 2\alpha + 2\beta, \\
\alpha\delta_1 = -2\alpha,\ \beta\delta_1 = -2\beta + \delta_{2^{m-1}-1} - \delta_1, \\
\delta_1^2 = \begin{cases} -4\delta_1 + \gamma, & \text{if } m = 2 \\ -4\delta_1 + \delta_2 + \alpha, & \text{if } m \geq 3, \end{cases} \\
\delta_{r+1} = \delta_1\delta_r + 2\delta_1 + 2\delta_r - \delta_{r-1}, \quad \text{if } 2 \leq r \leq 2^{m-1} - 2.
\end{gathered}$$

(*) Theorem 4.2 of [189].

Now recall that

$$\sum_{j=o}^{n+1} (-1)\Lambda^j\rho = \prod_{i=o}^{n} \left(2-\xi_{r_i+3}\right)$$

and that the cohomology groups of Q_m are

$$H^o(Q_m;\mathbb{Z}) \cong \mathbb{Z}, \; H^{2k+1}(Q_m;\mathbb{Z}) = 0$$

$$H^{4m}(Q_m;\mathbb{Z}) \cong Z_{2^{m+1}}, \quad \text{if } k > 0,$$

$$H^{4k+2}(Q_m;\mathbb{Z}) \cong \mathbb{Z}_2 \oplus \mathbb{Z}_2, \quad \text{if } k > 0 \quad [62; \text{page } 253]$$

These results, together with Atiyah's exact sequence (0.9) and the Atiyah-Hirzebruch Spectral Sequence (0.10) lead us immediately to the following statements.

Proposition 5.2. (i) *The rings* $KU(N^n(m;r_o,r_1,\ldots,r_n))$ *and* $RU(Q_m)/\langle \prod_{i=o}^{n} (2-\xi_{r_i+3})\rangle$ *are isomorphic. More precisely,* $\widetilde{KU}(N^n(m;r_o,r_1,\ldots,r_n))$, *as an additive group, is generated by* α,β,γ *and* $\{\delta_r\}_{1\leq r\leq 2^{m-1}-1}$, *while multiplicatively, is generated by* α,β *and* δ_1.

(ii) *The order of* $\widetilde{KU}(N^n(m;r_o,r_1,\ldots,r_n))$ *is* $2^{n(m+3)+2}$ *and* $\widetilde{KU}^{-1}(N^n(m;r_o,r_1,\ldots,r_n)) \cong \mathbb{Z}$.

The next Lemma is required for the comparison of the KU-theories of the generalized and ordinary Q_m-spherical forms:

Lemma 5.3. *For every integer* r, $4 \leq r \leq 2^{m-1}-1$, *there exist a polynomial* $p_r \in \mathbb{Z}[x]$ *of degree* $r-4$ *and an integer* α_r *such that:*

(i) $\xi_r = \xi_4 p_r(\xi_4) + \alpha_r(1+\xi_1)$, $p_r(2) = 1-\alpha_r$ *with* $\alpha_r = 0$ *or* $(-1)^{[(r+1)/2]}$ *according to* r *being even or odd;*

(ii) $p'_{r+1}(2) = \frac{1}{2}(r-3)(r-2) - V_r$, *where*

$$V_r = \begin{cases} -2k+1 & \textit{if} \quad r = 4k+1 , \\ -2k & \textit{if} \quad r = 4k+2 , \\ -2k+2 & \textit{if} \quad r = 4k \ \textit{or} \ 4k-1 \end{cases} \quad (*)$$

(Statement (i) is proved by induction; (ii) will be used later on.)

The reader should notice that the first part of the previous Lemma actually means that the ideal $\langle (2-\xi_4)^{n+1} \rangle$ of the ring $RU(Q_m)$ is canonically included in $\langle \prod_{i=o}^{n} \left(2-\xi_{r_i+3}\right) \rangle$. Then, because of 5.2 (ii), the following is true.

Proposition 5.4. *For every* $(m;r_o,r_1,\ldots,r_n)$ *there exists a canonical ring isomorphism of* $KU(N^n(m;r_o,r_1,\ldots,r_n))$ *onto* $KU(N^n(m))$.

As for the lens spaces, it is clear that this isomorphism can be carried over to real K - theory. Speaking about lens spaces, let us indicate how their K - theory can be used to give some information on the K - theory of the Q_m - spherical forms. Since the generators x and y of Q_m (see Definition 1.8, Chapter 5) generate cyclic groups of order 2^m and 4 respectively, define

$$i_m^! : KU\left(S^{4n+3}/Q_m\right) \longrightarrow KU\left(S^{4n+3}/\mathbb{Z}_{2^m}\right)$$

and

$$i_2^! : KU\left(S^{4n+3}/Q_m\right) \longrightarrow KU\left(S^{4n+3}/\mathbb{Z}_4\right)$$

to be the ring homomorphisms induced by the inclusions of $\mathbb{Z}_{2^m}$ and $\mathbb{Z}_4$ into Q_m, respectively. Let σ_m be the *stable class* of the canonical complex line bundle over $L^{2n+1}(2^m)$. The following is a consequence of Propositions 4.1 and 5.2.

Proposition 5.5. - [90] - $i_m^!(\alpha) = 0$, $i_m^!(\beta) = (\sigma_m+1)^{2^{m-1}} - 1$, $i_m^!(\delta_1) = \sigma_m + \overline{\sigma}_m$, $i_2^!(\alpha) = (\sigma_2+1)^2 - 1$, $i_m^!(\beta) = 0$, $i_m^!(\delta_1) = \sigma_2+\overline{\sigma}_2$.

Let us now switch back to the KO - theory of the Q_m-spherical forms. It is completely determined by the following

(*) p_r' is the derivative of p_r.

result due to D.Pitt, for ordinary forms.

Proposition 5.6. *The ring* $KO(N^n(m;r_o,r_1,\dots,r_n))$ *is isomorphic to:*

(i) *the ring* $RO(Q_m)/\langle \prod_{i=o}^{n} \left(2-\xi_{r_i+3}\right) \rangle$ *if* n *is odd;*

(ii) *the ring* $RO(Q_m)/\langle \prod_{i=o}^{n} \left(2-\xi_{r_i+3}\right) \rangle\, RSp(Q_m)$ *if* n *is even.*

Proof. The following procedure is standard. First of all, the ring (resp. group) homomorphism $c : RO(Q_m) \to RU(Q_m)$ (resp. $h : RSp(Q_m) \to RU(Q_m)$) allows us to consider $RO(Q_m)$ (resp. $RSp(Q_m)$) as a subring (resp. subgroup) of $RU(Q_m)$. These identifications and the well-known properties of the Bott homomorphisms r,c,q,h and t (namely, $rc = 2 = qh$, $cr = 1+t = hq$) (*) show that:

(i) the ring $RO(Q_m)$ of all orthogonal representations of Q_m is generated by $\xi_o = 1, \xi_1,\xi_2,\xi_3,\xi_{r+3}$ if $2 \le r$ even $\le 2^{m-1} - 2$ or $2\xi_{r+3}$ if $1 \le r$ odd $\le 2^{m-1} - 1$;

(ii) the group $RSp(Q_m)$ of all quaternionic representations of Q_m is the free abelian group generated by $2\xi_o = 2, 2\xi_1$, $2\xi_2$, $2\xi_3$, $2\xi_{r+3}$ if $2 \le r$ even $\le 2^{m-1} - 2$ or ξ_{r+3} if $1 \le r$ odd $\le 2^{m-1} - 1$.

Notice that the virtual representation $\rho = \prod_{i=o}^{n} \xi_{r_i+3}$ is quaternionic.

We now apply (0.9) : if n is *odd*, there is a commutative diagram with exact rows

$$\begin{array}{ccccccc}
RO(Q_m) & \xrightarrow{\Phi_o} & RO(Q_m) & \xrightarrow{\Theta_o} & KO(N^n(m;r_o,r_1,\dots,r_n)) & \longrightarrow & 0 \\
r\uparrow\downarrow c & & r\uparrow\downarrow c & & r\uparrow\downarrow c & & \\
RU(Q_m) & \xrightarrow{\Phi_U} & RU(Q_m) & \xrightarrow{\Theta_U} & KU(N^n(m;r_o,r_1,\dots,r_n)) & \longrightarrow & 0
\end{array}$$

(*) Let us recall that $q : RU(Q_m) \to RSp(Q_m)$ associates to each unitary representation ξ of Q_m the quaternionified representation $\xi \otimes 1_{\mathbb{H}}$, while h associates to each quaternionic representation of Q_m its underlying unitary representation.

and if n is *even* we obtain likewise a diagram

$$\begin{array}{ccccccc}
RSp(Q_m) & \xrightarrow{\Phi_o} & RO(Q_m) & \xrightarrow{\Theta_o} & KO(N^n(m;r_o,r_1,\dots,r_n)) & \longrightarrow & 0 \\
q \uparrow \downarrow h & & r \uparrow \downarrow c & & r \uparrow \downarrow c & & \\
RU(Q_m) & \xrightarrow{\Phi_U} & RU(Q_m) & \xrightarrow{\Theta_U} & KU(N^n(m;r_o,r_1,\dots,r_n)) & \longrightarrow & 0.
\end{array}$$

Proposition 5.6 now follows from these two diagrams, since Φ_U is nothing but multiplication by $\prod_{n=o}^{n}(2-\xi_{r_i+3})$ in $RU(Q_m)$.

Finally, using the Atiyah-Hirzebruch Spectral Sequence and the cellular structure of the Q_m - spherical forms, it is possible to determine the order of the groups $\widetilde{KO}(N^n(m;r_o,r_1,\dots,r_n))$:

<u>Proposition 5.7.</u> *The order of* $\widetilde{KO}(N^n(m;r_o,r_1,\dots,r_n))$ *is* $2^{n(m+3)+2}$ *if* n *is odd, or* $2^{n(m+3)+4}$ *if* n *is even.*

(For more details see [169]).

We shall limit ourselves to $N^n(m,r)$ *in studying the immersions and embeddings of* Q_m *- spherical forms into an euclidean space.* According to Corollary 1.6 (iii) and because $\det(\xi_{r+3}) = 1$ (see the definition of the representation ξ_{r+3} in Chapter 5), we obtain for

$$\tau_o = [\tau(N^n(m;r))] - (4n+3) \in \widetilde{KO}(N^n(m;r))$$

that

$$\gamma_t(-\tau_o) = \Big(1 + ([\xi_{r+3}]-2)\;(t-t^2)\Big)^{-2(n+1)} \quad ,$$

that is to say

$$\gamma^j(-\tau_o) = \sum_{i=[(j+1)/2]}^{j} (-1)^j \binom{2n+1+i}{i}\binom{i}{j-i}\,([\xi_{r+3}]-2)^i \;.$$

Hence, as an application of Atiyah's Criterion, we obtain:

<u>Proposition 5.8.</u> (i) *If* $N^{2p+1}(m;r)$ *can be immersed (resp. embedded) in* $\mathbb{R}^{8p+7+k}$ *then, for every* $j > k$ *(resp.* $j \geq k$*), there exists* $\Theta_j \in RO(Q_m)$ *such that*

$$\sum_{i=[(j+1)/2]}^{j} \binom{4p+3+i}{i}\binom{i}{j-i} \delta_r^i = \Theta_j \delta_r^{2p+2} ;$$

(ii) *if* $N^{2p}(m;r)$ *can be immersed (resp. embedded) in* $\mathbb{R}^{8p+3+k}$ *then, for every* $j > k$ *(resp.* $j \geq k$*), there exists* $\Theta_j \in RSp(Q_m)$ *such that*

$$\sum_{i=[(j+1)/2]}^{j} \binom{4p+1+i}{i}\binom{i}{j-i} \delta_r^i = \Theta_j \delta_r^{2p+1} .$$

For want of a precise knowledge of the orders and the generators of the groups which make up $KO(N^n(m;r))$, easily applicable criteria of immersion and embedding into euclidean spaces have been stated only for $m = 2$ or 3 (to the best of our knowledge). Meanwhile, Pitt has established that the order of $\delta_r^n = ([\xi_{r+3}]-2)^n \in \widetilde{KO}(N^n(m;r))$ is 2^{m+1} or 2^{m+2} according to n being even or odd; on the other hand, if we consider δ_r^n as an element of $\widetilde{KU}(N^n(m;r))$, its order is 2^{m+1}. Moreover, $\delta_r^{n+1} = 0$, so, if $k = 4p + 1$, the relation 5.8 (i) is reduced to

$$\binom{6p+4}{2p+1} \delta_r^{2p+1} = 0 ,$$

and if $k = 4p - 1$, 5.8 (ii) reduces to

$$\binom{6p+1}{2p} \delta_r^{2p} = 0 .$$

From these two relations we conclude that in $\widetilde{KU}(N^n(m;r))$, δ_r^n satisfies the equation

$$\binom{3n+1}{n} \delta_r^n = 0 .$$

Hence,

<u>Corollary 5.9.</u> *If* $\binom{3n+1}{n} \not\equiv 0 \pmod{2^{m+1}}$ *the* Q_m *- spherical form* $N^n(m;r)$ *cannot be immersed (resp. embedded) in* $\mathbb{R}^{6n+2}$ *(resp.* $\mathbb{R}^{6n+3}$*). In particular,* $N^1(m;r)$ *is never parallelisable.*

Using (4.17) the reader will be able to determine some Q_m - spherical forms for which the previous arithmetical condition is satisfied.

Let M^n be an arbitrary spherical form with fundamental group G ; if one of its associated p - spherical forms cannot be immersed in a given euclidean space, the same holds for M (cf. Chapter 5, § 1). This remark essentially motivated our concern we had about projective spaces, lens spaces and Q_m - spherical forms. Finally, we point out an example of a particularly interesting situation, namely, the case in which the group G contains a Sylow subgroup with a *unique* orthogonal representation without fixed point. In this case, we can state some non-immersion (non-embedding) criteria for *all* spherical forms S^n/G , that is to say, independently of the (free) action of G on S^n . Indeed, an illustration of this fact is given by Q_2 : this group has only one free representation (namely ξ_4) and is a 2 - Sylow subgroup of the generalized tetrahedral group T^*_m . Using (5.9) we obtain: if $\binom{3n+1}{n} \not\equiv 0$ (mod.8), no tetrahedral spherical form S^{4n+3}/T^*_m can be immersed (resp. embedded) in $\mathbb{R}^{6n+2}$ (resp. $\mathbb{R}^{6n+3}$). It follows, in particular, that no spherical form S^7/T^*_m is parallelizable!

6. Parallelizability of the Spherical Forms

In this section we shall be concerned with spherical forms of dimension 7 ; the results we describe (presented in [169]) complete those of Chapter 5.

For reasons alluded to before we shall examine in the following the lens spaces $L^3(p^m;a_o,a_1,a_2,a_3)$ with p *prime* and the Q_m - spherical forms $N^1(m;r_o,r_1)$.

We have already shown that for the generalized lens spaces,

$$\tau_o = \sum_{i=o}^{3} r\left([\xi]^{a_i} - 1\right) .$$

We are going to show that in $KO(L^3(p^m;a_o,a_1,a_2,a_3))$ this relation can actually be written as

$$\tau_o = \left(a_o^2 + a_1^2 + a_2^2 + a_3^2\right) r\sigma \quad . \tag{6.1}$$

If $p \neq 2$, (6.1) is an easy consequence of 4.7 (iii) and

4.8 (i) because in the ring $\widetilde{KO}(L^3(p^m;a_o,a_1,a_2,a_3))$

(6.2) $(r\sigma)^2 = 0$ and $r([\xi]^k-1) = k^2 r\sigma$, for every integer $k \geq 1$.

Actually the relations (6.2) are still true if $p = 2$; to see this fact, we must go back to the structure of the groups $\widetilde{K\mathbb{F}}(L^3(2^m))$. If $\mathbb{F} = \mathbb{C}$, using notably 4.3 and 4.4 , we show that

Lemma 6.3. *If* $m \geq 2$, *the group* $\widetilde{KU}(L^3(2^m))$ *is isomorphic to*

$$\mathbb{Z}_{2^{m+1}} \oplus \mathbb{Z}_{2^{m-1}} \oplus \mathbb{Z}_{2^{m-1}} ,$$

where each factor is generated respectively by $x_1 = \sigma$, $x_2 = \omega + 4\sigma$ *and* $x_3 = \omega\sigma$, *with* $\omega = \sigma^2 + 2\sigma$. *If* $m = 1$, $\widetilde{KU}(L^3(2^m)) = KU(\mathbb{R}P^7)$ *is isomorphic to* $\mathbb{Z}_8$, *generated by* σ.

For the real case we obtain:

Proposition 6.4. *The group* $\widetilde{KO}(L^3(2^m))$ *is described by the following table:*

$m = 1$	$\mathbb{Z}_8$	*generator:*	$y_1 = r\sigma$
$m = 2$	$\mathbb{Z}_8 \oplus \mathbb{Z}_2$	*generators:*	$y_1 = r\sigma$, $y_2 = \kappa + 2r\sigma$
$m \geq 3$	$\mathbb{Z}_{2^{m+1}} \oplus \mathbb{Z}_2$	*generators:*	$y_1 = r\sigma$, $y_2 = \kappa$.

Proof. The case $m = 1$ is well-known; let us then assume that $m \geq 2$. Using 4.9 and the notation and results of Lemma 6.3, we conclude that the images of $y_1, \kappa \in \widetilde{KO}(L^3(2^m))$ by the homomorphism c are given by

$$c(y_1) = -18x_1 + 3x_2 - x_3 ,$$

$$c(\kappa) = -(2^{2m} + 2^{2m-2})x_1 - 2^{m-2}x_2 .$$

It follows that the orders of $c(y_1)$ and $c(y_2)$ in $\widetilde{KU}(L^3(2^m))$ are 2^{m+1} and 2 , respectively. Indeed, these elements generate the subgroup $\mathbb{Z}_{2^{m+1}} \oplus \mathbb{Z}_2$. Since, on the one hand,

$$2^{m+2} \leq |\mathrm{im}\ c| \leq |\widetilde{KO}(L^3(2^m))|$$

and on the other hand (see Proposition 4.6)

$$|\widetilde{KO}(L^3(2^m))| = 2^{m+2} ,$$

we have that $\widetilde{KO}(L^3(2^m))$ is isomorphic to the image of c, namely, to the direct sum $\mathbb{Z}_{2^{m+1}} \oplus \mathbb{Z}_2$.

A consequence of this result is that

$$(6.5) \qquad c : \widetilde{KO}(L^3(2^m)) \longrightarrow \widetilde{KU}(L^3(2^m))$$

is an injection. Thus, relation 4.7 (iii) holds for $p = 2$. From (4.3) we obtain that $\sigma^4 = 0$ in $\widetilde{KU}(L^3(2^m;a_o,a_1,a_2,a_3))$; thus,

$$c((r\sigma)^2) = (cr\sigma)^2 = \sigma^4/(1+\sigma)^2 = 0$$

(see 4.7 (i)). The injectivity of c shows also that $(r\sigma)^2 = 0$ in $\widetilde{KO}(L^3(2^m;a_o,a_1,a_2,a_3))$. In other words, (6.2) holds for every prime number p.

<u>Theorem 6.6.</u> (i) *For every* $m \geq 1$ *and every odd prime* p, *the lens space* $L^3(p^m;a_o,a_1,a_2,a_3)$ *is parallelizable if, and only if,*

$$a_o^2 + a_1^2 + a_2^2 + a_3^2 \equiv 0 \quad (\mathit{mod.}\ p^m) ;$$

(ii) *the only parallelizable space amongst the lens spaces* $L^3(2^m;a_o,a_1,a_2,a_3)$ *is the real projective space* $\mathbb{R}P^7 = L^3(2)$.

<u>Proof.</u> The necessary conditions are evident (compare to [178]) because the order of $r\sigma \in \widetilde{KO}(L^3(p^m;a_o,a_1,a_2,a_3))$ is either p^m if $p \neq 2$ and $m \geq 1$, or 2^{m+1} if $p = 2$ and $m \geq 2$. Statement (ii) follows trivially, since the congruence

$$a_o^2 + a_1^2 + a_2^2 + a_3^2 \equiv 0 \quad (\mathrm{mod.}\ 8)$$

has no solution (a_o,a_1,a_2,a_3) with $(a_i,8) = 1$, $0 \leq i \leq 3$.

As for (i), it follows from Theorem 4.4 [222] stated for $p \geq 5$, because the congruence has no solutions for $p = 3$.

At this point we observe that because the forms $N^1(m,r)$ are never parallelizable (Corollary 5.9), it remains to examine the Q_m - generalized spherical forms $N^1(m;r_o,r_1)$ with $r_o \neq r_1$. Let d_r be the real representation of Q_m associated to δ_r; then,

$$\tau_o = d_{r_o} + d_{r_1} \quad . \tag{6.7}$$

As in the case of the lens spaces, the isomorphisms

$$\widetilde{K\mathbb{F}}(N^1(m,r_o,r_1)) \cong \widetilde{K\mathbb{F}}(N^1(m))$$

(cf. Proposition 5.4) lead us to work with $N^1(m)$.

First of all, Lemma 5.3 and the relation $\delta_1^2 = 0$ in $\widetilde{KU}(N^1(M))$ (Proposition 5.2 (i)) allow us to show that in $\widetilde{KU}(N^1(M))$ the δ_{odd}'s are integral multiples of δ_1 and the δ_{even}'s are integral linear combinations of α and δ_1. More precisely, for $\delta_r = a_r\delta_1$ (r odd) we have

$$a_r = 1 + 2P'_{r+3}(2) \quad . \tag{6.8}$$

Next we compute directly the groups $\widetilde{K\mathbb{F}}(N^1(M))$; since $\delta_1^2 = 0$, the relations

$$\alpha\delta_1 = -2\alpha, \ \beta\delta_1 + 2\beta = \delta_{2^{m-1}-1} - \delta_1 = \left(a_{2^{m-1}-1} - 1\right)\delta_1 = 0$$

obtained from (5.1) show that $4\alpha = 0$ and $2\beta\delta_1 = 0$, that is to say, $4\beta = 0$ because $\beta^2 = -2\beta$; it follows that

$$2(a_{2^{m-1}-1} - 1)\delta_1 = 4P'_{2^{m-1}+2}(2)\ \delta_1 = 0 \ .$$

But Lemma 5.3 (ii) tells us that

$$P'_{2^{m-1}+2}(2) = 2^{m-1}(2^{m-2} - 1)$$

and thus, $2^{m+1}\delta_1 = 0$. In fact, we claim that the orders of α, β and δ_1 are respectively, 4, 4 and 2^{m+1}, otherwise from Proposition 5.5 we would obtain that

a) in $\widetilde{KU}(L^3(4))$, $i_2^!(2\alpha) = \sigma_2^2 + 2\sigma_2 = 0$;

b) in $\widetilde{KU}(L^3(2^m))$, $i_m^!(2\beta) = 2\left((\sigma_m+1)^{2^{m-1}} - 1\right) = 0$, and

$$i_m^!(2^m\delta_1) = 2^m(\sigma_m + \bar{\sigma}_m) = 0 ,$$

and these equalities are impossible, according to Lemma 6.3. Since the order of the group $\widetilde{KU}(N^1(M)$ is 2^{m+5} (Proposition 5.2 (ii)), we have:

Proposition 6.9. *The group* $\widetilde{KU}(N^1(M)) \cong \mathbb{Z}_4 \oplus \mathbb{Z}_4 \oplus \mathbb{Z}_{2^{m+1}}$, *each factor generated respectively by* α, β *and* δ_1 .

If we regard α and β as the complexifications of the real representations a and b of Q_m , 5.6 (i), 5.7 and the previous proposition imply the

Corollary 6.10. *The group* $\widetilde{KO}(N^1(m))$ *is isomorphic to* $\mathbb{Z}_4 \oplus \mathbb{Z}_4 \oplus \mathbb{Z}_{2^{m+1}}$, *each factor being generated respectively by* a, b *and* d_1.

It follows now from (6.7) that a necessary condition for the immersibility of $N^1(m;r_o,r_1)$ in $\mathbb{R}^8$ is:

$$a_{r_o} + a_{r_1} \equiv 0 \quad (\text{mod.}\, 2^{m+1}) .$$

But, Lemma 5.3 (ii) and (6.8) show that the above congruence never holds; hence

Proposition 6.11. *The spaces* S^3/Q_m *are the only parallelizable* Q_m *- spherical forms.*

In Chapter 5 we subdivided the spherical forms in two classes: those of cyclic type and those of quaternionic type (Definition 5.2.9). Proposition 6.11 has an important consequence: no 7 - spherical form of quaternionic type is parallelizable. Even better, because no lens space of order 2^m with $m \geq 2$ is parallelizable (see Theorem 6.6), the only spherical forms which are succeptible of being parallelizable are those having fundamental group G with $v_2(|G|) = 0$ or 1.

In his classification of the spherical forms, Wolf [260] has listed all groups which can act freely on S^7 : if one excludes the groups which contain Q_m amongst their 2-Sylow subgroups, the only groups G left on the list are those with presentation

$$x^m = 1,\ y^n = 1,\ yxy^{-1} = x^r$$

where

$$mn > 0,\ ((r-1)m,n) = 1,\ r^n \equiv 1 \ (\mathrm{mod.}\, m),\ d = 1,2 \text{ or } 4,$$

with d equal to the order of r in the multiplicative subgroup of the invertible elements of $\mathbb{Z}_m$. One can check that necessarily m is odd and that n is a multiple of d (besides , d corresponds to the order of the free representations of G). Since the order of G is mn , only the groups for which $d = 1$ can give rise to parallelizable spherical forms; G is therefore cyclic. Hence, the only parallelizable spherical forms S^7/G are the lens spaces (G non-trivial)! These are completely determined by Theorem 6.6.

<u>Remark.</u> Working with Grothendieck operations on lens spaces and Q_m-spherical forms and applying Propositions 6.2.8 and 3.2, we can show that $\mathrm{span}(S^7/\mathbb{Z}_{2^m}) = 5$ $(m \geq 2)$ and $\mathrm{span}(S^7/Q_m) = 5$.

7. Immersions of Complex Projective Spaces

In the previous sections we have shown how we can take advantage of Atiyah's Theorem (3.1) to obtain information about immersions and embeddings of Dold Manifolds and Spherical Forms into euclidean spaces. Furthermore, as we have already indicated in § 2, K-theory gives other techniques leading to interesting results on the subject; for example, the reader is directed to the work of Atiyah and Hirzebruch [27] and to the methods developped by Mayer [173], which employ the famous Index Theorem. Actually, without recurring to primary operations in KO-theory, excellent results can be obtained with the use of just complex K - theory and the Adams ψ^k - Operations. To illustrate this last remark we shall present in this section some criteria for non-immersion of complex projective spaces (the reader should compare them with those deduced from (3.21) for $m = 0$). We investigate immersions of $\mathbb{C}P^n$ in $\mathbb{R}^{4n-2\alpha(n)}$, where $\alpha(n)$ is the number of 1's in the dyadic decomposition of the integer n (see Conjecture 2.3).

From this point and to the end of the chapter, we shall denote the Complex Projective space $\mathbb{C}P^n$ simply by P^n and the truncated projective space P^n/P^{n-k} by P^n_k. Furthermore, we shall write

$$(7.1) \qquad s(n,k) = \frac{(2n-\alpha(n)+1)!}{(2n-\alpha(n)+1+k)!} \cdot S(2n-\alpha(n)+1+k, 2n-\alpha(n)+1)$$

where the integers $S(m+q,m)$ are the Stirling numbers of the first kind defined as it is well known, by

$$(7.2) \qquad (\log(1+t)/t)^m = \sum_{q>0} (m!/(m+q)!)\, S(m+q,m)\, t^q .$$

Regarding the parity of these numbers, notice that

$$(7.3) \qquad S(m+q,m) \equiv \binom{m+q}{2q} \quad (\text{mod. } 2).$$

The result we have in mind is the following.

<u>Theorem 7.4.</u> - [220] - *If there is an immersion of* P^n *into* $\mathbb{R}^{4n-2\alpha(n)}$ *then, there exists an integer* e_o *such that:*

$$e_o s(n,k) \text{ is an } \begin{cases} \text{even integer, if } 0 \le k \le \alpha(n) - 2 \\ \text{odd integer, if } k = \alpha(n) - 1 \text{ or } k = \alpha(n). \end{cases}$$

The proof of this Theorem is essentially based on the study of the multiplicative structure of $\widetilde{KU}(T(\nu))$, where $T(\nu)$ is the Thom space of the normal bundle ν associated to the considered immersion. This structure can be determined working with a convenient suspension of $T(\nu)$, which is homeomorphic to a truncated projective space. We shall go over the details of the proof for (7.4) later on; presently, we prove the following.

Lemma 7.5. *If* P^n *can be immersed in* $\mathbb{R}^{2n+k}$, *there is an integer* b *such that the* (2b-k-2n-2) - *suspension of* $T(\nu)$ *is homeomorphic to* P_{n+1}^{b-1}.

Proof. Let η be the Hopf vector bundle over P^n. The order of $\tilde{J}(\eta)$ in the groupe $\tilde{J}(P^n)$ is equal to the James' number b_n (*); taking b as a multiple of b_n, the bundle $b\eta$ will have the same fibre homotopy type as the trivial real 2b - vector bundle. Then the usual relation

$$[\nu] = k - (n+1)([\eta] - 2)$$

in $\widetilde{KO}(P^n)$ becomes, in terms of stable fibre-homotopy equivalence,

$$\nu \oplus (2b-k-2n-2)\varepsilon \overset{J}{\cong} (b-n-1)\eta \quad .$$

In the language of Atiyah's Theory of Thom Complexes [12], the last relation means that $\Sigma^{2b-k-2n-2}\, T(\nu) \simeq P_{n+1}^{b-1}$.

Proposition 7.6. *If there is an immersion of* P^n *in* $R^{4n-2\alpha(n)}$, *the multiplicative structure of* $\widetilde{KU}(T(\nu))$ *is not trivial.*

Proof. Since $\psi^2(x) \equiv x^2$ (mod.2) (see 0.12) it suffices to show that on $\widetilde{KU}(T(\nu))$ the Adams operation ψ^2 is non-trivial modulo 2.

We set $q = b - 2n + \alpha(n) - 1$. In view of the suspension isomorphism

$$(\Sigma^{2q})^! : \widetilde{KU}(T(\nu)) \cong \widetilde{KU}(\Sigma^{2q}T(\nu))$$

and the relation

$$\psi^k \circ (\Sigma^{2q})^! = k^q (\Sigma^{2q})^! \circ \psi^k ,$$

it follows from Lemma 7.5 that the ψ - operations on $\widetilde{KU}(T(\nu))$ can be determined from those of the truncated projective spaces.

Let us recall that $\widetilde{KU}(P^{b-1})$ is generated by $\sigma = [\xi] - 1$ subject to the relation $\sigma^b = 0$, where ξ is the canonical complex line bundle over P^{b-1}, moreover, the

(*) See Theorem 4.5.8.

Adams operations are such that $\psi^{\ell}(\sigma) = (\sigma+1)^{\ell} - 1$. (see Chapter 4, § 3). The cofibration

$$P^{b-n-2} \longrightarrow P^{b-1} \longrightarrow P^{b-1}_{n+1}$$

gives rise to an exact sequence

$$0 \longrightarrow \widetilde{KU}(P^{b-1}_{n+1}) \longrightarrow \widetilde{KU}(P^{b-1}) \longrightarrow \widetilde{KU}(P^{b-n-2}) \longrightarrow 0$$

Hence, $\widetilde{KU}(P^{b-1}_{n+1})$ can be identified with the ideal of $\widetilde{KU}(P^{b-1})$ generated by σ^{b-n-1}. Let g_i $(0 \leq i \leq n)$ be the element of $\widetilde{KU}(P^{b-1}_{n+1})$ that corresponds to $\sigma^{b-n-1+i}$ under this identification.

From the relations

$$\Psi^2(\sigma^i) = (\sigma^2+2\sigma)^i = \sum_{k=0}^{i} \binom{i}{k} 2^{i-k} \sigma^{i+k}$$

in $\widetilde{KU}(P^{b-1})$ we obtain

$$\Psi^2(g_i) = \sum_{k=0}^{n-i} 2^{b-n-1+i-k} \binom{b-n-1+i}{k} g_{i+k} . \tag{7.7}$$

We claim that b can be chosen so that the latter relations reduce to

$$\begin{aligned} \Psi^2(g_0) &\equiv 2^{b-2n+\alpha(n)-1}(g_{n-1}+g_n) \quad (\text{mod}.\, 2^{b-2n+\alpha(n)}) , \\ \Psi^2(g_i) &\equiv 0 \quad (\text{mod}.\, 2^{b-2n+\alpha(n)}) , \end{aligned} \tag{7.8}$$

if $1 \leq i \leq n$.

To this end (assuming that b is already a multiple of b_n), it is sufficient to adjust the power of 2 dividing b in such a way that

$$\alpha(b-a) + \alpha(a-1) = \alpha(b-1), \text{ for every } a \in \{1,2,\ldots,n\} .$$

In fact, utilizing (4.17), such hypothesis allows us to guarantee that for every i ,

$$v_2\left(\binom{b-n-1+i}{k}\right) = \sum_{j\geq 1}\left[\frac{b-n-1+i}{2^j}\right] - \sum_{j\geq 1}\left[\frac{b-n-1+i-k}{2^j}\right] - \sum_{j\geq 1}\left[\frac{k}{2^j}\right]$$

$$= \sum_{j\geq 1}\left[\frac{n-i+k}{2^j}\right] - \sum_{j\geq 1}\left[\frac{n-i}{2^j}\right] - \sum_{j\geq 1}\left[\frac{k}{2^j}\right]$$

$$= v_2\left(\binom{n-i+k}{k}\right) .$$

Since for every integer N we have that $v_2(N!) = N - \alpha(N)$ and consequently,

$$N = v_2((2N)!) - v_2(N!) ,$$

it follows that the coefficient of g_{i+k} in (7.7) has the same dyadic valuation as

$$2^{b-2n+\alpha(n)-1} \binom{n-i+k}{k} \cdot n! \cdot \frac{(2i)!}{i!} \cdot \frac{k!}{(2k)!} .$$

The latter number is equal to

$$2^{b-2n+\alpha(n)-1} \cdot \frac{(2i)!(n-i+k)!}{(2k)!} \cdot \binom{n}{i} .$$

It is clear that $\frac{(2i)!(n-i+k)!}{(2k)!} \cdot \binom{n}{i}$ is odd only for $i = 0$ and $k = n - 1$ or n ; this establishes (7.8).

Let $\mu_o, \mu_1, \ldots, \mu_n$ be the basis of $\widetilde{KU}(T(\nu))$ which corresponds to $g_o, g_1, \ldots, g_n$ under the isomorphisms $\widetilde{KU}(T(\nu)) \cong \widetilde{KU}(\Sigma^{2q}T(\nu)) \cong \widetilde{KU}(P^{b-1}_{n+1})$, $q = b - 2n + \alpha(n) - 1$. The relation (7.8) implies

$$(7.9) \qquad \mu_o^2 \equiv \Psi^2(\mu_o) \equiv \mu_{n+1} + \mu_n \quad (\text{mod.}2)$$

and Proposition (7.6) is proved.

Remark. Using Lemma 7.5 and the description of $\Psi^2(g_i)$, $0 \leq i \leq n$, given in (7.8), the same kind of argument shows the well-known result that P^n cannot be immersed in $\mathbb{R}^{4n-2\alpha(n)-1}$ [173], [203] (resp. cannot be embedded in $\mathbb{R}^{4n-2\alpha(n)}$ [27]).

Proof of Theorem 7.4. Let us first note that the J-equivalence $\nu \oplus 2(b-2n+\alpha(n)-1) \overset{J}{\sim} (b-n-1)\eta$ and the behavior of the Thom class with respect to Whitney sums (see [134; 16.8.1] give rise to the following commutative diagram of isomorphisms

$$\begin{array}{ccc} H^*(P^n;R) & \xrightarrow[\cong]{\emptyset} & \tilde{H}^*(T(\nu);R) \\ \emptyset' \downarrow \cong & & \cong \downarrow (\Sigma^{2q})^* \\ \tilde{H}^*(P^{b-1}_{n+1};R) & \cong & \tilde{H}^*(\Sigma^{2q}T(\nu);R) \end{array}$$

$q = b - 2n + \alpha(n) - 1$, $R = \mathbb{Z}$ or $\mathbb{Q}$; (here, $\emptyset$ and $\emptyset'$ denote the Thom isomorphism of the appropriate vector bundle). In the following we will identify the two groups in the bottom line.

Let $a = c_1(\xi)$ be the canonical generator of $H^*(P^n;\mathbb{Z})$; then the elements

$$z^{n-\alpha(n)+i} = \emptyset(a^i) \ , \quad i = 0,\ldots,n \ ,$$

form a basis of $\tilde{H}^*(T(\nu);\mathbb{Z})$. The cup-product structure of $T(\nu)$ is completely controlled by the Euler class $e_o \in H^{2n-2\alpha(n)}(\mathbb{C}P^n;\mathbb{Z}) \cong \mathbb{Z}$, which we interpret as an integer. We have

$$z^s \cup z^r = e_o z^{s+r} .$$

By (7.6) the integer e_o is *non-zero*. (The Chern character $ch : \widetilde{KU}(T(\nu)) \to \tilde{H}^*(T(\nu);\mathbb{Q})$ is injective, since $T(\nu)$ is torsion free.).

As in the proof of (7.6) let μ_i be the KU-theory generator of $T(\nu)$ obtained desuspending the generator g_i of $\widetilde{KU}(P^{b-1}_{n+1})$. We intend to determine the integers a_k in

$$(7.10) \qquad \mu_o^2 = \sum_{k=o}^{\alpha(n)} a_k \cdot \mu_{n-\alpha(n)+k} .$$

This will be achieved using the Chern character, which is multiplicative and compatible with suspension.

Let

$$y^{b-n-1+i} = \emptyset'(a^i) = (\Sigma^{2q})^* z^{n-\alpha(n)+i}, \quad 0 \le i \le n ,$$

be the canonical basis of $\widetilde{H}^*(P_{n+1}^{b-1};\mathbb{Z})$. Desuspending the equation $ch\ g_i = (e^y-1)^{b-n-1+i}$ we get

$$ch\ \mu_i = \left(\frac{e^z-1}{z}\right)^{b-2n+\alpha(n)-1} (e^z-1)^{n-\alpha(n)+i} .$$

We then compute

$$ch\ \mu_o \cup ch\ \mu_o = e_o\left(\frac{e^z-1}{z}\right)^{2b-4n+2\alpha(n)-2} \cdot (e^z-1)^{2n-2\alpha(n)}$$

and

$$ch\ \mu_o^2 = \left(\frac{e^z-1}{z}\right)^{b-2n+\alpha(n)-1} \cdot \sum_k a_k(e^z-1)^{2n-2\alpha(n)+k}$$

Substituting $e^z = 1 + t$ in the equation $ch\ \mu_o \cup ch\ \mu_o = ch\ \mu_o^2$ we obtain

$$(7.11) \qquad e_o \cdot \left(\frac{t}{\log(1+t)}\right)^{b-2n+\alpha(n)-1} = \sum_k a_k t^k .$$

From this latter relation and in view of (7.9) and (7.10) we conclude: If P^n immerses into $\mathbb{R}^{4n-2\alpha(n)}$ there exists an integer e_o such that in the power series of (7.11) the coefficients $a_o,\ldots,a_{\alpha(n)-2}$ are even integers, whereas $a_{\alpha(n)-1}$ and $a_{\alpha(n)}$ are odd integers.

Since we may chose b such that the relevant coefficients of the power series $\left(\frac{\log(1+t)}{t}\right)^b$ are even integers (except for the constant term, which is of course 1), multiplication of (7.11) by $\left(\frac{\log(1+t)}{t}\right)^b$ does not change the integrality nor the parity of the first $\alpha(n) + 1$ coefficients of the power series involved and theorem 7.4 is proved.

Finally, congruence (7.3) allows us to establish the following non-immersion criterion

<u>Corollary 7.12.</u> *If* $\binom{n+1}{\alpha(n)}$ *is odd* P^n *cannot be immersed in* $\mathbb{R}^{4n-2\alpha(n)}$.

CHAPTER 7

GROUP HOMOMORPHISMS AND MAPS BETWEEN CLASSIFYING SPACES; VECTOR BUNDLES OVER SUSPENSIONS

It is well known that, for any two topological groups G and H, there exists a function $\alpha_{G,H}$ from the set $\mathrm{Hom}(G,H)$ of all classes of conjugate continuous homomorphisms from G to H into the set of free homotopy classes from BG into BH. In the first part of this chapter we give an example in which $\alpha_{G,H}$ is injective but not surjective; for this example, we take G to be compact, connected Lie group and H to be the infinite unitary group. This work is developed in sections 1 to 4; it follows the argument of [163].

The second part of the chapter presents some results about embeddings of vector bundles into trivial bundles; it is bases on work of Chan and Hoffman [64], [65].

1. Generalities

We begin with a quick review of the Milnor construction of classifying spaces; details can be read in [134]. Let G be a topological group and let SG be the set of all formal sequences $(t_o x_o,\dots,t_i x_i,\dots)$ where each $x_i \in G$ and $t_i \in [0,1]$, with $\Sigma t_i = 1$ and $t_i = 0$ except possibly for a finite number of indices. We define an equivalence relation in SG by saying that $(t_o x_o,\dots,t_i x_i,\dots) \equiv (t'_o x'_o,\dots, t'_o x'_i,\dots)$ if $t_i = t'_i$ and $x_i = x'_i$ whenever $t_i = t'_i > 0$. Give to $EG = SG/\equiv$ the initial topology with respect to the functions $t_i : EG \to [0,1]$ and $x_i : t_i^{-1}(0,1] \subset EG \to G$ which take the classes $\{(t_o x_o,\dots,t_i x_i,\dots)\}$ into t_i and x_i, respectively. It is easy to see that there is a continuous action of G on the right of EG; hence define BG to be the orbit space EG/G with the final topology relative to the quotient function $EG \to EG/G$.

A continuous homomorphism $h : G \to H$ between two topological groups induces a continuous function $Eh : EG \to EH$: $Eh\{(t_o x_o,\dots,t_i x_i,\dots)\} = \{(t_o h(x_o),\dots,t_i h(x_i),\dots)\}$ for every $\{(t_o x_o,\dots,t_i x_i,\dots)\} \in EG$. By passing to the quotient we define a map $Bh : BG \to BH$. One can easily show that if $h,h' : G \to H$ are *conjugate*, i.e., if there is an element $z \in H$ such that $h' = zhz^{-1}$, then Bh and Bh' are homotopic. In this way we obtain a function $\alpha_{G,H} : \mathrm{Hom}(G,H) \to [BG,BH]$. Let us study $\alpha_{G,H}$ in some particular cases.

<u>Example 1.1</u> G = finite abelian group, $H = S^1$; in this case, $\alpha_{G,H}$ is an isomorphism.

In fact, the resolution $Z \to R \twoheadrightarrow S^1$ shows that

$$H^1(G;\mathbb{R}) \longrightarrow H^1(G;S^1) \longrightarrow H^2(G;\mathbb{Z}) \longrightarrow H^2(G;\mathbb{R})$$

is exact [165; IV, 5.5]. Since $H^1(G;\mathbb{R}) = H^2(G,\mathbb{R}) = 0$ [165; IV, 5.4], $H^1(G,S^1) \simeq H^2(G;\mathbb{Z})$. Now $H^2(G;\mathbb{Z}) \simeq [BG,K(\mathbb{Z},2)]_* \simeq [BG,K(\mathbb{Z},2)]$ (the last isomorphism follows because $K(\mathbb{Z},2)$ is a path-connected H-space [78; 4.11]) and since $K(\mathbb{Z},2) \simeq BS^1$ it follows that $H^2(G;\mathbb{Z}) \simeq [BG,BS^1]$. On the other hand regarding S^1 as a trivial $\mathbb{Z}(G)$-module, $H^1(G;S^1) \simeq \mathrm{Hom}(G,S^1)$ [165; page 106]. Hence $\mathrm{Hom}(G,S^1) \simeq [BG,BS^1]$.

<u>Example 1.2</u> $G = \mathbb{Z}$, $H = \mathbb{R}$; then $\alpha_{G,H}$ is surjective but not injective. In fact, $B\mathbb{Z} = S^1$ and $B\mathbb{R} = *$.

<u>Example 1.3</u> G compact, connected Lie group, $H = U$, the infinite unitary group; in this case, $\alpha_{G,H}$ is injective but *not* surjective. The proof of this result requires some preparation; this will be done in sections 2 and 3.

<u>2. Cartan-Serre-Whitehead Towers and H - Spaces</u>

Let Y be a based path-connected CW-complex, Y^1 be its universal cover and $p_1 : Y^1 \to Y$ be the covering map; it is well known that Y^1 is a CW-complex [253; § 5 (N)], is 1-connected and that the map p_1 induces isomorphisms $p_{1\#}$: $\pi_j(Y^1) \simeq \pi_j(Y)$, $j \geq 2$. Suppose that we have contructed a sequence of based spaces and fibrations

$$Y^i \xrightarrow[p_i]{} Y^{i-1} \longrightarrow \cdots \xrightarrow[p_2]{} Y^1 \xrightarrow[p_1]{} Y$$

such that: (i) Y^1 is the universal cover of Y; (ii) $Y^2,\ldots,Y^i$ have the homotopy type of CW - complexes; (iii) for every k, $1 \leq k \leq i$, Y^k is k - connected and $(p_1\cdots p_k)_\# : \pi_j(Y^k) \simeq \pi_j(Y)$, $j \geq k+1$. We are going to construct a $(i+1)$-connected space Y^{i+1} of the homotopy type of a CW - complex, together with a map $p_{i+1} : Y^{i+1} \to Y^i$ such that $(p_1\cdots p_{i+1})_\# : \pi_j(Y^{i+1}) \simeq \pi_j(Y)$, $j \geq i+2$. In this way we obtain, by induction, an infinite sequence of based spaces progressively more connected starting at Y; this sequence is called a *Cartan - Serre - Whitehead tower* of Y.

The construction of Y^{i+1} goes as follows. Take $\pi = \pi_{i+1}(Y^i) \simeq \pi_{i+1}(Y)$ and recall that

$$[Y^i,K(\pi,i+1)]_* \simeq H^{i+1}(Y^i;\pi) \simeq \mathrm{Hom}\Big(H_{i+1}(Y^i;\mathbb{Z}),\pi\Big)\ ;$$

let $[g] \in [Y^i,K(\pi,i+1)]_*$ be the image under the above isomorphisms of the inverse of the Hurewicz isomorphism $\pi_{i+1}(Y^i)\simeq H_{i+1}(Y^i)$. Notice that $[g]$ is also the image of $[1_{K(\pi,i+1)}] \in [K(\pi,i+1),K(\pi,i+1)]_*$ under the homomorphism

$$[K(\pi,i+1),K(\pi,i+1)]_* \longrightarrow [Y^i,K(\pi,i+1)]_*$$

induced by g. On the other hand, $[1_{K(\pi,i+1)}]$ corresponds to the inverse of the Hurewicz isomorphism $\pi_{i+1}(K(\pi,i+1)) \overset{\rightarrow}{\simeq} H_{i+1}(K(\pi,i+1);\mathbb{Z})$ and thus, by naturality, $\mathrm{Hom}(g_*,1)$ takes an isomorphism into an isomorphism. It follows that $g_* : H_{i+1}(Y^i;\mathbb{Z}) \to H_{i+1}(K(\pi,i+1);\mathbb{Z})$ is an isomorphism and therefore,

$$\text{(2.1)} \qquad g_\# : \pi_{i+1}(Y^i) \longrightarrow \pi_{i+1}(K(\pi,i+1))$$

is an isomorphism. Now consider the pull-back diagram

$$\begin{array}{ccc}
Y^{i+1} = E_g & \xrightarrow{\ \ \bar{g}\ \ } & PK(\pi,i+1) \\
\Big\downarrow p_{i+1} & & \Big\downarrow p \\
Y^i & \xrightarrow[\ \ g\ \ ]{} & K(\pi,i+1)
\end{array}$$

where $PK(\pi,i+1)$ is the space of paths starting at the base point of $K(\pi,i+1)$ and $p(\lambda) = \lambda(1)$, for every path $\lambda \in PK(\pi,i+1)$. The Theorem of [205] shows that Y^{i+1} is of the homotopy type of a CW-complex; furthermore, Y^{i+1} is $(i+1)$-connected and $(p_1 \cdots p_{i+1})_\# : \pi_j(Y^{i+1}) \simeq \pi_j(Y)$, $j \geq i + 2$, as one can see comparing the exact sequences of homotopy groups of the appropriate fibrations and applying (2.1).

The tower over Y we have described is precisely the Moore-Postnikov factorization of the inclusion of the base point in Y given by Spanier in [225; 8.3.8]; this means that, for every $n \geq 0$ there is a map $f'_n : \{y_o\} \to Y^n$ such that $p_n f'_n = f'_{n-1}$. Furthermore, there is a map $f' : \{y_o\} \to Y^\infty = \varprojlim Y^n$ which is a weak homotopy equivalence [225; 8.3.2] .

We shall see next that if Y is an H-space, then so are the spaces at every stage of the tower. We begin with the following.

<u>Lemma 2.2.</u> *Let* Y *be a path-connected* CW-*complex and let* Y^1 *be its universal cover. If* Y *is an* H-*space, then so is* Y^1.

<u>Proof.</u> Let e be the unit of Y, which is also viewed as its base point, and let PY be the space of paths over Y, starting at e. We say that $\lambda, \lambda' \in PY$ are equivalent if $\lambda(1) = \lambda'(1)$ and $\lambda * \lambda'^{-1} \simeq \lambda_e$, where λ_e is the constant path at e and $\lambda * \lambda'^{-1}$ is the path defined by

$$\lambda * \lambda'^{-1}(t) = \begin{cases} \lambda(2t) & 0 \leq t \leq \frac{1}{2} \\ \lambda'(2-2t) & \frac{1}{2} \leq t \leq 1 \end{cases} .$$

Then $PY/_\sim = Y^1$; now, for every pair of classes $\{\lambda\},\{\mu\}$ of Y^i, define the multiplication $\{\lambda\}\{\mu\} = \{\lambda\mu\}$, where $\lambda\mu(t) = \lambda(t)\mu(t)$. This multiplication is well-defined: say $L : \lambda * \lambda'^{-1} \simeq \lambda_e$ and $M : \mu * \mu'^{-1} \simeq \lambda_e$ are homotopies and $m_Y : Y \times Y \to Y$ is the multiplication in Y. The reader can easily verify that

$$LM : I \times I \xrightarrow{(L,M)} Y \times Y \xrightarrow{m_Y} Y$$

is a homotopy $\lambda\mu\,(\lambda'\mu')^{-1} \simeq \lambda_e$. Finally, the multiplication on Y^1 is continuous and has a two-sided unit, namely $\{\lambda_e\}$.

Recall that if (Y,m_Y) and (Z,m_Z) are H - spaces, a map $g : Y \to Z$ is an *H - map* if $m_Z \circ (g \times g) \simeq g \circ m_Y$.

<u>Lemma 2.3.</u> *Let* Y *be an* (n-1) - *connected* H - *space, with* $n - 1 > 0$. *Then any map* $g : Y \to K(\pi,n)$ *is an* H - *map.*

<u>Proof.</u> The obstructions to g being an H - map lie in $H^i(Y \wedge Y;\pi_i(K(\pi,n)))$; now $Y \wedge Y$ is (2n-1) - connected and $\pi_i(K(\pi,n)) = 0$ for all $i \neq n$.

<u>Theorem 2.4.</u> (Stasheff, [226]) - *Let* Y,Z *be* H - *spaces and let* $\Omega Z \to E_g \to Y$ *be the fibration induced from* $\Omega Z \to PZ \to Z$ *by an* H - *map* $g : Y \to Z$. *Then* E_g *is an* H - *space.*

<u>Proof.</u> Recall that $E_g = \{(y,\lambda) \in Y \times PZ | g(y) = \lambda(1)\}$. An obvious candidate for a multiplication on E_g is $(y,\lambda)(y',\lambda') = (yy',\lambda\lambda')$, where $\lambda\lambda'(t) = \lambda(t)\lambda'(t)$. However, $g(yy')$ in general is not equal to $g(y)g(y')$ and thus $(yy',\lambda\lambda')$ is not necessarily in E_g . We correct this inconvenience using the fact that g is an H - map. Let

$$h : Y \times Y \times I \longrightarrow Z$$

be a homotopy with $h(y,y',0) = g(yy')$, $h(y,y',1) = g(y)g(y')$; now define a multiplication on E_g by

$$(y,\lambda)(y',\lambda') = (yy',\lambda\lambda' * h^{-1}(y,y', \ )).$$

Lemmas 2.2, 2.3 and Theorem 2.4 show the following.

<u>Theorem 2.5.</u> *Let* Y *be a path-connected* CW - *complex with an* H - *space structure. Then all the spaces of a Cartan-Serre-Whitehead tower over* Y *are* H - *spaces.*

<u>3. Remarks about the KU - Theory of certain Classifying Spaces</u>

Let X be a based CW - complex filtered by its skeleta X_n, $n \geq 0$. For every H - space Y there is an obvious group homomorphism

$$\Phi : [X,Y]_* \longrightarrow \varprojlim [X_n,Y]_* \quad .$$

It is easy to show that Φ is an epimorphism: if $\{[f_0],[f_1],[f_2],\ldots\}$ is an arbitrary element of the inverse limit, $f_n : X_n \to Y \simeq f_{n+1}|X_n$, for every $n \geq 0$. Because of the Homotopy

Extension Property for the pair (X_{n+1}, X_n), there is a homotopy $H_{n+1} : X_{n+1} \times I \to Y$ such that $H_{n+1}(\ ,0) = f_{n+1}$, $H_{n+1}(\ ,1)|X_n = f_n$; writing f'_{n+1} for $H_{n+1}(\ ,1)$ we see that $f'_{n+1}|X_n = f_n$. In this way we obtain a map $f : X \to Y$ such that, for every $n \geq 0$, $f|X_n \simeq f_n$.

The homomorphism Φ is not necessarily a monomorphism; for example, in [7] Adams and Walker construct two CW-complexes X,Y and an essential map $f : X \to Y$ whose restrictions to all skeleta X_n are inessential. In this section we prove that if $X = BG$, where G is a compact, connected Lie group, and $Y = BU$, then Φ is an isomorphism.

Lemma 3.1. *Let*

$$\cdots \longrightarrow Y^n \xrightarrow[p_n]{} Y^{n-1} \longrightarrow \cdots \longrightarrow Y^1 \xrightarrow[p_1]{} Y_o = Y$$

be a Cartan-Serre-Whitehead tower over Y . *Let* $f_n : X \to Y^n$ $n \geq 0$ *be a sequence of maps such that, for every* $n \geq 1$, $p_n f_n \simeq f_{n-1}$. *Then, all mappings* f_n *are inessential.*

Proof. Because of the Covering Homotopy Property we can deform the maps f_n into maps $f'_n : X \to Y^n$ so that, for every $n \geq 1$, $p_n f'_n = f'_{n-1}$. Hence there is a unique map $f' : X \to Y^\infty = \varprojlim Y^n$ such that $f'_n = q_n f'$ for every $n \geq 0$, where $q_n : Y^\infty \to Y^n$ is the canonical map. Hence, by [225; 7.6.23] we conclude that all the maps f'_n are inessential, and so are the maps f_n .

Lemma 3.2. *Let* X *and* Y *be based* CW-*complexes such that:*

(i) *for every* $n \geq 0, X_n$ *is finite;*

(ii) Y *is a path-connected* H-*space;*

(iii) *for every* $n \geq 1$, $\pi_n(Y)$ *is a finitely generated abelian group;*

(iv) *for every* $n \geq 0$, $H^n(X;Q) \otimes \pi_{n+1}(Y) = 0$. *Then,*

$$\Phi : [X,Y]_* \longrightarrow \varprojlim [X_n, Y]_*$$

is an isomorphism. (Compare with [60; Theorem 1.1]).

Proof. Let $f : X \to Y$ be such that $f|X_n \simeq$ constant map, for every $n \geq 0$; consider a tower over Y (notation of 3.1). We are going to construct - by induction - a sequence of maps $\{f_n : X \to Y^n | n \geq 0\}$ such that: (a) $f_o = f$; (b) for every $k \geq 0$ and every $n \geq 0$, $f_n|X_k \simeq$ constant map; c) for every $n \geq 1$, $p_n f_n \simeq f_{n-1}$. Then, by 3.1, all maps f_n (in particular, $f_o = f$) will be inessential, thereby proving the Lemma. Suppose that $f_n : X \to Y^n$ has been constructed and is such that for every $k \geq 0$, $f_n|X_k \simeq$ constant map and $p_n f_n \simeq f_{n-1}$.

We have seen that the space Y^{n+1} has been defined by an appropriate pull-back diagram

$$\begin{array}{ccc} Y^{n+1} & \longrightarrow & PK(\pi_{n+1}(Y),n+1) \\ {\scriptstyle p_{n+1}}\downarrow & & \downarrow \\ Y^n & \longrightarrow & K(\pi_{n+1}(Y),n+1) \end{array} \quad ;$$

clearly $f_n : X \to Y^n$ can be lifted to Y^{n+1} if, and only if, $gf_n \simeq$ constant map. But $[gf_n] \in H^{n+1}(X;\pi_{n+1}(Y)) \simeq H^{n+1}(X_{n+2};\pi_{n+1}(Y))$, where the isomorphism is induced by the inclusion $i_{n+2} : X_{n+2} \longrightarrow X$; thus, $(i_{n+2})_*[gf_n] = [gf_n|X_{n+2}] = 0$ because $f_n|X_{n+2} \simeq$ constant map, and therefore, $[gf_n] = 0$. Let $h : X \to Y^{n+1}$ be such that $p_{n+1}h = f_n$. Recall now that the base-homotopy classes of liftings of f_n to Y^{n+1} are in a one-to-one correspondence with the elements of $H^n(X;\pi_{n+1}(Y))$ [225; 8.2]. Also

$$H^n(X;\pi_{n+1}(Y)) \otimes \mathbb{Q} \simeq H^n(X;\mathbb{Q}) \otimes \pi_{n+1}(Y) \oplus \mathrm{Tor}\left(H^{n+1}(X),\pi_{n+1}(Y)\right) \otimes \mathbb{Q} = 0$$

because of the hypothesis and thus, $H^n(X;\pi_{n+1}(Y))$ is a finite group. Consider the following diagram of groups (use 2.5):

$$\begin{array}{ccccc} \varprojlim[X_k,K(\pi_{n+1}(Y),n)]_* & \longrightarrow & \varprojlim[X_k,Y^{n+1}]_* & \longrightarrow & \varprojlim[X_k,Y^n]_* \\ \uparrow{\scriptstyle \Phi_\pi} & & \uparrow{\scriptstyle \Phi_{n+1}} & & \uparrow{\scriptstyle \Phi_n} \\ H^n(X;\pi_{n+1}(Y)) \simeq [X,K(\pi_{n+1}(Y),n)]_* & \longrightarrow & [X,Y^{n+1}]_* & \longrightarrow & [X,Y^n]_* ; \end{array}$$

it shows that $K = \ker\left\{\varprojlim[X_k,Y^{n+1}]_* \to \varprojlim[X_k,Y^n]_*\right\}$ is a finite group and that $\Phi_{n+1}[h] \in K$. Indeed, the homotopy class of any lift of f_n to Y^{n+1} is taken by Φ_{n+1} into K . Since K is finite and Φ_{n+1} is onto, there is a lift f_{n+1} : $X \to Y^{n+1}$ such that $\Phi_{n+1}[f_{n+1}] = 0$, that is to say, $f_{n+1}|X_k \simeq$ constant map for all $k \geq 0$. Notice that $p_{n+1}f_{n+1} \simeq f_n$.

The reader should recall that if G is a compact Lie group then BG can be filtered by compact spaces $BG_1 \subset BG_2 \subset \dots$. Indeed, it is possible to construct a classifying space BG of G which is a CW-complex and whose skeleta $(BG)_n$ are finite CW-complexes; furthermore, if G is connected, BG and all $(BG)_n$'s are connected.

Theorem 3.3. *For every compact connected Lie group* G, $[BG,BU]_* \simeq \varprojlim[(BG)_n,BU]_*$.

Proof. The hypothesis of 3.2 hold: BU is a path-connected H-space (see, for example, [236; page 405]); (iii) is true because of the Bott Periodicity Theorem and (iv) holds because $H^{2n+1}(BG;\mathbb{Q}) = 0$ [43; 7.2 and 19.1] and $\pi_{2n+1}(BU) = 0$.

Remark. If we also require the Lie group G to have torsion free integral cohomology, Theorem 3.3 follows from an obstruction theory argument [195] .

4. A Theorem of Non-Surjectivity for $\alpha_{G,H}$

In this section we prove the following result.

Theorem 4.1. *If* G *is a compact, connected Lie group and* H *is the infinite unitary group* U, $\alpha_{G,U} : \mathrm{Hom}(G,U) \to [BG,BU]$ *is an injection but not a surjection.*

Proof. Theorem 3.3, Satz 4.11 of [9] and the connectivity of $(BG)_n$ show that

$$[BG,BU] \simeq [BG,BU]_* \simeq \varprojlim[(BG)_n,BU]_* \simeq \varprojlim \widetilde{KU}((BG)_n).$$

According to Atiyah and Hirzebruch, $\varprojlim KU((BG)_n)$ is expressible in terms of the ring $R(G)$ of unitary representations of G . In fact, let $\varepsilon : R(G) \to Z$ be the homomorphism obtained by sending each representation of G into its dimension.

Define $I(G) = \ker \epsilon$ and give to $R(G)$ the $I(G)$-adic topology, that is to say, the topology which has the set $\{I(G)^n \; n \geq 1\}$ as a sub-basis for the system of neighborhoods of $0 \in R(G)$. Then

$$\varprojlim KU((BG)_n) \simeq \varprojlim R(G)/I(G)^n \simeq \hat{R}(G)$$

[28; § 4] .

Since G is compact and connected, it has a maximal torus T which can be written as the group of k-tuples of reals mod 1 ; every irreducible representation of T is 1-dimensional and given by a homomorphism $\phi : T \to U(1)$,

$$\phi(x_1,\ldots,x_n) = \exp[2\pi i(a_1x_1 + \ldots + a_kx_k)] \quad ,$$

$a_i \in Z$. Hence $R(T) \simeq Z[\exp(2\pi ix_1),\exp(-2\pi ix_1),\ldots,\exp(2\pi ix_k),\exp(-2\pi ix_k)]$; writing z_j for $\exp(2\pi ix_j) - 1$, we see that $I(T)$ is the maximal ideal generated by $z_1,\ldots,z_k,(1+z_1)^{-1} - 1,\ldots,(1+z_k)^{-1} - 1)$. Hence, $I(T)^n = 0$ because an element of this intersection is a power series whose lowest term has arbitrarily high degree. Since $I(G)^n \subseteq I(T)^n$, it follows that $I(G)^n = 0$ and so, $R(G)$ is Hausdorff; now [51; chap. III, § 5, n.4] shows that $R(G)$ is isomorphic to a totally dense proper sub-ring of $\hat{R}(G)$. We are going to show that there is an injection $Hom(G,U) \to R(G)$; with this, the proof of 4.1 is complete: in fact, it follows from the commutative diagram

$$\begin{array}{ccccc}
Hom(G,U) & \rightarrowtail & R(G) & \underset{\text{not a surjection}}{\rightarrowtail} & \hat{R}(G) \\
\downarrow{\scriptstyle \alpha_{G,U}} & & & & \downarrow{\scriptstyle \simeq} \\
[BG,BU] \simeq \varprojlim[(BG)_n,BU]_* \simeq \varprojlim \widetilde{KU}((BG)_n) & & \rightarrowtail & & \varprojlim KU((BG)_n)
\end{array} \quad .$$

To obtain the injection $Hom(G,U) \rightarrowtail R(G)$ we first notice that because G is compact, $Hom(G,U) \simeq \varinjlim Hom(G,U(n))$. On the other hand, if $\rho,\rho' : G \to U(n)$ are two conjugate homomorphisms (actually, unitary representations of G) they have the same character and so, are identified in $R(G)$. This defines an injection of $Hom(G,U(n))$ into $R(G)$ and ultimately, due to the compactness of G, an injection of $\mathrm{Hom}(G,U)$ into $R(G)$.

5. Vector Bundles over Suspensions.

In this section we shall consider finite dimensional complex vector bundles over a compact path-connected Hausdorff space X. It is well known that if ξ is a bundle of this type, there is a bundle η over X such that $\xi \oplus \eta$ is trivial. We shalll prove that if X is the suspension of a space Y there exists such a bundle η which has the same dimension as ξ. Furthermore, K-theory will be used to show that the result obtained is the best possible.

Theorem 5.1. *Let $\xi = (E,p,SY)$ be a complex n-bundle, with Y a based, compact path-connected Hausdorff space. Then there is a complex n-bundle η over SY such that $\xi \oplus \eta$ is trivial.*

Proof. First we construct a vector bundle which is isomorphic to ξ but which allows a certain degree of manipulation. Let us regard SY as the union of two cones C^-Y and C^+Y, where

$$C^-Y = Y \times [-1,0]/(-1 \times Y \cup [-1,0] \times \{y_0\})$$

and

$$C^+Y = Y \times [0,1]/(1 \times Y \cup [0,1] \times \{y_0\}) \quad .$$

Since both cones are contractible the restrictions of to C^-Y and C^+Y are isomorphic to trivial vector bundles; let $\alpha_- : E|C^-Y \to C^-Y \times \mathbb{C}^n$ and $\alpha_+ : E|C^+Y \to C^+Y \times \mathbb{C}^n$ be bundle isomorphisms. Notice that $Y = C^-Y \cap C^+Y$ and consider the isomorphism of trivial vector bundles $\beta : Y \times \mathbb{C}^n \to Y \times \mathbb{C}^n$ defined by $\beta = (\alpha_-|(\xi|Y))(\alpha_+^{-1}|(\xi|Y))$. Now form the vector bundle $\beta(\xi)$ over SY having for total space $(C^+Y \times \mathbb{C}^n) \cup_\beta (C^-Y \times \mathbb{C}^n)$ the adjunction space of $C^+Y \times \mathbb{C}^n$ and $C^-Y \times \mathbb{C}^n$ by the function β. Next, define the bundle map

$$\alpha : \xi \longrightarrow \beta(\xi)$$

as follows: for every $e \in E$,

$$\alpha(e) = \begin{cases} \alpha_+(e) & , \text{ if } e \in E|C^+Y \\ \alpha_-(e) & , \text{ if } e \in E|C^-Y \quad . \end{cases}$$

Since α_+ and α_- are isomorphisms on the fibres, so is α and hence, ξ and $\beta(\xi)$ are isomorphic.

Now consider the vector bundle $\eta = \beta^{-1}(\xi)$ over SY whose total space is the adjunction space $(C^+Y \times \mathbb{C}^n) \cup_{\beta^{-1}} (C^-Y \times \mathbb{C}^n)$, and whose bundle map is the projection on the first component. We are going to show that $\beta(\xi) \oplus \beta^{-1}(\xi)$ is trivial, so proving the theorem. The isomorphism β and its inverse β^{-1} correspond respectively to maps Φ and Φ^{-1} from Y into $GL(n,\mathbb{C})$; one should notice that for every $y \in Y, \Phi(y)^{-1} = \Phi^{-1}(y)$. Furthermore, one observes that

$$(C^+Y \times \mathbb{C}^n) \cup_\beta (C^-Y \times \mathbb{C}^n) \oplus (C^+Y \times \mathbb{C}^n) \cup_{\beta^{-1}} (C^-Y \times \mathbb{C}^n)$$

$$\simeq (C^+Y \times \mathbb{C}^{2n}) \cup_{\beta \oplus \beta^{-1}} (C^-Y \times \mathbb{C}^{2n}) \quad ,$$

with $\beta \oplus \beta^{-1}$ corresponding to a map $\Psi : Y \to GL(2n,\mathbb{C})$ given by the matrix

$$\Psi(y) = \begin{pmatrix} \Phi(y) & 0 \\ 0 & \Phi^{-1}(y) \end{pmatrix}$$

for every $y \in Y$. Following [17; Theorem 2.4.6] we define a homotopy $\Theta : \Psi \simeq I_{2n}$, the $2n \times 2n$ identity matrix:

$$\Theta(y,t) = \begin{pmatrix} \Phi(y) & 0 \\ 0 & I_n \end{pmatrix} \begin{pmatrix} I_n \cos t\,\pi/2 & I_n \sin t\,\pi/2 \\ -I_n \sin t\,\pi/2 & I_n \cos t\,\pi/2 \end{pmatrix}$$

$$\begin{pmatrix} I_n & 0 \\ 0 & \Phi^{-1}(y) \end{pmatrix} \begin{pmatrix} I_n \cos t\,\pi/2 & -I_n \sin t\,\pi/2 \\ I_n \sin t\,\pi/2 & I_n \cos t\,\pi/2 \end{pmatrix} .$$

This, of course, completes the proof.

The previous theorem is saying, in particular, that because SY is the union of two contractible spaces, any n-dimensional vector bundle over SY can be embedded in a $2n$-dimensional trivial vector bundle. This result can be extended.

<u>Theorem 5.2.</u> *If* X *is also normal and has an open covering*

$\{U_1,\ldots,U_k\}$ *of contractible spaces, every* n-*dimensional vector bundle* ξ *over* X *can be embedded in a* kn-*dimensional trivial vector bundle.*

Proof. If $\xi = (E,p,X)$, for every integer j between 1 and k, there is a homeomorphism $\phi_j : E|U_j \to U_j \times \mathbb{C}^n$ whose restrictions to the fibres are isomorphisms of vector spaces. Let $\{q_j \mid 1 < j < k\}$ be a partition of unity subordinated to $\{U_j \mid 1 < j < k\}$; define, for each j,

$$\psi_j : E \longrightarrow X \times \mathbb{C}^n$$

by

$$\psi_j(e) = \begin{cases} q_j(p(e))\phi_j(e), & \text{if } e \in E|U_j, \\ 0, & \text{otherweise.} \end{cases}$$

Then set $\psi = \Sigma\, i_j \psi_j$, where $i_j : X \times \mathbb{C}^n \to X \times (\mathbb{C}^n)^k$ takes $(x,\vec{v})$ into $(x,(0,\ldots,\vec{v},\ldots,0))$ with $\vec{v}$ sitting on the jth-component of the k-tuple. The morphism $\psi : E \to X \times (\mathbb{C}^n)^k$ gives the embedding.

We now comment on the previous two theorems and for this, we need K-theory. First, we show that in Theorem 5.1, the condition that ξ is a vector bundle *over a suspension* is necessary; our forthcoming Theorem will also show that the result of (5.2) is the best possible for $n = 1$, $k = 3$.

Theorem 5.3. *Let* $X = \mathbb{C}P^k$, $k \geq 2$, *and* ξ *be the canonical line bundle over* $\mathbb{C}P^k$. *Then every* $(r-1)$-*dimensional vector bundle* η *over* X *with* $r - 1 < k$ *is such that* $\xi \oplus \eta$ *is not trivial.*

Note: This result shows that there is no line bundle η over $\mathbb{C}P^2$ such that $\xi \oplus \eta$ is trivial.

Proof. Suppose that $\xi \oplus \eta$ is trivial and apply λ-operations to $[\xi] + [\eta]$:

$$\binom{r}{i} = \lambda^i([\xi] + [\eta]) = \lambda^i[\eta] + [\xi]\lambda^{i-1}[\eta]$$

and thus,

$$\lambda^i[\eta] = \sum_{j=0}^{i} (-1)^j \binom{r}{i-j} [\xi]^j$$

Since $KU(\mathbb{C}P^k) = \mathbb{Z}[v]/\langle v^{k+1}\rangle$ (see Chapter 4, § 3), where $v = [\xi] - 1$, $\lambda^r[\eta] \neq 0$. This contradicts the fact that $\lambda^r[\eta] = 0$ (the triviality of $\lambda^r[\eta]$ comes from $\dim \eta = r-1$).

Next, we show that in (5.1), n is the minimum dimension for η so that $\xi \oplus \eta$ is trivial; our argument will also show that (5.2) is the best result if $k = 2$.

Theorem 5.4. *If* $X = S^{2n}$ *, there exists an* n *- vector bundle* ξ *over* X *such that, for every* r *- dimensional vector bundle* η *over* X *with* $r < n$, $\xi \oplus \eta$ *is not trivial.*

Proof. Let α be a generator of $H^{2n}(S^{2n},Z)$. As a consequence of the Integrality Theorem (O.II) There is an element $z \in \ker\left[KU(S^{2n}) \to KU((S^{2n})_{2n-1})\right] \subseteq \widetilde{KU}(S^{2n})$ such that $ch(z) = \alpha$. But $\widetilde{KU}(S^{2n}) \simeq [S^{2n},G_n(\mathbb{C}^{2n})]_*$ (where $G_n(\mathbb{C}^{2n})$ is the Grassman manifold of n spaces in $\mathbb{C}^{2n}$) and so there is an n - dimensional vector bundle ξ over S^{2n} whose image in $\widetilde{KU}(S^{2n})$ is z . Therefore $ch[\xi] = \alpha \neq 0$ showing that the nth - Chern class $c_n(\xi)$ is non-trivial. Suppose that η is a vector bundle over S^{2n} such that $\xi \oplus \eta$ is trivial. Then,

$$0 = c_n(\xi \oplus \eta) = \sum_{i+j=n} c_i(\xi)c_j(\eta) = c_n(\xi) + c_n(\eta) ;$$

hence $c_n(\eta) \neq 0$ and then, $\dim \eta \geq n$.

We complete this section with an example which answers in the negative the following question: if X is a normal space which is covered by k contractible open sets and η is an n - dimensional vector bundle over X embedded in a trivial vector bundle of dimension less than kn , can we decompose η as a Whitney sum of non-zero bundles?

Example 5.5. Let $X = \mathbb{C}P^2$, ξ be the canonical line bundle over X and η be a 2-dimensional vector bundle over X such that $\xi \oplus \eta \cong (X \times \mathbb{C}^3, pr_1, X)$. Then η cannot be decomposed into the Whitney sum of two line bundles.

In fact, suppose that η is the sum of two line bundles η_1 and η_2 .

Then,

$$[\eta_1] - 1 = a([\xi] - 1) + b([\xi] - 1)^2 \quad \text{for some} \quad a,b \in \mathbb{Z} ,$$
$$[\eta_2] - 1 = [\eta] - a([\xi] - 1) - b([\xi] - 1)^2 - 2 ,$$

and because $[\eta] + [\xi] = 3$,

$$[\eta_2] - 1 = -(a+1)([\xi] - 1) - b([\xi] - 1)^2 .$$

Recall that for any line bundle τ , $\gamma_t([\tau] - 1) = 1 + ([\tau] - 1)t$ (see Chapter O) and hence,

$$\gamma_t([\eta] - 2) = (1+([\eta]_1 - 1)t)\ (1 + ([\eta]_2 - 1)t)$$
$$= 1 - ([\xi] - 1)t - a(a+1)([\xi] - 1)^2t^2 ;$$

on the other hand,

$$\gamma_t([\eta] - 2) = \gamma_t(1 - [\xi]) = (\gamma_t([\xi] - 1))^{-1}$$
$$= 1 - ([\xi] - 1)t + ([\xi] - 1)^2t^2$$

(in both cases one must remember that $KU(CP^2) = \mathbb{Z}[[\xi] - 1]/\langle[\xi]-1)^3\rangle$
Thus, $-a(a+1) = 1$, absurd.

CHAPTER 8

ON THE INDEX THEOREM OF ELLIPTIC OPERATORS

1. Introduction.

It would be inconceivable to present a certain number of geometric and topological problems in which K - theory appears, either as their adequate theoretical framework or as a powerful tool capable of simplifying greatly their solution, without mentioning the deep connections between the Theory of Differential Operators and K - theory. One should recall moreover that this relationship has played a large role in the development of K - theory itself and that it has produced some of the most important results obtained in the mathematics of the last fifteen years. This motivation and some other good reasons, easily convinced us to present in this book - even if sketchily - some aspects of the Atiyah-Singer Theorem and of its several formulations.

The purpose of this chapter is two-fold: we first wish to direct the attention of the reader to this important area of mathematics we mentioned above and second, to present enough bibliographical material as to permit a more profound independent study of the subject. It should therefore be understood more as a presentation of the Index Theorem and some of its classical consequences, rather than a detailed and complete exposition of the subject. At any rate, one must recognize that the large number of works relative to that delicate and complex subject would not leave us with any other alternative. It is our belief that we shall have reached part of our objectives if the study of the following pages will help in vulgarizing this beautiful mathematical Theory.

It is neither new nor surprising that there are tight relations between Analysis and Algebraic Topology. For example, let γ be a closed curve of $\mathbb{R}^2$ which does not contain the origin $O \in \mathbb{R}^2$ and consider the number of times γ turns around O ; this numerical "invariant" is the index of γ with respect to O , or the degree $d(f)$ of a continuous function

$$f : S^1 \longrightarrow \mathbb{C}^*$$

such that $f(S^1) = \gamma$. The number $d(f)$ can be computed analytically, for example, via Cauchy's Integral Formula

$$d(f) = \frac{1}{2\pi i} \int_{\gamma} \frac{dg}{g} ,$$

where g is an holomorphic function approximating f . We recall that $d(f)$ does not change when f is deformed with continuity (homotopically); furthermore, given an integer d , we can always find a function f such that $d(f) = d$. This is not a fortuitous example. In fact, let $C(S^{n-1}, GL(m,\mathbb{C}))$ be the set of all maps from S^{n-1} into $GL(m,\mathbb{C})$; then, if $2m \geq n \geq 2$, it was proved by Bott in [46] that

(i) if n = even, there exists a surjection

$$d : C(S^{n-1}, GL(m,\mathbb{C})) \longrightarrow \mathbb{Z}$$

such that $d(f) = d(g)$ if, and only if, f and g are homotopic;

(ii) if n = odd, all maps $f \in C(S^{n-1}, GL(m,\mathbb{C}))$ are homotopically trivial.

Incidentally, we wish to observe that the periodicity of the stable homotopy groups of the general linear groups - the most remarkable property of K - theory - was established for the first time in that same paper. The proof of the periodicity given subsequently by Atiyah and Bott in [22] is another example of the connections between K - theory and Analysis - in this case, speciffically with the theory of Complex Functions.

Let U be an open subset of $\mathbb{R}^n$ and let $C^\infty(U)$ (resp. $C_o^\infty(U)$) be the space of all C^∞ - differentiable maps on U (resp. with compact support). Then, a pseudo-differential operator D of order r on U is of the form

$$Du(x) = \frac{1}{(2\pi)^n} \int_{\mathbb{R}^n} p(x,y)\hat{u}(y)e^{i(x|y)}dy, \quad u \in C_o^\infty(U), x \in U$$

where $\hat{u}(y) = \int_U u(x)e^{-i(x|y)}dx$ is the Fourier transform of u (here $(x|y)$ is the scalar product of x and y in $\mathbb{R}^n$) and p is a C^∞-differentiable function on $U \times \mathbb{R}^n$ with convenient properties over each compact subspace of U (these are necessary conditions for $Du \in C^\infty(U)$ and which allow differentiation under the integral - see [126]); furthermore, we require that for every $x \in U$ and every $y \in \mathbb{R}^n - \{0\}$,

$$\sigma_r(x,y) = \lim_{a\to\infty} \frac{p(x,ay)}{a^r}$$

exists. This limit is the symbol of order r of D. Whenever D is a differentiable operator, σ_r is a polynomial in y of degree r and with coefficients in $C^\infty(U)$; moreover, if $\sigma_r(x,y) \in GL(m,\mathbb{C})$ for every $x \in U$ and every $y \in \mathbb{R}^n - \{0\}$, then D is said to be an elliptic differential operator. In this case, for a fixed x and by restriction, we obtain a map

$$\sigma_r(x,-) : S^{n-1} \longrightarrow GL(m;\mathbb{C}) ;$$

if n is even, this map has a degree, called the degree of the elliptic operator D. As expected, in view of the local chart structure of a differentiable manifold, all these notions can be extended globally to a compact manifold. With this, we now make a crucial observation relative to an elliptic operator D defined on a compact manifold: the solution space of the equation

$$Du = 0$$

is finite dimensional; the same happens to D^*, the adjoint of D. We then define the index of the elliptic operator D as the integer

$$i(D) = \dim(\ker D) - \dim(\ker D^*) .$$

Let us finish this paragraph with the observation that Atiyah, at the end of his interesting paper [16], after commenting upon the (real) Bott periodicity and the idea of defining the index of an elliptic operator as a function of its degree was lead to say that "analysis and topology are now inextricably mixed".

One can probably say that the first study about the index of an elliptic operator appeared in F.Noether's paper on tidal movements, published in 1921 [190]. Then, except for a few sparse results (e.g., those obtained by S.G.Mihlin in 1948), very little was done in this context and it was only during the sixties that the question attracted considerable interest. In fact, in 1960 Gelfand clearly formulated the problem of computing explicitly the index of an elliptic operator in pure topological terms; this problem was taken up by the analysts - among whom we quote M.S.Agranovich, A.S.Dymin, R.T.Seeley and A.I.Vol'pert - who established some particular results. It was only in 1963 that Atiyah and Singer obtained the first general answer (see Section 4).

The strength of the Index Theorem lies partly in the fact that it unifies in one same formulation, results apparently as different as the Chern-Gauss-Bonnet formula (cf. [97]), the Hirzebruch-Riemann-Roch Theorem and the Hirzebruch Signature Theorem (cf. [116]), which we presently recall. For a compact riemannian surface M with gaussian curvature k, the classical Gauss-Bonnet formula tells us that

$$\chi(M) = \frac{1}{2\pi} \int_M k \, dv ,$$

where $\chi(M)$ is the Euler-Poincaré Characteristic of M and dv is the density defined by the riemannian metric. The original Riemann-Roch problem consisted in computing the number of holomorphic cross-sections of a 1 - dimensional holomorphic bundle η over a riemannian compact surface M, that is to say, find $H^o(M;\Omega(\eta))$ where $\Omega(\eta)$ is the sheaf defined by the germs of the holomorphic cross-sections of η; we conclude that

$$\dim H^o(M;\Omega(\eta)) - \dim H^1(M;\Omega(\eta)) = \chi(\eta) + 1 - g$$

where $\chi(\eta)$ is the Euler characteristic of η and g is the genus of M. Finally, Hirzebruch's Signature Theorem shows that if M is a 4ℓ-dimensional, compact, oriented manifold, the signature of the quadratic form

$$H^{2\ell}(M;\mathbb{R}) \times H^{2\ell}(M;\mathbb{R}) \longrightarrow H^{4\ell}(M;\mathbb{R}) \simeq \mathbb{R}$$

defined by the cup-product is the L-genus of M (see Section 3). It is important to observe that the Atiyah-Singer Theorem is much more than a useful unifying proposition: indeed, several interesting generalizations of the results we quoted were derived from it. For example, the Theorem of Hirzebruch-Riemann-Roch was extended to every compact complex manifold following this route. In this context, we quote Bott in saying that the Index Theorem "may also be thought of as a beautiful and far reaching generalization of Hirzebruch's Riemann-Roch Theorem - both in statement and in the spirit of the proof". The reader is invited to get acquainted with the different proofs and multiple applications of the Index Theorem by reading the following sections and browsing through the standard bibliography (a recent introductory study of the Atiyah-Singer Index Theorem written by Shanahan [216] contains a good bibliography). In the meantime, we observe that Atiyah and Bott proposed in [23] a proof of the Index Theorem using the ζ-function

$$\zeta(s) = \Sigma\lambda^{-s} \quad ,$$

usually introduced for the determination of the proper values of an operator; the index formula obtained in this way is actually too complicated. Nevertheless, the idea of approaching the Index problem to the study of the proper values of a differential operator proved itself to be an efficient and rewarding technique. We bring to our testimony the papers of MacKean and Singer [164], Patodi [194] and Gilkey [96]; these papers actually suggested a new proof of the Index Theorem to Atiyah, Bott and Patodi [25] , using the Heat Equation's method.

Recently the Atiyah-Singer Theorem crossed its pure mathematical boundaries to make its entrance in the physics of elementary particles, specially in problems related to Gauge

Theories. Such theories have excited the interest of theoretical physicists and owe their present success to the experimental proof of some of their predictions: the discovery of neutrius's diffusion reactions on matter (1973) and the detection of charmed particles (1974). We observe also that the right mathematical framework for Gauge Theories is precisely the geometry of principal bundles; for example, the result obtained by Atiyah, Hitchin and Ward about certain solutions (instantons) of the Yang-Mills equations (these are, roughly speaking, generalizations of the Maxwell equations of Electro-Magnetism) concerns SU(n) - bundles [31].(*) Finally, Atiyah and Ward were lead towards Algebraic Geometry problems, by interpreting the equations defining the instantons as equations in the 3 - dimensional complex projective space [38].

2. The Index of an Elliptic Differential Operator

Given $x = (x, x_2, \ldots, x_n) \in \mathbb{R}^n$ and $p = (p_1, p_2, \ldots, p_n) \in \mathbb{N}^n$, define

$$D^p = \left(\frac{1}{i}\frac{\partial}{\partial x_1}\right)^{p_1}\left(\frac{1}{i}\frac{\partial}{\partial x_2}\right)^{p_2}\cdots\left(\frac{1}{i}\frac{\partial}{\partial x_n}\right)^{p_n}, \quad |p| = \sum_{i=1}^{n} p_i$$

with $i^2 = -1$. On the other hand, for any open set $U \subseteq \mathbb{R}^n$ and any finite dimensional complex vector space V, we denote by $C^\infty(U,V)$ the space of all C^∞ - differentiable functions from U into V. Then, if A and B are finite dimensional vector spaces over $\mathbb{C}$, a linear function

$$D : C^\infty(U,A) \longrightarrow C^\infty(U,B)$$

is a *differential operator of order* r if there exist functions $g_p \in C^\infty(U, \mathrm{Hom}(A,B))$ such that

$$\text{(2.1)} \qquad Df = \sum_{|p| \leq r} g_p(x) D^p f .$$

To such an operator we associate a function

$$\sigma_r(D) : U \times \mathbb{R}^n \longrightarrow \mathrm{Hom}(A,B)$$

defined by

(*) It is worth noticing that this result was obtained using the Index formula.

$$(2.2) \qquad \sigma_r(D)(x,y) = \sum_{|p|=r} g_p(x)y^p ,$$

where $y = (y_1,\dots,y_n) \in \mathbb{R}^n$ and $y^p = y_1^{p_1} y_2^{p_2} \cdots y_n^{p_n}$;

$\sigma_r(D)$ is called the *symbol* of D (it depends on r). If for every $x \in U$ and every $y \in \mathbb{R}^n - \{0\}$, $\sigma_r(D)(x,y)$ is an isomorphism (thus, we are admitting that dim A = dim B), we say that D is an *elliptic* differential operator of order r . As an example we quote the Laplacian

$$\Delta = \sum_{i=1}^{n} \frac{\partial^2}{\partial x_i^2} ;$$

indeed, $\sigma_2(\Delta)(x,y) = -\left(y_1^2 + y_2^2 + \cdots + y_n^2\right) \neq 0$ if $y \neq 0$. We finally observe that any operator D of order r is also of order $r + 1, r + 2, \dots,$ and $\sigma_{r+i}(D) = 0$ for every $i \geq 1$; furthermore, σ_r is additive for operators of order $\leq r$.

We now transpose all the previous notions to differentiable manifolds; we shall assume, from this point to the end of the chapter that *all manifolds considered are finite dimensional,* C^∞ *- differentiable (smooth), compact and provided with a riemannian metric.* Let ξ and η be two C^∞ - differentiable vector bundles over an n - dimensional manifold M . We denote by $\Gamma(\xi)$ and $\Gamma(\eta)$ the spaces of cross-sections of ξ and η , respectively. Then, a linear function

$$D : \Gamma(\xi) \longrightarrow \Gamma(\eta)$$

is said to be a *linear differential operator of order* r if it is locally of type (2.1) ; in other words, we shall assume that:

(i) there is an open covering $\{U_j\}_{j\in J}$ of M so that each U_j is a local chart of M (and so, it is C^∞ - diffeomorphic to an open subset of $\mathbb{R}^n$) and moreover, both ξ and η are trivial over U_j ;

(ii) for every $j \in J$, the restriction D_j of D to U_j is a differential operator of order r (in the sense of (2.1)),

$$D_j : C^\infty(U_j,A) \longrightarrow C^\infty(U_j,B)$$

where A and B are typical fibres of ξ and η, respectively. We also want to define the symbol $\sigma_r(D)$ so to retrieve locally the description given by (2.2). We assume that $\sigma_r(D_j)$ is given (over the local chart U_j) by $\sigma_r(D_j)(x,y) = \sum_{|p|=r} g_p(x)y^p$, where y is a cotangent vector of M with origin $x \in U_j$. This approach permits us to give an actual intrinsic definition of the symbol (*): let $\tau^*(M)$ be the cotangent bundle of M; then, the *symbol* of D is a morphism of vector bundles

$$(2.3) \qquad \tilde{\sigma}_r(D) : \tau^*(M) \longrightarrow \mathrm{Hom}(\xi,\eta)$$

defined by

$$(2.4) \qquad \tilde{\sigma}_r(D)(x,y)(z) = \frac{i^r}{r!} D(f^r s)(x)$$

for every $x \in M$, $y \in T_x^*(M)$, $z \in \xi_x$ and where f is a C^∞-map such that $f(x) = 0$, $df(x) = y$ and s is a cross-section of $\xi|U$ (U is a local chart containing x) such that $s(x) = z$. Of course, we must show that (2.4) is independent of the choice of f and s; to this end, it suffices to express the righthand side of (2.4) in local coordinates. In fact,

$$D(f^r s)(x) = \sum_{|p|\leq r} g_p(x) D^p(f^r s)(x) = \sum_{|p|\leq r} g_p(x)(D^p f^r)(x) z$$

$$= \frac{r!}{i^r} \sum_{|p|=r} g_p(x) y^p z .$$

Here again, we say that D is an *elliptic operator of order* r if, and only if, $\tilde{\sigma}_r(D)(x,y)$ is an isomorphism for every $x \in M$ and every non-zero cotangent vector $y \in T_x^*(M)$; in other words, the symbol must be an isomorphism on the complement of the trivial cross-section of $\tau^*(M)$ for D to be elliptic. This suggests an alternative definition of ellipticity. Let $(D(M),\pi,M)$ be the disc-bundle associated to $\tau^*(M)$; then, the existence of the symbol (2.3) is equivalent to the exist-

(*) For more details over the motivation conducting to the definition of the symbol, please consult the excellent lecture notes of L.Schwartz [206]; for other historical notes, see [209].

ence of a fibre-preserving homomorphism

(2.5) $$\sigma_r(D) : \pi^*\xi \longrightarrow \pi^*\eta$$

given by $\sigma_r(D)(\omega,z) = (\omega,\tilde{\sigma}_r(x,\omega)z)$ for every $\omega \in D(M)$ and $z \in \xi_x$, where $x = \pi(\omega)$ (after conveniently identifying the elements of $D(M)_x$ with the vectors of $T^*_x(M)$ of length ≤ 1). Then, if $S(M)$ is the total space of the sphere bundle associated to $\tau^*(M)$ we have the following.

Definition 2.6. *A linear differential operator* D *of order* r *is elliptic if, and only if, the restriction of* $\sigma_r(D)$ *to* $\pi^*\xi|S(M)$ *is an isomorphism.*

The latter definition of ellipticity has the advantage of setting that notion in a convenient Algebraic Topology framework; indeed, one can see that the correct procedure is to interpret it with the aid of the group $KU(D(M),S(M))$.

We recall that if ξ is a C^∞ - differentiable complex vector bundle over a manifold M (compact, riemannian, according to our convention), then ξ is endowed with an hermitian metric $(\ ,\ )_\xi$. Hence, the space $\Gamma(\xi)$ of cross-sections has automatically the structure of a real pre-Hilbert space; the scalar product is defined by

$$(s|t)_\xi = \int_M (s,t)_\xi \, v_g$$

for every $s,t \in \Gamma(\xi)$, where v_g is the canonical measure associated to the riemannian metric g of M. Thus, for every differential operator $D : \Gamma(\xi) \rightarrow \Gamma(\eta)$, there exists a *formal adjoint* $D^* : \Gamma(\eta) \rightarrow \Gamma(\xi)$ defined by

(2.7) $$(Ds|t)_\eta = (s|D^*t)_\xi$$

for every $s \in \Gamma(\xi)$ and $t \in \Gamma(\eta)$. Since the hermitian structures of ξ and η can be transmitted to $\pi^*\xi$ and $\pi^*\eta$ respectively, it is possible to define the adjoint $\sigma^*_r(D) : \pi^*\eta \rightarrow \pi^*\xi$ of the symbol $\sigma_r(D)$. The crucial results concerning us at this moment are the following (for their proof, consult [193], pages 70-73):

1) the differential operator D has one and only one adjoint D^* ;

(2.8) 2) $\sigma_r(D^*) = (-1)^r \sigma_r^*(D)$.

From relations (2.7) and (2.8) we obtain the following two conclusions

(I) ker D and im D^* are orthogonal; more precisely, if s is orthogonal to im D^* , from $(s|D^*Ds)_\xi = 0$ we deduce that $(Ds|Ds)_\eta = 0$ and hence, $Ds = 0$, showing thereby that ker D is the complement of im D^* in $\Gamma(\xi)$;

(II) if D is elliptic, its adjoint D^* is also such. On the other hand, as shown by Gelfand [95], ker D *is a finite dimensional complex vector space whenever* D *is elliptic*. This fundamental result plus (I) and (II) justify the notion of index.

Definition 2.9. *The analytic index of an elliptic differential operator* D *is the integer*

$$\begin{aligned} i_a(D) &= \dim(\ker D) - \dim(\ker D^*) \\ &= \dim(\ker D) - \dim(\operatorname{coker} D) . \end{aligned}$$

At that point, mathematicians started to contend with the verification of Gelfand's conjecture (see [95]) namely, that $i_a(D)$ could be expressed uniquely in terms of topological invariants. In the remainder of this section we shall express the analytic index of an elliptic operator in different ways; thus, we hope to expose some facets of the deepness and richness of the integer $i_a(D)$. To begin with, since the ellipticity of D implies that of D^* , it is clear that the operators $\Delta_\xi = D^*D$ and $\Delta_\eta = DD^*$ are elliptic; moreover, they are positive-definite and self-adjoint. Then, because $\ker \Delta_\xi = \ker D$ and $\ker \Delta_\eta = \ker D^*$,

(2.10) $i_a(D) = \dim(\ker \Delta_\xi) - \dim(\ker \Delta_\eta)$.

This is a convenient formula to have since it allows the use of some fundamental results concerning the eigenvalues of operators like Δ_ξ . For every real number λ , consider the set

$$\Gamma_\lambda(\xi) = \{s \in \Gamma(\xi) \,|\, \Delta_\xi(s) = \lambda s\} \quad ;$$

then, Hodge's Theory [118] implies the following.

Theorem 2.11. - (i) *The space* $\Gamma_\lambda(\xi)$ *is finite dimensional, for every* $\lambda \in \mathbb{R}$ *; furthermore,* $\Gamma_\lambda(\xi) = 0$ *for all but a countable number of* λ's ;

(ii) *the sum of the spaces* $\Gamma_\lambda(\xi)$ *is an orthogonal direct sum; moreover, the completion of* $\bigoplus_\lambda \Gamma_\lambda(\xi)$ *is the Hilbert space* $L^2(\Gamma(\xi))$ *(here* $L^2(\Gamma(\xi))$ *is just the completion of* $\Gamma(\xi)$ *with respect to the metric* $(\ |\ )_\xi$) ;

(iii) *for every* $\lambda \neq 0$, *the restriction of* D *to* $\Gamma_\lambda(\xi)$ *is an isomorphism of* $\Gamma_\lambda(\xi)$ *onto* $\Gamma_\lambda(\eta)$.

The cornerstone in the proof of Theorem 2.11 is the construction of a *parametrix*, that is to say a differential operator $P : \Gamma(\eta) \to \Gamma(\xi)$ such that $PD - 1_{\Gamma(\xi)}$ and $DP - 1_{\Gamma(\eta)}$ are integral operators with C^∞ - differentiable kernels. Since the operators of that kind are compact, one can use Riesz' classical theory on compact operators over Banach spaces and the conclusions stated in (2.11) follow from results on Fredholm operators. Details can be found in the standard references [63] and [193] .

The most concrete situation where to interpret the previous Theorem is obtained assuming that $M = S^1 = \{z = e^{i\Theta}\}$, $\xi = \eta$ = trivial line bundle and $D = \frac{1}{i}\frac{d}{d\Theta}$. Then, $\Delta = D \cdot D$ is elliptic and self-adjoint; its eigenvalues are the positive integers $\lambda_m = m^2$, so that $\left\{e^{im\Theta}\right\}_{m\in\mathbb{Z}}$ is a complete orthogonal system of $L^2(S^1)$. Hence, for every function φ of $L^2(S^1)$, we can write

$$\varphi(\Theta) = \sum_m a_m e^{im\Theta}$$

where the coefficients a_m are given by the classical Theory of Fourier Series.

Theorem 2.11 shows that $i_a(D) = \dim \Gamma_o(\xi) - \dim \Gamma_o(\eta)$ can be written in the form

$$(2.12) \qquad i_a(D) = \sum_\lambda h(\lambda) \dim \Gamma_\lambda(\xi) - \sum_\lambda h(\lambda) \dim \Gamma_\lambda(\eta)$$

for every function $h : \mathbb{R} \to \mathbb{R}$ such that $h(0) = 1$, provided that the series introduced converge. The use of the Heat Equation for the study of the index of an elliptic operator can be justified by (2.12). In fact, whenever discussing the spectrum of an operator defined on a riemannian manifold (*) it is useful to consider the function

$$h_\xi(t) = \sum_\lambda e^{-\lambda t} \dim \Gamma_\lambda(\xi) \tag{2.13}$$

(well-defined for $t > 0$). If the operator considered is Δ_ξ (or Δ_η) there is an assymptotic development for $t \to 0$ given by

$$h_\xi(t) \sim \sum_{k \geq -n} \mu_k(\Delta_\xi) t^{k/2r} , \tag{2.14}$$

where $n = \dim M$ and the coefficients $\mu_k(\Delta_\xi)$ can be constructed locally from the coefficients of Δ_ξ. This important result, first proved for the Laplacian by Minakshisundaram and Pleijel [183], was shown in the general case by Seeley [208]. With its aid, the index can be expressed by the formula

$$i_a(D) = \mu_0(\Delta_\xi) - \mu_0(\Delta_\eta) . \tag{2.15}$$

More precisely, let $\{\varphi_m\}$ be a complete orthogonal set of eigenfunctions associated to the eigenvalues $\{\lambda_m\}$ of Δ (here Δ means either Δ_ξ or Δ_η). For every $t > 0$, let H_t be the differential operator

$$H_t = e^{-\Delta t} ;$$

notice that, on the one hand, H_t satisfies the Heat Equation

$$\left(\frac{d}{dt} + \Delta\right) H_t = 0 ,$$

and on the other hand, the eigenvalues of H_t are given by $e^{-\lambda_m t}$. Then, writing an arbitrary $\varphi \in L^2(\Gamma(\xi))$ in the form $\varphi = \Sigma a_m \varphi_m$ (Theorem 2.11) we have:

$$(H_t\varphi)(x) = \sum_m a_m (H_t\varphi_m)(x) = \sum_m a_m e^{-\lambda_m t} \varphi_m(x) ;$$

because

$$a_m = \int_M (\varphi, \varphi_m)_y \, v_g(y) ,$$

it follows that

(*) The book of Berger et al. [42] is a good reference about this subject.

$$H_t\varphi)(x) = \sum_m e^{-\lambda_m t} \int_M (\varphi,\varphi_m)_y \, v_g(y)\varphi_m(x)$$

$$= \int_M \left(\sum_m e^{-\lambda_m t} \varphi_m(x) \otimes \overline{\varphi_m(y)}\right) \varphi(y) v_g(y)$$

or, by setting

$$h(t;x,y) = \sum_m e^{-\lambda_m t} \varphi_m(x) \otimes \overline{\varphi_m(y)} \quad ,$$

$$(H_t\varphi)(x) = \int_M h(t;x,y)\varphi(y) v_g(y) \quad .$$

In other words, $h(t;x,y)$ is the kernel of H_t.(*) Considering the vector bundle $\xi \otimes \xi^* \cong \mathrm{Hom}(\xi,\xi)$ over $M \times M$ (the hermitian product on ξ induces an isomorphism between its conjugate $\overline{\xi}$ and its dual ξ^*), the series $h(t;x,y)$ - which converge for $t > 0$ - can be viewed as elements of the fibre $(\xi \otimes \xi^*)_{(x,y)}$. Hence, the trace of $h(t,x) = h(t;x,x)$ is given by

$$\mathrm{tr}(h(t,x)) = \sum_m e^{-\lambda_m t} (\varphi_m,\varphi_m)_x$$

and so

$$\int_M \mathrm{tr}(h(t,x)) v_g(x) = \sum_m e^{-\lambda_m t} \int_M (\varphi_m,\varphi_m)_x v_g(x) \tag{2.16}$$

$$= \sum_m e^{-\lambda_m t} = h(t) \quad .$$

(*) The reader can easily verified that if $M = S^1$, $\xi = \eta =$ trivial line bundle and $\Delta = -\dfrac{d^2}{d\Theta^2}$, the kernel of H_t is the C^∞ - function

$$h(t;\Theta,\Theta') = \sum_m e^{-m^2 t} e^{im(\Theta-\Theta')} \quad ;$$

this function converges and $h(t,\Theta,\Theta) = \sum_m e^{-m^2 t} \sim \dfrac{1}{2\sqrt{\pi t}}$ when $t \to 0$.

Since $i_a(D) = h_\xi(t) - h_\eta(t)$ (this equality is a consequence of (2.12) and (2.13)) we obtain the following integral expression for the index:

$$(2.17) \qquad i_a(D) = \int_M \Big(\mathrm{tr}(h_\xi(t,x)) - \mathrm{tr}(h_\eta(t,x))\Big) v_g(x) \ .$$

When $t \to 0$, $h(t,x)$ has an assymptotic development

$$(2.18) \qquad h(t,x) \sim \sum_{k \geq -n} \mu_k(x,\Delta) t^{k/2r} \ ;$$

then, the format of formula (2.17) can be approximated to that of (2.15). Utilizing (2.14) and (2.16), it follows that

$$\mu_k(\Delta) = \int_M \mathrm{tr}(\mu_k(x,\Delta)) v_g(x)$$

and hence, (2.17) can also be written in the form

$$(2.19) \qquad i_a(D) = \int_M \Big(\mathrm{tr}(\mu_o(x,\Delta_\xi)) - \mathrm{tr}(\mu_o(x,\Delta_\eta))\Big) v_g(x) \ .$$

The trouble with this approach to the analytic index via the Heat Equation (as initially tried by Atiyah and Bott) was the difficulty in obtaining a good description of the coefficients $\mu_k(x,\Delta)$ of (2.18). Meanwhile, Patodi's decisive contribution [194] gave an inkling of the simplified formula (2.19). For a characterization of the $\mu_k(x,\Delta)$'s , the reader may consult Gilkey's papers [96] and [97] , and of course, the work of Atiyah, Bott and Patodi [25] .

Next, we describe certain arithmetic considerations which yield information about the coefficients $\mu_k(\Delta_\xi)$. The central idea is to replace $h_\xi(t) = \sum_\lambda e^{-\lambda t} \dim \Gamma_\lambda(\xi)$ by a kind of Riemann's ζ - function, namely

$$(2.20) \qquad \zeta_\xi(s) = \sum_\lambda \lambda^{-s} \dim \Gamma_\lambda(\xi) \ ,$$

with s complex. On the other hand, it is known that the gamma function is given by

$$\Gamma(s) = \int_0^{+\infty} u^{s-1} e^{-u} du$$

with $\mathrm{Re}(s) > 0$. Writing $u = \lambda t$ we obtain

$$\Gamma(s) = \int_0^{+\infty} \lambda^s t^{s-1} e^{-\lambda t} dt$$

and therefore,

$$\lambda^{-s} = \frac{1}{\Gamma(s)} \int_0^{+\infty} t^{s-1} e^{-\lambda t} dt \quad .$$

Because of (2.20) ,

$$\zeta_\xi(s)\Gamma(s) = \int_0^{+\infty} t^{s-1} h_\xi(t) dt \quad .$$

Now we can show that the points $s_k = -k/2r,\ k \geq -n$ are the simple poles of $\zeta_\xi(s)\Gamma(s)$; indeed, at these points the residues are precisely $\mu_k(\Delta_\xi)$. In particular,

$$\zeta_\xi(0) = \mu_o(\Delta_\xi) - 1$$

and we obtain the equation

$$\text{(2.21)} \qquad i_a(D) = \zeta_\xi(0) - \zeta_\eta(0) \ .$$

The previous considerations are intrinsecally connected to the idea of assymptotic development; remarks of this kind were already made in the work of Minakshisundaram and Pleijel. The reader who desires to find out how extensive are the connections between Number Theory and the Index Theorem should consult the excellent book of Hirzebruch and Zagier [117].

To conclude this section we shall indicate how the definition of the index of an elliptic differential operator can be generalized to a family of differential operators. A differentiable complex $(\Gamma(\xi),d)$ over a manifold M is a finite sequence $\{\xi_o,\xi_1,\ldots,\xi_n\}$ of smooth complex vector bundles over M and linear differential operators $d_i : \Gamma(\xi_i) \to \Gamma(\xi_{i+1})$ such that $d_{i+1} \cdot d_i = 0$ for every $i \in \{0,1,\ldots,n-1\}$. (Notice that we do not exclude the possibility of the d_i's having different orders r_i).

Definition 2.22. *An elliptic complex over* M *is a differentiable complex* $(\Gamma(\xi),d)$ *such that the sequence*

$$0 \longrightarrow \xi_o \xrightarrow{\sigma(d_o)|S(M)} \pi^*\xi_1 \xrightarrow{\sigma(d_1)|S(M)} \cdots$$

$$\cdots \xrightarrow{\sigma(d_{n-1})|S(M)} \pi^*\xi_n \longrightarrow 0$$

is exact.

Given an elliptic complex $(\Gamma(\xi),d)$ over M, it is natural to study its cohomology, namely, the spaces $H^i(\Gamma(\xi)) = \ker d_i/\mathrm{im}\, d_{i-1}$. These spaces happen to be finite dimensional; this suggests us to view the index of such a complex as its Euler-Poincaré characteristic.

Definition 2.23. *The analytic index of an elliptic complex* $(\Gamma(\xi),d)$ *over a manifold* M *is the integer*

$$i_a(\xi,d) = \Sigma(-1)^i \dim H^i(\Gamma(\xi)) .$$

In order to show that the spaces $H^i(\Gamma(\xi))$ are finite dimensional we proceed in two steps (see [24, Part I, § 6]). First, we construct for each $i \in \{1,\ldots,n-1\}$ a continuous linear operator (parametrix)

$$P_i : \Gamma(\zeta_{i+1}) \longrightarrow \Gamma(\xi_i)$$

such that

$$P_i d_i + d_{i-1} P_{i-1} = 1_{\Gamma(\xi_i)} - S_i$$

where S_i is a C^∞-operator on $\Gamma(\xi_i)$; next, we use the finiteness properties of the Fredholm operators $1_{\Gamma(\xi_i)} - S_i$.

3. Four Standard Complexes

(I) Let M be a compact, oriented, riemannian manifold of dimension n and let $\tau^*(M)$ be its cotangent bundle. Consider the complex vector bundle

$$\xi = \bigoplus_{p=o}^{n} \xi^p = \bigoplus_{p=o}^{n} \Lambda^p(\tau^*(M)) \otimes \epsilon$$

where ϵ is the trivial complex line bundle over M. The elements of the set $\Gamma(\xi)$ of all cross-sections of ξ will be called (complex) *exterior differential forms over* M ; in particular, a cross-section of the fibration ξ^p will be an *ex-*

terior differential p - *form* or just a p - *form*, for short. We denote the set of all p - forms over M by $\Omega^p(M)$. with the aid of the differential operators

$$d : \Omega^p(M) \longrightarrow \Omega^{p+1}(M)$$

we form the De Rham Complex $(\Omega(M), d)$:

$$0 \longrightarrow \Omega^0(M) \xrightarrow{d} \Omega^1(M) \xrightarrow{d} \cdots \xrightarrow{d} \Omega^n(M) \longrightarrow 0 ;$$

because of De Rham's Theorems, its cohomology groups are naturally isomorphic to the ordinary cohomology groups of M with complex coefficients.

As we have seen in the previous section $\Gamma(\xi)$ is automatically endowed with an inner product $(\ |\ )_\xi$; furthermore, let us recall that there is a well-known automorphism $*$ of $\Omega(M)$ - the *Hodge star operator* - characterized by the following properties:

(3.0) for every integer p , $0 \leq p \leq n$,

$$* : \Omega^p(M) \longrightarrow \Omega^{n-p}(M) ;$$

(3.1) for every $\alpha, \beta \in \Omega^p(M)$, $\alpha \wedge *\beta = (\alpha|\beta)_\xi {}^* 1$, $1 \in \Omega^0(M)$;

(3.2) for every $\alpha \in \Omega^p(M)$, $**\alpha = (-1)^{p(n-p)}\alpha$.

Notice that (3.1) implies $(\alpha|\beta)_\xi = \int_M \alpha \wedge *\beta$. We use Hodge's star operator to define the exterior codifferentiation

$$\delta : \Omega^p(M) \longrightarrow \Omega^{p-1}(M) :$$

$$\delta = (-1)^{np+n+1} *d* .$$

One should observe that given any $\alpha \in \Omega^{p-1}(M)$ and $\beta \in \Omega^p(M)$, since

$$d(\alpha \wedge *\beta) = d\alpha \wedge *\beta + (-1)^{p-1}\alpha \wedge d*\beta$$

and $\int_M d(\alpha \wedge *\beta) = 0$, it follows that $(d\alpha|\beta)_\xi = (\alpha|\delta\beta)_\xi$.

Hence, δ is just the adjoint of d . An element $s \in \Gamma(\xi)$ is *closed* (resp. *coclosed*) if $ds = 0$ (resp. $\delta s = 0$).

We now define the *Hodge-Laplace operator*

$$\Delta = (d+\delta)^2 = d\delta + \delta d$$

and claim it to be elliptic. In order to prove this assertion it suffices to study the 1 - symbols of d and δ. To this end, let $x \in M$, $y \in T_x^*M$ and s like in (2.4); if f is a smooth function such that $f(x) = 0$ and $df(x) = y$,

$$\begin{aligned}\tilde{\sigma}_1(d)(x,y)s(x) &= i(df \wedge s + f \wedge ds)(x) \\ &= i(y \wedge s(x)).\end{aligned}$$

Thus, the symbol of d is just left exterior multiplication by iy and therefore, because of (2.8), the symbol of δ is equal to the internal product ω_y over ξ_y, up to a factor i. (Recall that $\omega_y : \xi_x \quad \xi_x$ is defined from

$$\omega_y(e_1\wedge \cdots \wedge e_p) = \sum_{k=1}^{p} (-1)^{k+1} (y,e_k)_\xi \, e_1\wedge \cdots \wedge \hat{e}_k \wedge \cdots \wedge e_p$$

- where $\{e_1,\ldots,e_n\}$ is an orthonormal basis for T_x^*M - and moreover, for every $a,b \in \xi_x$, $(y \wedge a,b)_\xi = (a,\omega_y(b))_\xi$). We conclude that

$$\begin{aligned}\tilde{\sigma}_2(\Delta)(x,y) &= (\tilde{\sigma}_1(d)\tilde{\sigma}_1(\delta) + \tilde{\sigma}_1(\delta)\tilde{\sigma}_1(d))(x,y) \\ &= \tilde{\sigma}_1(d)(-i\omega_y) + \tilde{\sigma}_1(\delta)(iy \wedge -) \\ &= y \wedge \omega_y + \omega_y y \wedge - \\ &= \|y\|^2 .\end{aligned}$$

In other words, $\tilde{\sigma}_2(\Delta)(x,y)$ is scalar multiplication by $\|y\|^2$ in ξ_x and so, Δ is elliptic. The self-adjointness of Δ implies that its analytic index is 0; this and other considerations severely reduce its usefulness. However, from the ellipticity of Δ, the decomposition $\Delta = (d+\delta)(d+\delta)$ and (2.8), we deduce that $D = d + \delta$ *is elliptic*. Let us now assume that, besides our usual requirements on M, $\partial M = \emptyset$. Moreover, let

$$b_p = \dim H^p(M;\mathbb{C}) , \qquad 0 \leq p \leq n ,$$

be the Betti numbers of the manifold M and set

$$\xi^{\text{even}} = \bigoplus_{p \geq 0} \xi^{2p} \quad , \quad \xi^{\text{odd}} = \bigoplus_{p \geq 0} \xi^{2p+1} \ .$$

Proposition 3.3. *The differential operator*

$$D = d + \delta : \Gamma(\xi^{\text{even}}) \longrightarrow \Gamma(\xi^{\text{odd}})$$

is elliptic and

$$i_a(D) = \chi(M) = \sum_{p=o}^{n} (-1)^p b_p \ .$$

Proof. A differential form $s \in \Gamma(\xi)$ such that $\Delta s = 0$ is said to be *harmonic*; in that case, because

$$(\Delta s|s)_\xi = (d\delta s|s)_\xi + (\delta ds|s)_\xi = (\delta s|\delta s)_\xi + (ds|ds)_\xi \ ,$$

$ds = 0$ and $\delta s = 0$. It is clear from the definition of Δ that if $s \in \Gamma(\xi)$ is such that $ds = \delta s = 0$ then s is harmonic; thus, in other words, a differential form over M is harmonic if, and only if, it is closed and coclosed. Notice that every harmonic form over M defines a cohomology class in $H^*(M;\mathbb{C})$; but, as one learns in Hodge Theory, each cohomology class of M contains a unique harmonic representative (see [98]). Hence,

$$\ker D = \left\{ s \in \Gamma(\xi^{\text{even}}) \,|\, ds + \delta s = 0 \right\} = \ker(\Delta|\Gamma(\xi^{\text{even}}))$$

$$\cong \bigoplus_{p \geq 0} H^{2p}(M;\mathbb{C})$$

and

$$\ker D^* = \ker(\Delta|\Gamma(\xi^{\text{odd}})) \cong \bigoplus_{p \geq 0} H^{2p+1}(M;\mathbb{C}) \ .$$

Definition (2.9) then shows

$$i_a(D) = \sum_{p=o}^{[n/2]} b_{2p} - \sum_{p=o}^{[n/2]} b_{2p+1} = \chi(M) \ .$$

From the previous Proposition and the Poincaré Duality Theorem we conclude that if M (oriented) is odd-dimensional, $i_a(D) = 0$.

Remark. If $n \equiv 1 \pmod 4$ one can use Hodge Theory to prove that the differential operator K, defined in the algebra of even differential forms from the formula

$$K\alpha = (-1)^p d^*\alpha + {}^*d\alpha$$

for every $\alpha \in \Omega^{2p}(M)$, $0 \leq p \leq [n/2]$, is elliptic and furthermore, $i_a(K)$ (mod 2) is the (real) Kervaire semi-characteristic

$$k(M) = \sum_{p=0}^{[n/2]} \dim H^{2p}(M;\mathbb{R}) \quad (\text{mod } 2) \quad .$$

Finally, Proposition 3.3 and Gauss-Bonnet Theorem show that the total curvature of a closed riemannian manifold is $i_a(D)$. In fact, the Chern-Gauss-Bonnet Formula can be obtained through a direct treatment (for more details, please consult [97, Chapter 4]).

(II) For our second example, M will be a *closed, oriented, riemannian manifold of even dimension* $n = 2m$. We use the same notation as in the previous example.

Let τ be the endomorphism of $\xi = \bigoplus_{p=0}^{n} \xi^p$, defined on ξ^p by the formula

(3.4) $$\tau = i^{p(p-1)+m} * \qquad (i^2 = -1) \quad .$$

Because of (3.2) and the definitions of d and δ,

(3.5) $$\tau^2 = 1_\xi \; , \; (d + \delta)\tau = -\tau(d + \delta) \quad .$$

Since τ is an involution, we can decompose ξ as $\xi^+ \oplus \xi^-$ so that

$$\tau|\xi^+ = 1_{\xi^+} \quad \text{and} \quad \tau|\xi^- = -1_{\xi^-} \quad .$$

Thus, with the usual notational abuses, $\Gamma(\xi)$ is decomposed in two proper subspaces Γ^+ and Γ^-, associated respectively to the eigenvalues $+1$ and -1 of τ, so that $\Gamma^\pm = \Gamma(\xi^\pm)$. Moreover, the second equality of (3.5) shows that $D = d + \delta$ sends Γ^+ (resp. Γ^-) onto Γ^- (resp. Γ^+). Writing $D^+ = D|\Gamma^+$ and $D^- = D|\Gamma^-$, the operators

$$D^+ : \Gamma^+ \longrightarrow \Gamma^-$$
$$D^- : \Gamma^- \longrightarrow \Gamma^+$$

are adjoint to each other (because D is self-adjoint) *and elliptic* (this follows from the ellipticity of D and the functoriality of the symbol). Let us study the analytic index of D^+, namely,

$$i_a(D^+) = \dim(\ker D^+) - \dim(\ker D^-) .$$

We begin by observing that $\Delta\tau = \tau\Delta$; hence, identifying $\ker \Delta$ (space of the harmonic forms of $\Gamma(\xi)$) with $H^*(M;\mathbb{C})$, the involutivity of τ implies the decomposition

$$H^*(M;\mathbb{C}) = H^*_+(M;\mathbb{C}) \oplus H^*_-(M;\mathbb{C})$$

where $H^*_\pm(M;\mathbb{C})$ are the eigenspaces associated to the eigenvalues ± 1 of τ, respectively. Hence,

$$(3.6) \qquad i_a(D^+) = \dim H^*_+(M;\mathbb{C}) - \dim H^*_-(M;\mathbb{C}) ,$$

because the solutions of $D^+s = 0$ (resp. $D^-s = 0$) are precisely the harmonic forms of Γ^+ (resp. Γ^-). The subspaces $E_p = H^p(M;\mathbb{C}) \oplus H^{2m-p}(M;\mathbb{C})$ $0 \leq p < m$ and $H^m(M;\mathbb{C})$ of $H^*(M;\mathbb{C})$ are stable under the action of τ on $H^*(M;\mathbb{C})$; actually, Definition 3.4 shows that τ exchanges the two factors of E_p; hence, the subspace of E_p associated to the eigenvalue $+1$ of τ equals the one associated to the eigenvalue -1. Consequently, (3.6) becomes

$$i_a(D^+) = \dim H^m_+(M;\mathbb{C}) - \dim H^m_-(M;\mathbb{C}) .$$

If m *is odd,* (3.4) shows that $\tau = i^*$. As * is real and $^{**} = -1$, $\tau(\bar{s}) = -\overline{\tau(s)}$ for every $s \in H^m(M;\mathbb{C})$; hence τ is an isomorphism of $H^m_+(M;\mathbb{C})$ onto $H^m_-(M;\mathbb{C})$ and therefore

$$i_a(D^+) = 0 .$$

If m *is even,* $\tau = {}^*$ is real and therefore, we can write

$$i_a(D^+) = \dim H^m_+(M;\mathbb{R}) - \dim H^m_-(M;\mathbb{R}) .$$

But, for every non-trivial $s \in H^m_+(M;\mathbb{R})$ (resp. $t \in H^m_-(M;\mathbb{R})$)

$$\int_M s \wedge s = \int_M s \wedge {}^*s = (s|s)_\xi > 0$$

$$\int_M t \wedge t = -\int_M t \wedge {}^*t = -(t|t)_\xi < 0 \quad ;$$

this means that $i_a(D^+)$ *is the signature of the quadratic form* $s \longmapsto \int_M s \wedge s$ *defined by cup product on* M. This topological invariant $\mathrm{Sign}(M)$ is called the *Hirzebruch Signature* of M. If we make the convention that $\mathrm{Sign}(M) = 0$ whenever $\dim M = 2m$ with m odd, we can state the following.

<u>Proposition 3.7.</u> *The analytic index of the operator* D^+ *is the Hirzebruch Signature of* M.

This result explains why the elliptic operator D^+ is usually called the *Signature Operator* of M. We suggest that the reader searches in the "ad hoc" literature to evaluate the fundamental rôle this operator plays in the general theory of the Index.

(III) Our third example is supplied by Complex Analytic Geometry. In this case, M will be a (compact) complex manifold of complex dimension m. We begin by noticing that an exterior differential form over M can be written as

$$a\,dz^{k_1} \wedge \ldots \wedge dz^{k_p} \wedge d\bar{z}^{\ell_1} \wedge \ldots \wedge d\bar{z}^{\ell_q}$$

(form of type (p,q)) where $z^1,\ldots,z^m$ and $\bar{z}^1,\ldots,\bar{z}^m$ are holomorphic local coordinates of M. Because of this, the complex vector bundle $\xi = \Lambda(\tau^*(M)) \otimes \epsilon$ decomposes into the direct sum

$$\xi = \bigoplus_{r=0}^{m} \xi^r \quad , \quad \xi^r = \bigoplus_{p+q=r} \xi^{p,q}$$

and $\Gamma(\xi^{p,q})$ is the set of all forms of type (p,q). Hence, the exterior differentiation d defined on $\Gamma(\xi) = \bigoplus_{p,q} \Gamma(\xi^{p,q})$ decomposes into two operators ∂ and $\bar{\partial}$; more precisely, ∂ and $\bar{\partial}$ are defined by linearity from the formulas

$$\partial = \sum_k dz^k \wedge \frac{\partial}{\partial z^k} \quad , \quad \bar{\partial} = \sum_k d\bar{z}^k \wedge \frac{\partial}{\partial \bar{z}^k}$$

so that $\partial^{p,q}$ (resp. $\bar{\partial}^{p,q}$) takes $\Gamma(\xi^{p,q})$ onto $\Gamma(\xi^{p+1,q})$ (resp. $\Gamma(\xi^{p,q+1})$). In this situation, the complex analogue to De Rham's is the *Dolbeault's Complex*

$$0 \longrightarrow \Gamma(\xi^{0,0}) \xrightarrow{\bar{\partial}} \Gamma(\xi^{0,1}) \xrightarrow{\bar{\partial}} \quad \xrightarrow{\bar{\partial}} \Gamma(\xi^{0,m}) \longrightarrow 0$$

(it is easy to see that $\bar{\partial}\,\bar{\partial} = 0$) . We now call the riemannian metric of M into the play. More precisely, the underlying 2m-dimensional real differentiable structure of M has a riemannian metric such that the canonical involution $J : \tau(M) \to \tau(M)$ defined by

$$J\left(\frac{\partial}{\partial x^k}\right) = \frac{\partial}{\partial y^k} \quad , \quad J\left(\frac{\partial}{\partial y^k}\right) = -\frac{\partial}{\partial x^k}$$

(for local coordinates $x^1, y^1, \ldots, x^m, y^m$ such that $z^k = x^k + iy^k$, $i^2 = -1$, $1 \leq k \leq m$) is an isometry. Then, we can talk about the adjoint $\bar{\partial}^*$ of $\bar{\partial}$ (we have the equality $\bar{\partial}^* = -{*}\bar{\partial}{*}$) and one easily verifies that

$$\square = \bar{\partial}\,\bar{\partial}^* + \bar{\partial}^*\bar{\partial}$$

is just the complex Laplacian over M . (*) It follows that $\bar{\partial}$ *is elliptic and so is Dolbeault's Complex.*

More generally, let η be an holomorphic vector bundle over M and let $\Omega(\eta)$ be the sheaf of germs of all holomorphic cross-sections of η . Suppose that η is endowed with an hermitian metric; then one can extend $\bar{\partial}$ to $\Gamma(\xi \otimes \eta)$ so that

$$0 \longrightarrow \Gamma(\xi^{0,0} \otimes \eta) \xrightarrow{\bar{\partial}} \Gamma(\xi^{0,1} \otimes \eta) \xrightarrow{\bar{\partial}} \cdots \xrightarrow{\bar{\partial}} \Gamma(\xi^{0,m} \otimes \eta) \to 0$$

becomes elliptic (for the details, the reader is referred to [116] , p. 187 and on). This complex is the so-called *Twisted Dolbeault Complex*. For two given integers p and q, $0 \leq p,q \leq m$, we denote by $\mathcal{H}^{p,q}$ the space of all complex harmonic forms of type (p,q) and with coefficient in η , that is to say

(*) In fact, if M is a Kähler manifold (cf. Chapter 3, § 1), $\square = 1/2\,\Delta$.

$$\mathcal{H}^{p,q} = \{s \in \Gamma(\xi^{p,q} \otimes \eta) \mid \square s = 0\} \quad .$$

The results of Dolbeault [73], Kodaira [158] and Serre [215] form the equivalent of Hodge's Theory in Complex Analytic Geometry. We should point out, in particular, that the complex vector spaces $\mathcal{H}^{p,q}$ are finite dimensional and that $\mathcal{H}^{o,q} \cong H^q(M;\Omega(\eta))$. Thus, if we write

$$\eta^{even} = \bigoplus_{q\geq o} \xi^{o,2q} \otimes \eta \ , \ \eta^{odd} = \bigoplus_{q\geq o} \xi^{o,2q+1} \otimes \eta \ ,$$

we obtain the following.

<u>Proposition 3.8.</u> *The differential operator*

$$\bar\partial + \bar\partial^* : \Gamma(\eta^{even}) \longrightarrow \Gamma(\eta^{odd})$$

is elliptic and

$$i_a(\bar\partial + \bar\partial^*) = \chi(M;\eta) = \sum_{q=o}^{m} (-1)^q \dim H^q(M;\Omega(\eta)) \ .$$

The characteristic $\chi(M;\eta) = \sum_{q=o}^{m} (-1)^q \dim \mathcal{H}^{o,q}$ is known usually as the *Riemann-Roch Characteristic* of M (if $\eta = \epsilon$, $\chi(M;\epsilon)$ is the *arithmetic genus* of M). Slight generalizations of the preceeding considerations are the departure point for extremely fruitful developments in the theory of kählerian manifolds, notably in questions connected to Algebraic Manifolds(*) and the Riemann-Roch Theorem, one of the most profound results of Algebraic Geometry.

(IV) In our last example we shall be concerned with the Dirac Operator. Since this operator can be defined only on Spin-manifolds, we shall give a brief account of those results on Clifford Algebras and Spin groups which are relevant to our work (for the details one can consult Chapters 11 and 13 of [134]

(*) We recall that a compact, complex manifold M is algebraic if and only if, M admits a kählerian metric whose fundamental class belongs to the image of the natural homomorphism $H^2(M;\mathbb{Z}) \to H^2(M;\mathbb{R})$ (cf. [159]).

or the basic paper of Atiyah, Bott and Shapiro [26]). Please notice that we have already made our acquaintance with Chifford Algebras, namely, during the proof of the Hurwitz-Radon-Eckmann Theorem (see Chapter 4).

Let $A = \mathbb{R}^n$ be a real vector space with the canonical base $\{e_1,\dots,e_n\}$ and let q be the quadratic form defined by

$$q(a) = - \sum_{k=1}^{n} a_k^2$$

for every $a = (a_1,\dots,a_n) \in A$. The *Clifford Algebra* C_n is the quotient of the tensor algebra $\otimes A$ by the ideal generated by all elements of the form $a \otimes a - q(a) \cdot 1$. Hence, C_n is an associative algebra with a unit element 1, generated by the elements $e_1,\dots,e_n$ with the relations

$$e_i^2 = -1,\ e_i e_j = -e_j e_i \quad \text{if} \quad i \neq j,\ i,j = 1,\dots,n \quad ,$$

(cf. Chapter 4). The multiplication on C_n is called *Clifford multiplication*. Notice that we can regard C_n as generated by the products

$$e_{k_1} e_{k_2} \dots e_{k_r} \ ,\ k_1 < k_2 < \dots < k_r,\ 1 \leq r \leq n$$

and thus, $\dim C_n = 2^n$; moreover, C_n and $\Lambda(A)$ are isomorphic as left $O(n)$-modules. Using this isomorphism we define a $\mathbb{Z}_2$-graduation on C_n, namely,

$$C_n = C_n^+ \oplus C_n^-$$

which corresponds to $\Lambda(A) = \Lambda^{\text{even}}(A) \oplus \Lambda^{\text{odd}}(A)$. Clearly A is naturally embedded in C_n^-. One of the readily visible facts about Clifford Algebras is that for $n = 2m$, the complexification $\tilde{C}_n = C_n \otimes_{\mathbb{R}} \mathbb{C}$ is a matrix algebra over the complex numbers; it follows that C_n is a matrix algebra over $\mathbb{R}$ or $\mathbb{H}$. On the other hand, if n is odd, $\tilde{C}_n$ is a direct sum of matrix algebras. Next, consider the automorphism $\alpha : C_n \to C_n$ defined from

$$\alpha\left(e_{k_1} e_{k_2} \dots e_{k_r}\right) = (-1)^r e_{k_r} \dots e_{k_2} e_{k_1}$$

and let Spin(n) be the set of all invertible elements $g \in C_n$ such that

$$(3.9) \quad \begin{cases} \text{(i)} & g \in C_n^+ , \ \alpha(g) \cdot g = 1 \ ; \\ \text{(ii)} & gag^{-1} \in A \ \text{ for every } \ a \in A \ . \end{cases}$$

This set, together with the multiplication induced from C_n, is a group. It is easy to verify that, for every element g of the group Spin(n), the function $\pi_g : A \to A$ defined by $\pi_g(a) = gag^{-1}$ is an orientation preserving isometry; hence, we can define a surjection

$$\pi : \mathrm{Spin}(n) \longrightarrow \mathrm{SO}(n)$$

by taking every $g \in \mathrm{Spin}(n)$ into π_g and moreover, $\ker \pi = \{-1,1\}$. This shows that we can regard Spin(n) as a double covering of SO(n). Indeed, if $n \geq 3$, the Lie group Spin(n) is the universal covering of SO(n), because $\pi_1(\mathrm{SO}(n)) = \mathbb{Z}_2$. Spin(n) has a complex representation of dimension 2^n, denoted usually by S and called the *Spin Representation*; its representation space (denoted also by S) is called the *Space of Spinors*. We wish to observe that Spin(n) acts on S and this action is induced by the Clifford multiplication

$$(3.10) \qquad \mu : A \otimes S \longrightarrow S \ ;$$

furthermore, since for every $g \in \mathrm{Spin}(n)$, $a \in A$ and $x \in S$,

$$g(ax) = (gag^{-1})(gx) = \pi_g(a)(gx) \ ,$$

the function μ is actually a Spin(n)-module homomorphism.

Now let M be a compact, connected, oriented, riemannian manifold of dimension n. We say that M is a *Spin-manifold* if the structural group SO(n) of $\tau(M)$ can be lifted to Spin(n). More generally,

Definition 3.11. *A Spin-structure on* M *is a fibre-preserving isomorphism between the oriented bundles* $\tau(M)$ *and* $P \times_{\mathrm{Spin}(n)} \mathbb{R}^n$, *where* P *is a principal* Spin(n)-*bundle over* M.

Each lifting of SO(n) to Spin(n) defines a Spin-structure on M and so, the Spin-structures of M are classified (up to isomorphism) by $H^1(M;\mathbb{Z}_2)$. If M is simply connected, M has essentially a unique Spin-structure. It is possible to show that M has a Spin structure if, and only if, its second Stiefel-Whitney class $w_2(M)$ is trivial [44]. Thus, $\mathbb{C}P^n$, n = odd, is an example of Spin-manifold.

From now to the end of the section we shall assume that M has a Spin-structure defined by a principal Spin-bundle P over M. Let ζ be the bundle with fibre S associated to P; its total space is given by $E = P \times_{\mathrm{Spin}(n)} S$. This bundle ζ is called the *Spinor Bundle* over M; the *Dirac Operator* $\not\partial$ over M is a differential operator of order 1 over $\Gamma(\zeta)$. More precisely, $\not\partial$ is defined as the composition

$$\begin{array}{ccc} \Gamma(\zeta) & \xrightarrow{\;\not\partial\;} & \Gamma(\zeta) \\ & {\scriptstyle\nabla}\searrow \quad \nearrow{\scriptstyle\mu} & \\ & \Gamma(\tau^*(M) \otimes \zeta) & \end{array}$$

where ∇ is the covariant derivative induced by the riemannian connection of M (Levi-Civita Connection) and μ^* is the morphism induced by the Clifford multiplication (3.10) (recall that we can identify $\tau(M)$ and $\tau^*(M)$ because M is riemannian). Thus, if $\{v_1,\dots,v_n\}$ is an orthonormal basis of $\tau(M)$, for every $s \in \Gamma(\zeta)$,

$$\not\partial(s) = (\mu^* \cdot \nabla)(s) = \mu^*\left(\sum_{k=1}^{n} v_k^* \otimes \nabla_{v_k} s\right)$$

$$= \sum_{k=1}^{n} \mu\left(v_k \otimes \nabla_{v_k} s\right)$$

where $\nabla_{v_k} s$ is the covariant derivative of s in the direction v_k. Since $\not\partial^2 = -\Delta$ = Laplacian of M, it follows that $\not\partial$ is elliptic.

A particularly interesting situation arises whenever the dimension of M is even, say $n = 2m$. In this case, the Spin Representation S becomes the direct sum of two irreducible (non-equivalent) representations S^+ and S^- of dimension 2^{n-1}. More precisely, the Space of Spinors, identified to a minimal left-ideal of $\tilde{C}_n$ (for example, to the left ideal generated by $(e_1 + ie_{m+1})(e_2 + ie_{m+2})\dots(e_m + ie_{2m})$) decomposes as

$$S = S^{+} \oplus S^{-} \text{ , with } S^{+} = \tilde{C}_n^{+} \cap S \text{ and } S^{-} = \tilde{C}_n^{-} \cap S \text{ .}$$

Because the Clifford multiplication exchanges the factors S^+ and S^- among themselves, one obtains two Spin(n) - module homomorphisms:

$$\mu^{+} : A \otimes S^{+} \longrightarrow S^{-} , \quad \mu^{-} : A \otimes S^{-} \longrightarrow S^{+} .$$

Setting $E^{\pm} = P \times_{Spin(n)} S^{\pm}$, ζ becomes the direct sum $\zeta = \zeta^{+} \oplus \zeta^{-}$ and the properties of the morphisms μ^+ and μ^- allow us to define the operators

$$\vartheta^{+} : \Gamma(\zeta^{+}) \longrightarrow \Gamma(\zeta^{-}), \quad \vartheta^{-} : \Gamma(\zeta^{-}) \longrightarrow \Gamma(\zeta^{+})$$

which are still *elliptic*. Let H be the space of the *Harmonic Spinors* over M , that is to say, $H = \{s \in \Gamma(\zeta) \,|\, \vartheta s = 0\}$. Since ϑ is self-adjoint (consider the usual inner product on $\Gamma(\zeta)$) it follows that ϑ^- is adjoint to ϑ^+ . Thus, $H = H^{+} \oplus H^{-}$ with

$$H^{+} = \ker \vartheta^{+}, \ H^{-} = \ker \vartheta^{-} \cong \operatorname{coker} \vartheta^{+}$$

and consequently

$$i_a(\vartheta^{+}) = \dim H^{+} - \dim H^{-} .$$

Definition 3.12. *The analytic index of* ϑ^+ *is called the Spin-index of* M *and is denoted by* Spin(M).

Whenever $n \equiv 0 \pmod 4$ Spin(M) coincides with the $\hat{A}$ - *genus of* M (see next Section); this is one of the interesting topological facts about Spin(M).

To conclude this section, we wish to observe that there are interesting connections between the Dirac Operator and the other elliptic operators we described before (the reader is referred in particular to Atiyah's lectures in [20]). Finally, we notice that the study of elliptic operators over Spin - manifolds allows one to retrieve certain classical results concerning compact Riemann surfaces [19].

4. The Index Theorem

As we have said at the beginning of this chapter, we do not intend to give here a proof of the Index Theorem, but rather, to present the various concepts which make up its statement, comment on the main ideas involved and finally, apply the Theorem to some of the situations discussed in the previous section.

Let $D : \Gamma(\xi) \to \Gamma(\eta)$ be an *elliptic differential operator* of order r on a manifold M (not necessarily oriented) but which we suppose to be riemannian, closed and of dimension n. In view of Definition 2.6, the ellipticity of D is equivalent to saying that the symbol $\sigma_r(D) : \pi^*\xi \to \pi^*\eta$ is a vector bundle isomorphism when the bundles involved are restricted to $S(M)$. The presence of the bundles $\pi^*\xi$, $\pi^*\eta$ and of the bundle isomorphism $\sigma_r(D)|S(M)$ indicates that we are in a classical situation of K-theory. We associate the element $d = d(\pi^*\xi, \pi^*\eta, \sigma_r(D)|S(M)) \in KU(D(M), S(M))$ to the triple $(\pi^*\xi, \pi^*\eta, \sigma_r(D)|S(M))$; the element d is called the *difference bundle* of the triple (the reader can find its actual construction and properties in [29, § 3]). The Chern character of the difference bundle is an element of $H^*(D(M), S(M); \mathbb{Q})$ which shall be denoted by ch D. Moreover, we know from Chapter 3, § 2, that the total space T^*M of the cotangent bundle $\tau^*(M)$ is an almost complex manifold; its almost complex structure is defined by the $GL(n;\mathbb{C})$-bundle $\alpha = \pi^*(\tau_{\mathbb{C}}(M))$ where $\tau_{\mathbb{C}}(M)$ is the complexification of $\tau(M)$ (cf. Definition 2.2, Chapter 3, and the lines which follow it). Hence, α induces an orientation - the α-*orientation* - of T^*M and consequently, of $D(M)$ and $S(M)$; if $[D(M), S(M)]_\alpha \in H_{2n}(D(M), S(M); \mathbb{Q})$ is the fundamental class defined by the α-orientation, for every $x \in H^*(D(M), S(M); \mathbb{Q})$, $x[D(M), S(M)]_\alpha$ denotes the value of the maximum rank component of x computed on the fundamental class.

Definition 4.1. *The topological index of the elliptic operator* D *is the rational number*

$$i_t(D) = (\text{ch } D.\text{Td}\alpha)[D(M), S(M)]_\alpha \quad ,$$

where Td α *is the total Todd class of* α .

Before continuing, let us recall that the total Chern class $c(\lambda) = \sum_{j=o}^{m} c_j(\lambda)$ of a complex vector m-bundle λ over X can be written as a product

$$c(\lambda) = \prod_{k=1}^{m} (1+x_k)$$

Where x_k is a polynomial in the symmetric elementary functions of the elements $c_k(\lambda) \subset H^{2k}(X;\mathbb{Q})$.(*) Moreover, we know that the Chern character of λ is given by

$$ch(\lambda) = \sum_{k=1}^{m} e^{x_k} .$$

Thus, the (total) Todd class of λ is defined by

(4.2) $$\mathrm{Td}\ \lambda = \prod_{k=1}^{m} \frac{x^k}{1-e^{-x_k}}$$

(Td $\lambda = 1$ if λ is trivial). This definition shows that for every pair of complex vector bundles λ and μ over X ,

(4.3) $$\mathrm{Td}(\lambda \oplus \mu) = \mathrm{Td}\ \lambda \cdot \mathrm{Td}\ \mu .$$

If λ is the complexification $\Theta_{\mathbb{C}}$ of a real vector bundle Θ , its Todd class can be expressed in terms of the rational Pontrjagin classes of Θ . More precisely, writing

$$c(\Theta_{\mathbb{C}}) = c(\lambda) = \prod_{k=1}^{[m/2]} (1-y_k)(1+y_k) ,$$

we shall have

(4.4) $$\mathrm{Td}(\Theta_{\mathbb{C}}) = \prod_{k=1}^{[m/2]} \left(\frac{y_k}{1-e^{-y_k}}\right) \left(\frac{-y_k}{1-e^{y_k}}\right)$$

$$= \prod_{k=1}^{[m/2]} \left(\frac{1/2\cdot y_k}{\sinh \frac{1}{2}\cdot y_k}\right)^2 ;$$

(*) Use the formal factorization

$$c_t(\lambda) = \sum_{j=o}^{m} c_j(\lambda)t^j = \prod_{k=1}^{m} (1+x_k t) .$$

but the total Pontrjagin class of Θ is given by

$$p(\Theta) = \prod_{k=1}^{[m/2]} \left(1+y_k^2\right) ,$$

so, our claim follows (the reader can find a justification for the formulas we presented here for $c(\Theta_{\mathbb{C}})$ and $p(\Theta)$ in Chapter 6, § 3; there one can also find the easy relations between the elements x_k and y_k employed before).

Going back to Definition 4.1, we notice explicitly that because of (4.4), $\mathrm{Td}\ \alpha$ can be written as a polynomial in the rational Pontrjagin classes of the manifold M. Also, if M is *oriented*, $i_t(D)$ can be expressed in a different way, via the Thom isomorphism. We make this claim more precise. Recall that the Thom isomorphism

$$\varphi_j : H^j(M;\mathbb{Q}) \longrightarrow H^{j+n}(D(M),\ S(M);\ \mathbb{Q})$$

is given by the cup product $\varphi_j(a) = \pi^*a \cup U$, where $\pi^* : H^j(M;\mathbb{Q}) \to H^j(D(M);\mathbb{Q})$ is the isomorphism induced by the projection $\pi : D(M) \to M$ (π is a homotopy equivalence) and $U \in H^n(D(M),S(M);\mathbb{Q})$ is the Thom class of $\tau^*(M)$ (*) corresponding to the (standard) orientation of $\tau^*(M)$ induced by that of M. If $[M]$ and $[D(M),S(M)]$ are the fundamental classes associated to the orientation of M,

$$[D(M),S(M)] = (-1)^{\frac{n(n-1)}{2}} [D(M),S(M)]_\alpha$$

(see Chapter 3, § 2) and

$$\text{(4.5)} \qquad i_t(D) = \varphi_*^{-1}\left((-1)^{\frac{n(n-1)}{2}} \mathrm{ch}\ D \cdot \mathrm{Td}\ \alpha\right)[M] .$$

This formula will actually be considered as an equivalent definition of the topological index of an elliptic differential operator over M.

At this point we wish to observe that the Thom isomorphism

(*) Observe that $H^*(D(M),S(M);\mathbb{Q})$ is a free $H^*(D(M);\mathbb{Q})$ - module generated by U (see [238]).

$$\varphi_! : KU(M) \xrightarrow{\cong} KU(D(M), S(M))$$

in KU - Theory (see (0.8)) is related to the cohomology Thom isomorphism φ_* by the formula

(4.6) $$\text{ch}\,\varphi_!(x) = (-1)^n \varphi_*(\text{ch}\,x \cdot (\text{Td}\,\alpha)^{-1})$$

for every $x \in KU(M)$, where n = dimension of α (see [49]). Moreover, the Chern character ch D of the difference bundle $d(\pi^*\xi, \pi^*\eta, \sigma_r(D) | S(M))$ is such that

(4.7) $$e(\tau^*(M)) \cdot \varphi_*^{-1}(\text{ch}\,D) = \text{ch}\,\xi - \text{ch}\,\eta$$

where $e(\tau^*(M))$ is the Euler class of the real $SO(2n)$ - bundle $\tau^*(M)$.

We are now ready to concentrate all our attention on the celebrated *Atiyah-Singer Index Theorem* [34].

<u>Theorem 4.8.</u> *Let* ξ *and* η *be differentiable vector bundles over a closed differentiable manifold* M *and let* $D : \Gamma(\xi) \to \Gamma(\eta)$ *be an elliptic differential operator. Then,*

$$i_a(D) = i_t(D) .$$

Note that from this Theorem it follows immediately that $i_t(D)$ is actually an integer. Furthermore, it is easy to see directly that $i_t(D) = 0$ *whenever* $\dim M = n$ *is odd* (it suffices to use the antipodal map of D(M)); this generalizes the analogous result given by Proposition 3.3 relative to the particular elliptic operator $D = d + \delta : \Gamma(\xi^{\text{even}}) \to \Gamma(\xi^{\text{odd}})$ and which yields the Euler-Poincaré characteristic of M.

The proof of the equality $i_a(D) = i_t(D)$ rests on the following key idea : extend the definitions of the analytic and topological indices so that they become homomorphisms from $KU(D(M), S(M)) \cong KU(T^*M)$ (*) into $\mathbb{Q}$ and then, show that these homomorphisms coincide. The proof of that equality is obtained after a careful analysis of the properties which characterize i_t ; indeed, from that analysis one constructs a system of

(*) If X is locally compact, we agree to define $KU(X) = \widetilde{KU}(X^+)$, where X^+ is the one-point compactification of X. Hence $KU(T^*M) \cong KU(D(M), S(M))$.

axioms which completely determine an "index function". At this point, one has to show that i_a satisfies the axioms obtained. Observe that the expression

$$i_t(x) = \varphi_*^{-1}((-1)^{\frac{n(n-1)}{2}} ch(x) Td\ \alpha)[M]$$

is well-defined for every $x \in KU(D(M),S(M)) \cong KU(T^*M)$ (cf. (4.5)); furthermore, if M' is another manifold, one can show that i_t is multiplicative with respect to tensor product, that is to say,

(4.9) $$i_t(x \otimes y) = i_t(x) \cdot i_t(y)$$

for every $x \in KU(T^*M)$ and $y \in KU(T^*M')$. Hence, i_t can be viewed as a homomorphism of $KU(D(M),S(M))$ into $\mathbb{Q}$.

The following construction is essential for a more systematic investigation of the properties satisfied by i_t. Let f be an embedding of M into $X = \mathbb{R}^{n+q}$ (q sufficiently large) and let $\nu_f(M)$ be the normal bundle associated to f. We have seen in Chapter 6 that the total space $D(\nu_f(M))$ of the disc-bundle associated to $\nu_f(M)$ is homeomorphic to a tubular neighborhood V of $f(M)$ in $\mathbb{R}^{n+q}$; an analogous observation can be made with respect to the tubular neighborhood T^*V obtained in the embedding T^*f of T^*M into T^*X (induced by f). We should also recall from Chapter 3, Section 2, that the $2n$-dimensional manifold T^*M has an almost complex structure induced by $\alpha = \pi^*(\tau_{\mathbb{C}}(M))$; since

$$\tau(T^*M) \oplus \nu_{T^*f}(T^*M) \cong \epsilon^{2(n+q)}$$

(cf. (1.3), Chapter 6), it follows that $\nu_{T^*f}(T^*M) \cong \pi^*(\nu_f(M) \oplus \nu_f(M))$ is endowed with an almost complex structure, namely that defined by $\nu = \pi^*(\nu_f(M) \otimes \mathbb{C})$. These facts let us define a homomorphism

$$\psi_! : KU(T^*M) \longrightarrow KU(T^*V)$$

via the Thom isomorphism $KU(T^*M) \cong KU(D(\nu_{T^*f}(T^*M)),S(\nu_{T^*f}(T^*M)))$ and the identification $KU(D(\nu_{T^*f}(T^*M))) = KU(T^*V)$. On the other hand, if U is an open set of a locally compact space

X and, as befores X^+ is the one-point compactification of X, the map $X^+ \to X^+/(X^+-U) \cong U^+$ gives rise to a natural homomorphism

$$KU(U) = \widetilde{KU}(U^+) \longrightarrow \widetilde{KU}(X^+) = KU(X) .$$

Hence, to each embedding $f : M \to X$ we associate a homomorphism

$$f_! : KU(T^*M) \longrightarrow KU(T^*X)$$

defined by $f_! = i_* \cdot \psi_!$, where i_* is the natural homomorphism induced by the open inclusion $i : T^*V \subset T^*X$. (*)

Let us now observe that for every $x \in KU(T^*M)$,

$$(\mathrm{ch}\ f_!(x))[T^*X]_\nu = (\mathrm{ch}\ i_*\psi_!(x))[T^*X]_\nu = (\mathrm{ch}\ \psi_!(x))[T^*V]_\nu \ ;$$

moreover, considering $\psi_* : H^*(T^*M;\mathbb{Q}) \to H^*(T^*V;\mathbb{Q})$ (ψ_* is defined analogously to $\psi_!$) and relation (4.6), we obtain

$$(\mathrm{ch}\ \psi_!(x))[T^*V]_\nu = (-1)^q\psi_*(\mathrm{ch}\ x \cdot (\mathrm{Td}\ \nu)^{-1})[T^*V]_\nu$$

$$(\mathrm{ch}\ \psi_!(x))[T^*V]_\nu = (-1)^{n+q}(\mathrm{ch}\ x \cdot (\mathrm{Td}\ \nu)^{-1})[T^*M]_\alpha$$

and therefore,

$$(4.10) \qquad (\mathrm{ch}\ f_!(x))[T^*X]_\nu = (-1)^{n+q}(\mathrm{ch}\ x \cdot (\mathrm{Td}\ \nu)^{-1})[T^*M]_\alpha .$$

In particular, consider the case in which the manifold M is just the origin x_o of $X = \mathbb{R}^{n+q}$ and the embedding f is the inclusion $j : \{x_o\} \subset X$. Because $j_!$ is an isomorphism, relation (4.10) is written simply as

$$(\mathrm{ch}\ j_!(y))[T^*X]_\nu = (-1)^{n+q}\mathrm{ch}\ y[T^*\{x_o\}]_\alpha$$

$$= (-1)^{n+q}\ y \ ,$$

for every $y \in KU(T^*\{x_o\}) \cong \mathbb{Z}$ (since, in this case, the normal bundle ν is trivial, $\mathrm{Td}\ \nu = 1$). In other words, for every $z \in KU(T^*X)$, $j_!^{-1}$ is defined by

$$j_!^{-1}(z) = (-1)^{n+q}\ \mathrm{ch}(z)[T^*X]_\nu .$$

(*) This construction is valid for any manifold X on which M is embedded.

This and (4.10) show that

$$j_!^{-1} \circ f_!(x) = j_!^{-1}(f_!(x)) = (-1)^{n+q}(\mathrm{ch}\, f_!(x))[T^*X]_\nu \ ,$$

$$j_!^{-1} \circ f_!(x) = (\mathrm{ch}\, x \cdot (\mathrm{Td}\ \nu)^{-1})[T^*M]_\alpha \ .$$

Finally, because $\alpha \oplus \nu$ is the trivial $(n+q)$ - dimensional complex vector bundle, we obtain from (4.3) that $(\mathrm{Td}\ \nu)^{-1} = \mathrm{Td}\alpha$ and hence,

(4.11) $$j_!^{-1} \circ f_!(x) = (\mathrm{ch}\, x \cdot (\mathrm{Td}\ \alpha)[T^*M]_\alpha \ ,$$

for every $x \in KU(T^*M) = KU(D(M), S(M))$. We now define

$$\mathrm{ind} : KU(T^*M) \longrightarrow KU(T^*\{x_o\}) \cong \mathbb{Z}$$

to be the homomorphism which makes the following diagram commutative:(*)

$$\begin{array}{ccc} KU(T^*M) & \xrightarrow{\ \mathrm{ind}\ } & KU(T^*\{x_o\}) \cong \mathbb{Z} \\ & {\scriptstyle f_!}\searrow \quad \swarrow{\scriptstyle j_!}\ \cong & \\ & KU(T^*X) & \end{array}$$

Using (4.1) and (4.11) we conclude that the function ind *coincides with the topological index* $i_t : KU(T^*M) \to \mathbb{Q}$. This coincidence shows immediately that i_t is a *function with integral values*. We wish to note that the homomorphism ind is *functorial* : if $h : M \to M'$ is a diffeomorphism, the following diagram commutes:

$$\begin{array}{ccc} KU(T^*M') & \xrightarrow{\ h^*\ } & KU(T^*M) \\ & {\scriptstyle \mathrm{ind}_{M'}}\searrow \quad \swarrow{\scriptstyle \mathrm{ind}_M} & \\ & \mathbb{Z} & \end{array} \quad ;$$

moreover, the index functions ind_M satisfy the following properties:

(A 1) if M is a point, ind_M = identity;

(*) The homomorphism ind is independent of the embedding $f : M \to X$ selected (cf. [35, Part I, page 498]).

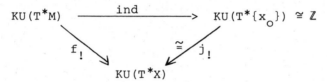

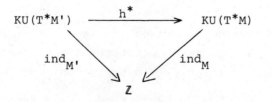

(A 2) if g is an embedding of M into another compact manifold M', the diagram

$$\begin{array}{ccc} KU(T^*M) & \xrightarrow{\;g_!\;} & KU(T^*M') \\ & {\scriptstyle ind_M} \searrow \quad \swarrow {\scriptstyle ind_{M'}} & \\ & \mathbb{Z} & \end{array}$$

commutes.

According to [35, Part I, Proposition 4.1], (A 1) and (A 2) characterize completely the index functions ind_M ; in other words, if for every compact manifold M there is a function $\gamma_M : KU(T^*M) \to \mathbb{Z}$ satisfying (A 1) and (A 2), then, $\gamma_M = ind_M$.

The uniqueness property of the index functions we alluded to suggests that a path one could follow in order to prove that $i_a(D) = i_t(D)$ would be to redefine $i_a(D)$ as a function on $KU(T^*M)$, with values on $\mathbb{Z}$, and satisfying conditions (A 1) and (A 2). However, when we attempt to give a meaning to $i_a(x)$ for every $x \in KU(D(M),S(M))$, we immediately run into two difficulties:

(i) one must make sure that if D_1 and D_2 are two elliptic operators whose symbols $\sigma_{r_1}(D_1)$ and $\sigma_{r_2}(D_2)$ define the same difference bundle in $KU(D(M),S(M))$, then $i_a(D_1) = i_a(D_2)$;

(ii) every $x \in KU(D(M),S(M))$ must be the class defined by the difference bundle of some elliptic operator.

But, two symbols which determine the same difference bundle are homotopic and two homotopic operators have the same analytic index; hence (i) is actually a consequence of

(i') a homotopy between $\sigma_{r_1}(D_1)$ and $\sigma_{r_2}(D_2)$ implies the existence of a homotopy between D_1 and D_2.

Unfortunately, (i') and (ii') are in general not true for elliptic differential operators!

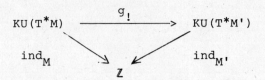

To circumvent these problems, Atiyah and Singer [35, §§ 5 and 6] considered a large class of pseudo-differential operators constructed from the singular integral Calderón-Zygmund operators extended by Seeley [207] to vector bundles over compact manifolds without boundary.(*) Since the notion of symbol still exists in such class one can consider the elliptic operators within the class; interestingly enough, the notions of analytic and topological index still have a meaning for these elliptic operators and moreover, conditions (i') and (ii) hold. Because i_a is an integral valued function, (A 1) follows easily. However, the verification of (A 2) appears to be a delicate process; the rub is that for a given operator D over M it is necessary to construct an operator D' over M' with symbol $\sigma(D') = g_!(\sigma(D))$ and such that $i_a(D') = i_a(D)$. Having observed that (A 2) can be proved whenever the manifolds involved are spheres, the idea is then to make a reduction to such case. With this in mind, Atiyah and Singer introduced three new axioms which imply (A 2) for any index function; these are the excision, normalization and multiplicative axioms. The excision axiom is the following. Let V be a non-compact manifold and $i : V \to Y$, $i' : V \to Y'$ be two *open* embeddings into the compact manifolds Y and Y'; then, the diagram

$$\begin{array}{ccccc} & & KU(T^*Y) & & \\ & \overset{i_*}{\nearrow} & & \overset{ind_Y}{\searrow} & \\ KU(T^*V) & & & & \mathbb{Z} \\ & \underset{i'_*}{\searrow} & & \underset{ind_{Y'}}{\nearrow} & \\ & & KU(T^*Y') & & \end{array}$$

commutes. The multiplicative axiom - "the most significant one" according to Atiyah and Singer - makes it possible to compute the index function of a product manifold, while the normalization axiom is concerned with certain operators over spheres. Let us describe, very briefly, how these axioms lead towards (A 2). Let V be a tubular neighborhood of $M \cong g(M)$ in M' and let $k : M \to Z$ be the inclusion of M into the double of

(*) We refer the reader to [193] (particularly to Chapters XI and XVI of that paper) for the detailed construction and properties of the "Seeley Algebra" thus obtained.

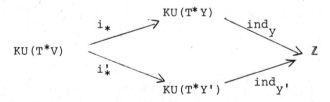

V , indicated here by Z . Then, the excision axiom applied to $Y = M'$ and $Y' = Z$ shows that

$$\text{ind}_{M'} \circ g_! = \text{ind}_{M'} \circ i_* \circ \psi_! = \text{ind}_Z \circ i_*' \circ \psi_! = \text{ind}_Z \circ k_! .$$

Since Z can be viewed as a sphere bundle over M , applying the multiplicative axiom and then, the normalization axiom, we obtain that $\text{ind}_Z \circ k_! = \text{ind}_M$ and so, $\text{ind}_{M'} \circ g_! = \text{ind}_M$ as we wanted.

The proof of the Index Theorem whose main steps we just retraced here is not the one given initially by Atiyah and Singer in [34] , but rather, that contained in [35] .(*) Indeed, Cobordism Theory was used in [34] to show that the analytic index satisfies the characteristic properties of the topological index. Furthermore, the earlier proof did not allow one to envisage certain generalizations, like the study of the index in the equivariant situation. We should point out that in [35] the Index Theorem is stated and proved within the framework of equivariant KU - theory; there the index functions are given as homomorphisms $KU_G(T^*M) \to RU(G)$ of $RU(G)$ - modules. In our previous description we supposed G to be trivial for the sake of simplicity; however, all the preceeding constructions can be transposed "mutatis mutandis" to the equivariant case (in the next section we shall describe an important application of the Index Theorem in the equivariant case). We also wish to observe that the theorem was extended to manifolds with boundary in [21], and in [35, IV], to families of elliptic operators parametrized by the points of a compact space (here the index of an operator is viewed as an element of the Grothendieck group of that given space); finally, the case for the real operators was discussed in [35, V] (the appropriate K - theoretical framework in this situation is given by the KR-theory studied in [15]). In 1973 , Atiyah, Bott and Patodi published a new proof of the Index Theorem (see [25]), based on the assymptotic developments of the Heat Equation (cf. Section 2). In its main lines, this proof follows that of [34] ,

(*) The remarks about cohomology we utilized are taken from [35, Part III] .

with local Riemannian Geometry replacing Cobordism. As a technical remark, we observe that in [25], the authors use systematically Invariant Theory (see Gilkey's paper [96]) for the orthogonal group.(*) Hence, the Riemann-Roch Theorem (for kählerian manifolds) and the generalized Hirzebruch Signature Theorem are totally proved within the framework of Riemannian Geometry. Furthermore, although the proof given in [25] is less general than that of [35,I] it emphasizes, for example, in a very clear fashion, the interesting connections between the eigenvalues of an operator and the geometric properties of manifolds (see [32], in particular).

Now, just as an example (sic !) we apply the Index Theorem to the *Signature Operator* $D^+ = d + \delta : \Gamma(\xi^+) \to \Gamma(\xi^-)$ over a closed oriented manifold M of dimension $n = 2m$ (see Section 3). Because of Proposition 3.7 , we know that

$$i_a(D^+) = \mathrm{Sign}(M) \;;$$

we also know that if m is odd, both sides of this relation are equal to zero. Using (4.5) we obtain

$$i_t(D^+) = \varphi_*^{-1}((-1)^m \operatorname{ch} D^+ \cdot \operatorname{Td} \alpha)[M] .$$

Now, because of (4.4) we have that

$$\operatorname{Td} \alpha = \prod_{k=1}^{m} \left(\frac{1/2\cdot y_k}{\sinh 1/2\cdot y_k}\right)^2 \;;$$

moreover, since $\operatorname{ch}(\xi^+) - \operatorname{ch}(\xi^-) = \prod_{k=1}^{m} \left(e^{-y_k} - e^{y_k}\right)$

and $e(\tau^*(M)) = \prod_{k=1}^{m} y_k$, we conclude from (4.7) that

$$\varphi_*^{-1}(\operatorname{ch} D^+) = \prod_{k=1}^{m} \frac{e^{-y_k} - e^{y_k}}{y_k}$$

$$= (-1)^m 2^m \prod_{k=1}^{m} \frac{\sinh 1/2\cdot y_k \cdot \cosh 1/2\cdot y_k}{1/2\cdot y_k}$$

(*) Please read Bott's lectures [50] .

(note that $1/2\left(e^{y_k} - e^{-y_k}\right) = \sinh\ y_k = 2\sinh 1/2\cdot y_k \cdot \cosh 1/2\cdot y_k$).
It follows that

$$(4.12 \qquad i_t(D^+) = \left(2^m \prod_{k=1}^{m} \frac{1/2\cdot y_k}{\tanh\ 1/2\cdot y_k}\right) [M] \ .$$

Thus, the topological index of D^+ is nothing else but the L - genus L(M) of M (observe that if m is odd, the right-hand side of (4.12) is trivial, because the m functions $y_k \longmapsto \dfrac{1/2\ y_k}{\tan h\ 1/2\ y_k}$ are even) . This shows that the Atiyah-Singer Index Theorem (4.8) contains, as a particular case, the *Hirzebruch Signature Theorem* (stated in 1953, [113], Theorem 3.1); the latter theorem represents Sign(M) as a polynomial in the Pontrjagin classes of the cotangent bundle of M . Putting together the previous remarks, we have:

Theorem 4.13. *If* M *is a compact, oriented manifold of even dimension,*

$$\mathrm{Sign}(M) = L(M) \ .$$

We have already seen that the analytic index of the operator

$$\bar\partial + \bar\partial^* : \Gamma(\eta^{\mathrm{even}}) \longrightarrow \Gamma(\eta^{\mathrm{odd}})$$

(where η is a holomorphic vector bundle over a complex compact manifold M of dimension m) is equal to the Riemann - Roch characteristic of M , namely to

$$\chi(M;\eta) = \sum_{q=o}^{m} (-1)^q \dim H^q(M;\Omega(\eta)) \quad ,$$

where $\Omega(\eta)$ is the sheaf defined by the germs of the holomorphic cross-sections of η (see Proposition 3.8). Let $\beta = \beta(M)$ be the *complex* tangent bundle of M . Using the notation introduced before, α is just the complexification of the real bundle $\tau(M)$ subjacent to $\beta(M)$; hence,

$$\alpha = cr(\beta(M)) \cong \beta(M) \oplus \overline{\beta(M)} \quad ,$$

where $\overline{\beta(M)}$ is conjugated to $\beta(M)$. It follows that

$$\text{Td}\ \alpha = \text{Td}(\beta + \overline{\beta}) = \text{Td}\ \beta \cdot \text{Td}\ \overline{\beta} \ .$$

Since M is endowed with an hermitian metric, $\beta \cong \overline{\beta}^*$ and so, we can identify $\xi^{o,q}$ to $\Lambda^q\beta$ (in this context $\overline{\beta(M)}$ is viewed as the set of all differential forms of type $(1,0)$). Then, the relation

$$c_m(\beta)\varphi_*^{-1}(\text{ch}(\overline{\partial} + \overline{\partial}^*)) = \text{ch}\ \eta^{\text{even}} - \text{ch}\ \eta^{\text{odd}}$$

$$= \text{ch}\ \eta \cdot \left(\sum_{q=o}^{m} (-1)^q\ \text{ch}\ \xi^{o,q}\right)$$

where $c_m(\beta)$ is the mth Chern class of β (cf. (4.7)), shows that

$$\varphi_*^{-1}(\text{ch}(\overline{\partial} + \overline{\partial}^*)) = (-1)^m\ \text{ch}\ \eta \cdot (\text{Td}\ \beta^*)^{-1} \ .$$

Because of (4.5) we have

$$i_t(\overline{\partial} + \overline{\partial}^*) = (\text{ch}\ \eta \cdot (\text{Td}\ \beta^*)^{-1} \cdot \text{Td}\ \beta \cdot \text{Td}\ \overline{\beta})[M]$$

that is to say,

$$i_t(\overline{\partial} + \overline{\partial}^*) = (\text{ch}\ \eta \cdot \text{Td}\ \beta)[M] \ .$$

Hence, the Index Theorem implies the following *Generalized Riemann - Roch Theorem:* (*)

Theorem 4.14. [34] *Let* M *be a complex, compact manifold and* η *be a holomorphic vector bundle over* M *then*

$$\chi(M;\eta) = (\text{ch}\ \eta \cdot \text{Td}\ \beta)[M] \ ,$$

where $\text{Td}\ \beta$ *is the Todd class of the complex tangent bundle over* M .

We wish to observe that Hirzebruch had established the general format (4.14) of Riemann - Roch's Theorem whenever M was an algebraic manifold [114]; it was only with the aid of the Index Theorem that the result could be extended to complex, compact manifolds. Observe also that Theorem 4.14 says, in particular, that the arithmetic genus of a complex manifold coincides with its Todd genus (just take η to be the trivial vector m - bundle).

(*) The Riemann - Roch problem is stated in the introduction to this Chapter.

The reader should have noticed that the application of the Index Theorem to the operators D^+ and $\bar{\partial} + \bar{\partial}^*$ consisted ultimately, in the evaluation of $ch(D^+)$ and $ch(\bar{\partial} + \bar{\partial}^*)$. Indeed, the computations one must make are taken out of a general result concerning manifolds with a G-structure [34; Theorem 2], which utilizes a characterization of the cohomology of the classifying space BG [44]. In the previous section we described the notion of Spin-structure on a manifold; now we generalize it. Let G be a Lie group and V an oriented real G-module; a G-*structure* on M is a fibre (and orientation) preserving isomorphism between the oriented bundles $\tau(M)$ and $P\times_G V$ where P is a principal G-bundle over M. Thus, for the Signature Operator D^+ (resp. the Dolbeault Operator $\bar{\partial} + \bar{\partial}^*$), $G = SO(2m)$ (resp. $G = U(m) \times U(\ell)$, $\ell = \dim \eta$) and $V = \mathbb{R}^{2m}$.

Therefore, if M is a $2m$-dimensional Spin-manifold (Def. 3.11) the same kind of computations show that the topological index of the Dirac Operator

$$\mathcal{D}^+ : \Gamma(\zeta^+) \longrightarrow \Gamma(\zeta^-)$$

is given by

$$i_t(\mathcal{D}^+) = \left(\left(\prod_{k=1}^{m} \frac{\sinh 1/2 \cdot y_k}{1/2 \cdot y_k}\right) \cdot \mathrm{Td}\ \alpha\right) [M]$$

or, using (4.4),

$$i_t(\mathcal{D}^+) = \left(\prod_{k=1}^{m} \frac{1/2 \cdot y_k}{\sinh 1/2 \cdot y_k}\right) [M] . \qquad (4.15)$$

This means that the topological index of $\mathcal{D}^+$ is precisely the $\hat{A}$-*genus* of M (notation: $\hat{A}(M)$) defined by Hirzebruch (see [115]) in case m = even (if m is odd, (4.15) implies that $i_t(\mathcal{D}^+) = 0$ because of the parity of the function $y_k \longmapsto \dfrac{1/2 \cdot y_k}{\sinh 1/2 \cdot y_k}$). Using Definition 3.12, one can thus obtain the following.

<u>Theorem 4.16.</u> *If M is a Spin-manifold of dimension 4ℓ,*

$$\mathrm{Spin}(M) = \hat{A}(M) .$$

5. The Generalized Lefschetz Fixed-Point Formula.

As we pointed out in the previous section, the Atiyah-Singer Index Theorem was stated for an equivariant situation. For every compact Lie group G, the topological index i_t is viewed as an index function from $KU_G(T^*M)$ into $RU(G)$, where M is a compact G-manifold (in what follows we shall identify T^*M with TM because we assume that M is endowed with a riemannian G-invariant metric). We construct i_t precisely as in Section 4: we begin by associating a homomorphism $f_! : KU_G(TM) \to KU_G(TX)$ to each G-embedding $f : M \to X$ (here X is a real representation space of G)(*) then we take the isomorphism $j_! : KU_G(T\{x_o\}) \to KU_G(TX)$ which corresponds to the G-embedding of the origin $x_o \in X$ into X and finally, we define

$$i_t : KU_G(TM) \longrightarrow RU(G) \cong KU_G(T\{x_o\})$$

as the composition $(j_!)^{-1} \circ f_!$. It goes without saying that i_t satisfies condition (A 1) and also, the excision, normalization and multiplicativity axioms of the index functions; hence, i_t satisfies (A 1) and (A 2). On the other hand, it is still possible to define the analytic index of an elliptic operator $D : \Gamma(\xi) \to \Gamma(\eta)$ over the G-manifold M, or more generally, of an elliptic complex $(\Gamma(\xi), d)$ over M (see Definition 2.22)

$$0 \longrightarrow \Gamma(\xi_o) \xrightarrow{d_o} \Gamma(\xi_1) \xrightarrow{d_1} \cdots \xrightarrow{d_{n-1}} \Gamma(\xi_n) \longrightarrow 0$$

where the ξ_i's $(0 \leq i \leq n)$ are complex vector G-bundles whose G-actions commute with the differentials d_i (in other words, $(\Gamma(\xi), d)$ is a G-module); in the latter case, the analytic index is defined by

$$i_a(\xi, d) = \sum_{i=o}^{n} (-1)^i \dim H^i(\Gamma(\xi)) .$$

Thus, $i_a(\xi, d)$ is itself viewed as an element of $RU(G)$. How-

---

(*) Such an embedding is allways possible, according to R. Palais [192].

ever, one should note that in order to talk about the adjoints d_i^* of the differentials d_i, it is necessary to assume that each bundle ξ_i is endowed with a G-invariant metric.

The first objective of this section is to study the topological index in relation to the fixed point set of the action of G on M; the result we are aiming at is Theorem 5.3, due to Atiyah and Segal [33]. Let us notice "en passant" that such result, when conveniently interpreted in Cohomology, allows us to obtain specific formulas representing the topological index in terms of the characteristic classes (compare with the remarks about G-structures made at the end of Section 4). The second objective is to describe briefly how the topological index leads to a generalized Lefschetz formula (relation (5.4)).

For a fixed $g \in G$, let M^g be the set of points in M which are invariant by the action of g on M or, in short,

$$M^g = \{x \in M | g \cdot x = x\} .$$

The canonical G-embedding $f : M^g \to M$ gives rise to $f_! : KU_G(TM^g) \to KU_G(TM)$ and induces $f^* : KU_G(TM) \to KU_G(TM^g)$; thus, composing these homomorphisms and using the Thom isomorphism in KU_G-theory, we obtain a homomorphism

$$f^* \circ f_! : KU_G(M^g) \longrightarrow KU_G(M^g)$$

which is just multiplication by the element

$$\lambda_{-1}(\nu_g) = \sum_{k \geq 0} (-1)^k \Lambda^k(\nu_g) \in KU_G(M^g), \text{ where } \nu_g = \nu_f(M^g) \otimes \mathbb{C}$$

is the complexification of the normal bundle $\nu_f(M^g)$ of $M^g \subset M$ (see [35,I,(3.1)]). Next, we describe an essential result established by Segal. Let G be a compact Lie group, γ be a conjugacy class in G and $M^\gamma = \bigcup_{g \in \gamma} M^g$. The set $\mathcal{P} = \mathcal{P}_\gamma$ of all characters vanishing on γ is a prime ideal of $RU(G)$; for every $RU(G)$-module A, let $A_{\mathcal{P}}$ be the module obtained from A by localizing at $\mathcal{P}$. We can now state the following localization theorem

<u>Proposition 5.1.</u> [210] - *Let* $i^! : KU_G(M) \to KU_G(M^\gamma)$ *be the homomorphism of* RU(G) - *modules defined by the canonical inclusion* $i : M^\gamma \to M$. *Then, the induced homomorphism*

$$(i^!)_{\mathcal{P}} : KU_G(M)_{\mathcal{P}} \longrightarrow KU_G(M^\gamma)_{\mathcal{P}}$$

is an isomorphism.

From now on *we shall suppose the existence of an element* $g \in G$ *whose powers form a dense subgroup of* G (then G is the product of a cyclic group and a torus). In this case G acts trivially on M^g : in fact, $M^g = M^G = \{x \in M | (\forall h \in G)\ hx = x\}$; moreover, if γ is the conjugacy class of this distinguished element g , $M^\gamma = M^g$. Then, using the homeomorphism $(TM)^g \cong TM^g$, Proposition 5.1 and the Thom Isomorphism Theorem, we obtain that

$$(f^*)_{\mathcal{P}} : KU_G(TM)_{\mathcal{P}} \longrightarrow KU_G(TM^g)_{\mathcal{P}}$$

is an isomorphism. Since $\lambda_{-1}(\nu_g)$ is invertible in $KU_G(M^g)_{\mathcal{P}}$ [33; Lemma 2.7], it follows that the RU(G) - module homomorphism $f_!$ induces an isomorphism

$$(f_!)_{\mathcal{P}} : KU_G(TM^g)_{\mathcal{P}} \longrightarrow KU_G(TM)_{\mathcal{P}} ;$$

its inverse is given by the isomorphism

$$(f^*)_{\mathcal{P}}/\lambda_{-1}(\nu_g) : KU_G(TM)_{\mathcal{P}} \longrightarrow KU_G(TM^g)_{\mathcal{P}} . \quad (5.2)$$

Consider the following commutative diagram

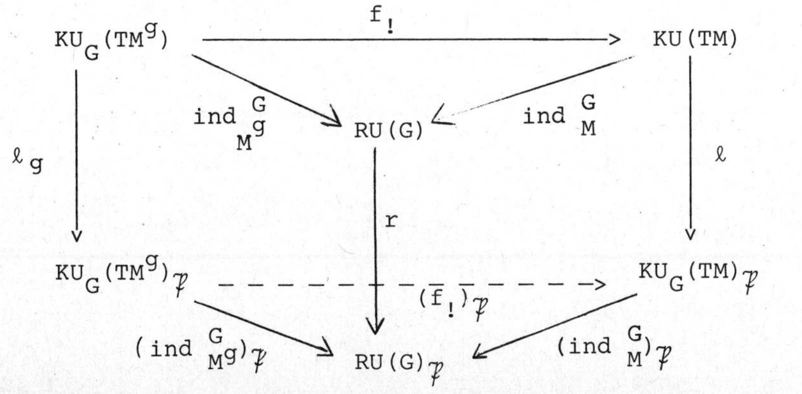

;

its top is given by Axiom (A 2) of § 4 , its bottom is the localization of the top and the vertical arrows are the localization homomorphisms. Next, consider the evaluation map

$$\epsilon : RU(G) \longrightarrow \mathbb{C}$$

given by $\epsilon(\chi) = \chi(g)$ for every $\chi \in RU(G)$. Defining $\epsilon_{\mathcal{P}} : RU(G)_{\mathcal{P}} \to \mathbb{C}$ so that $\epsilon_{\mathcal{P}} \circ r = \epsilon$, we have

$$\epsilon \circ \mathrm{ind}_M^G = \epsilon_{\mathcal{P}} \circ r \circ \mathrm{ind}_M^G = \epsilon_{\mathcal{P}} \circ \left(\mathrm{ind}_M^G\right)_{\mathcal{P}} \circ \ell$$

$$\epsilon \circ \mathrm{ind}_M^G = \epsilon_{\mathcal{P}} \circ \left(\mathrm{ind}_{M^g}^G\right)_{\mathcal{P}} \circ (f_!)_{\mathcal{P}}^{-1} \circ \ell$$

or, using the inverse (5.2) of $(f_!)_{\mathcal{P}}$,

$$\epsilon \circ \mathrm{ind}_M^G = \epsilon_{\mathcal{P}} \circ \left(\mathrm{ind}_{M^g}^G\right)_{\mathcal{P}} \circ \left\{(f^*)_{\mathcal{P}}/\lambda_{-1}(\nu_g)\right\} \circ \ell \quad .$$

In other words, for every $x \in KU_G(TM)$,

$$(5.3) \qquad \left(\epsilon \circ \mathrm{ind}_M^G\right)(x) = \epsilon_{\mathcal{P}} \circ \left(\mathrm{ind}_{M^g}^G\right)_{\mathcal{P}} \left\{ \frac{\ell_g(f^*(x))}{\lambda_{-1}(\nu_g)} \right\} \quad .$$

Since G acts trivially on TM^g ,

$$KU_G(TM^g) \cong KU(TM^g) \otimes RU(G)$$

and by localization,

$$KU_G(TM^g)_{\mathcal{P}} \cong KU(TM^g) \otimes RU(G)_{\mathcal{P}} \quad .$$

Because of these isomorphisms we make the following identifications:

(1) $\ell_g \equiv 1_{KU(TM^g)} \otimes r : KU(TM^g) \otimes RU(G) \to KU(TM^g) \otimes RU(G)_{\mathcal{P}}$,

so that $\ell_g(f^*(x))$ can be viewed as an element of $KU(TM^g) \otimes RU(G)_{\mathcal{P}}$;

(2) $\left(\mathrm{ind}_{M^g}^G\right)_{\mathcal{P}} \equiv \mathrm{ind}_{M^g} \otimes 1_{(RU(G)_{\mathcal{P}}} : KU(TM^g) \otimes RU(G)_{\mathcal{P}} \to \mathbb{Z} \otimes RU(G)_{\mathcal{P}}$,

Where ind_{M^g} is the topological index $i_t : KU(TM^g) \to \mathbb{Z}$ corresponding to the case in which G is trivial (just as in § 4);

(3) $$\epsilon_{\mathcal{P}} \equiv 1_{\mathbb{Z}} \otimes \epsilon_{\mathcal{P}} : \mathbb{Z} \otimes RU(G)_{\mathcal{P}} \longrightarrow \mathbb{Z} \otimes \mathbb{C} \cong \mathbb{C} .$$

The reader should also notice that the element $\lambda_{-1}(\nu_g)(g)$ - obtained evaluating $\lambda_{-1}(\nu_g) \in KU(TM^g)_{\mathcal{P}}$ at g - is, in this context, viewed as an invertible element of the ring $KU(TM^g) \otimes \mathbb{C}$. Finally, if

$$\text{index} = \text{ind}_{M^g} \otimes 1_{\mathbb{C}} : KU(TM^g) \otimes \mathbb{C} \longrightarrow \mathbb{Z} \otimes \mathbb{C} \cong \mathbb{C}$$

is the complexification of the topological index, the following diagram commutes:

$$\begin{array}{ccc}
1_{KU(TM^g)} \otimes RU(G)_{\mathcal{P}} & \xrightarrow{\ \text{ind}_{M^g} \otimes 1_{RU(G)_{\mathcal{P}}}\ } & \mathbb{Z} \otimes RU(G)_{\mathcal{P}} \\
\cong KU(TM^g)_{\mathcal{P}} & & \cong RU(G)_{\mathcal{P}} \\
{\scriptstyle 1_{KU(TM^g)} \otimes \epsilon_{\mathcal{P}}} \Big\downarrow & & \Big\downarrow {\scriptstyle 1_{\mathbb{Z}} \otimes \epsilon_{\mathcal{P}}} \\
KU(TM^g) \otimes \mathbb{C} & \xrightarrow[\ \text{index}\ ]{} & \mathbb{Z} \otimes \mathbb{C} \cong \mathbb{C} .
\end{array}$$

Hence, from the preceeding considerations, (5.3) gives the relation

$$(\epsilon \circ \text{ind}_M^G)(x) = (1_{\mathbb{Z}} \otimes \epsilon_{\mathcal{P}})\left(\text{ind}_{M^g} \otimes 1_{RU(G)_{\mathcal{P}}}\right) \circ$$

$$\circ \left\{ \frac{(1_{KU(TM^g)} \otimes r)(f^*(x))}{\lambda_{-1}(\nu_g)} \right\}$$

or

$$(\epsilon \circ \text{ind}_M^G)(x) = \text{index} \left\{ \left(1_{KU(TM^g)} \otimes \epsilon_{\mathcal{P}}\right) \left(\frac{f^*(x)}{\lambda_{-1}(\nu_g)}\right)\right\}$$

$$= \text{index} \left\{ \frac{(f^*(x))(g)}{\lambda_{-1}(\nu_g)(g)} \right\} .$$

With all these results one can show the following.

Theorem 5.4. [33] - *Let* M *be a compact* G-*manifold. Suppose that there is an element* $g \in G$ *whose powers form a dense subgroup of* G. *Let* M^g *be the set of points of* M *which are invariant by the action of* g; *let* ν_g *be the complexification of the normal bundle defined by the embedding* $M^g \subset M$. *Then, the element* $\lambda_{-1}(\nu_g)(g)$, *obtained by evaluating* $\lambda_{-1}(\nu_g) = \sum_{k\geq o} (-1)^k \Lambda^k(\nu_g)$ *at* g, *is invertible in the ring* $KU(TM^g) \otimes \mathbb{C}$. *Moreover, for every* $x \in KU_C(TM)$,

$$(\mathrm{ind}_M^G\, x)(g) = \mathrm{index}\left\{ \frac{(f^*(x))(g)}{\lambda_{-1}(\nu_g)(g)} \right\},$$

where $f^* : KU_G(TM) \to KU_G(TM^g)$ *is the map induced by the canonical inclusion* $f : M^g \to M$.

Let f be a finite rank endomorphism of a (finite) complex of R-modules; its *Lefschetz number* is the alternate sum of the traces of the homomorphisms induced by f in the cohomology of the complex. For example, the Lefschetz number of the identity homomorphism is the Euler-Poincaré characteristic of the complex. For a G-invariant elliptic complex $(\Gamma(\xi),d)$ defined over a compact G-manifold M, the Lefschetz number of an element $g \in G$ is given by

$$L(g;\xi) = \sum_{i\geq o} (-1)^i \mathrm{tr}(g|H^i(\Gamma(\xi)));$$

notice that $L(g;\xi)$ is just the evaluation of the analytic index of $(\Gamma(\xi),d)$ at g (see Definition 2.23). Let $x = [\sigma(\xi)] \in KU_G(TM)$ be the class of the symbol sequence $\sigma(\xi)$ for the complex $(\Gamma(\xi),d)$ (see [35;I,§ 2]); then, restricting the action of G to that defined by the closure of the cyclic group generated by an element $g \in G$, applying Theorem 5.4 and the Index Theorem, we obtain

$$L(g;\xi) = \mathrm{index}\left\{ \frac{(f^*(x))(g)}{\lambda_{-1}(\nu_g)(g)} \right\}. \tag{5.5}$$

This is the *generalized Lefschetz fixed point formula* obtained by Atiyah and Segal [33; Theorem 2.12]. Its discovery opened the way to several diversified results; unfortunately, due to their large number, we can only suggest that the reader consult the extensive "ad hoc" literature. At any rate, we wish to observe that if we apply the Lefschetz formula to the elliptic complexes described in § 2 , we obtain the equivariant version of the results described in § 3 ; thus, besides the case of the De Rham complex, for which (5.5) gives back the classical Lefschetz formula, we find an equivariant form of the Riemann-Roch Theorem and also, some G - signature theorems (see [33] and [35; III]). If G is a finite group, $L(g,\xi)$ is an algebraic integer; thus (5.5) becomes the source of several integrality theorems. If the action of G is such that M^g is finite, say $M^g = \{y_1,\dots,y_r\}$, the Atiyah-Segal formula (5.5) becomes particularly simple; this is due to the fact that in such case, the topological index is easily obtained. Let us explain our statement. Because

$$KU(TM^g) \cong \bigoplus_{k=1}^{r} KU(\{y_k\}) \quad ,$$

the map $i_t : KU(TM^g) \to \mathbb{Z}$ is the sum of the natural isomorphisms $KU(\{y_k\}) \cong \mathbb{Z}$. In other words, if f_k is the natural inclusion $\{y_k\} \subseteq M$ and $\nu_{g,k}$ is the complexification of the normal bundle induced by such an inclusion, then

$$i_t \left(\frac{f^*([\sigma(\xi)])}{\lambda_{-1}(\nu_g)} \right) = \sum_{k=1}^{r} \frac{f_k^*([\sigma(\xi)])}{\lambda_{-1}(\nu_{g,k})} \quad .$$

Notice that $f_k^*([\sigma(\xi)]) = \sum_{i \geq o} (-1)^i [\xi_{i,k}]$, where $\xi_{i,k}$ is the restriction of ξ_i to $\{y_k\}$ and hence,

$$f_k^*([\sigma(\xi)])(g) = \sum_{i \geq o} (-1)^i tr(g|\xi_{i,k}) \quad ;$$

on the other hand, because the total space of the normal bundle over $\{y_k\}$ coincides with $T_{y_k}(M)$, if we write $\tau_k = f_k^*[\tau(M)]$, we obtain

$$\lambda_{-1}(\nu_{g,k})(g) = \sum_{i\geq o} (-1)^i \mathrm{tr}\{\Lambda^i(g|\tau_k \otimes \mathbb{C})\}$$

$$= \det_{\mathbb{C}}(1-g|\tau_k \otimes \mathbb{C})$$

$$= \det_{\mathbb{R}}(1-g|\tau_k)$$

(here $\det_{\mathbb{C}}$ (resp. $\det_{\mathbb{R}}$) indicates the complex (resp. real) determinant). Finally, we obtain the formula

$$(5.6) \qquad L(g;\xi) = \sum_{k=1}^{r} \frac{\sum_{i\geq o}(-1)^i \mathrm{tr}(g|\xi_{i,k})}{\det_{\mathbb{R}}(1-g|\tau_k)} \ .$$

Formula (5.6) was announced by Atiyah and Bott in [23] and proved in [24; I (Theorem A)] . If $(\Gamma(\xi),d)$ is the De Rham's Complex, (5.6) reduces to the Lefschetz formula : $L(g;\xi) = L(g) =$ number of points kept fixed by g . Clearly, (5.6) is particularly adapted to the study of the fixed point set for the isometries of a group G acting on M .

If $(\Gamma(\xi),d)$ is the Dolbeault Complex, formula (5.6) gives rise to the following.

<u>Proposition 5.7.</u> [24; II] - *Let* M *be a complex compact manifold,* η *a holomorphic vector bundle over* M *and* $g: M \to M$, *a holomorphic function with simple fixed points. Take* $L(g,f)$ *to be the Lefschetz number given by the action of* (g,f) *on* $H^*(M;\Omega(\eta))$, *where* $f : g^*\eta \to \eta$ *is a morphism of holomorphic bundles. Then,*

$$L(g,f) = \sum_{k=1}^{r} \frac{\mathrm{tr}\, f_k}{\det_{\mathbb{C}}(1-dg_k)}$$

where dg_k *is the differential of* g *at* $y \in M^g$ *and* f_k *is the* $\mathbb{C}$ *- endomorphism induces by* f *on the fibre* η_{y_k} *of* η .

Proposition (5.7) has several applications to Representation Theory; for example, the Atiyah-Bott Formula is connected to the Hermann Weyl Formula, which evaluates the trace of certain representations over a maximal torus of a compact Lie group. Furthermore, we point out that whenever M is a curve and g is a transversal morphism of M into M (every fixed point y_k of g is simple, i.e., $\det_{\mathbb{C}}(1-dg_k) \neq 0$) we can use (5.7) to show that the degree d of g is related to the number r of points kept fixed by g ; the relation is given by the formula

$$\frac{r-d+1}{2} = \mathrm{Re}\left(\sum_{k=1}^{r} \frac{1}{1-g'(y_k)} \right)$$

($\mathrm{Re}(z)$ = real part of the complex number z).

Now we apply the Atiyah-Bott Formula (5.6) to the Signature Operator

$$D^+ = d + \delta : \Gamma(\xi^+) \longrightarrow \Gamma(\xi^-)$$

defined over a closed, oriented, riemannian manifold M of dimension $n = 2m$ (see § 3). We suppose that the G-*action preserves the orientation* of M (we shall continue to identify $\tau^*(M)$ to $\tau(M)$ via the G-invariant riemannian metric of M) and furthermore, we assume that there is a $g \in G$ whose action on M is an *isometry having only isolated fixed points* (since M is compact, M^g is finite). The action induced by g on $\Gamma(\xi^+)$ (resp. $\Gamma(\xi^-)$) will be denoted by g^+ (resp. g^-); hence, we have a commutative diagram

$$\begin{array}{ccc} \Gamma(\xi^+) & \xrightarrow{\ D^+\ } & \Gamma(\xi^-) \\ {\scriptstyle g^+}\downarrow & & \downarrow{\scriptstyle g^-} \\ \Gamma(\xi^+) & \xrightarrow[\ D^+\ ]{} & \Gamma(\xi^-) \end{array} .$$

The *Signature* of g (notation: $\mathrm{Sign}(g;M)$) is defined to be the Lefschetz number of g . To compute $\mathrm{Sign}(g;M)$ we shall determine its contribution $\mathrm{Sign}(g;M)_k$ at each point $y_k \in M^g$ ($1 \leq k \leq r$). Since g is an isometry its differential dg_k : $T_{y_k}(M) \to T_{y_k}(M)$ is also an isometry. Hence,

$$\mathrm{Sign}(g;M)_k = \frac{\mathrm{tr}(g^+|\Gamma(\xi^+)) - \mathrm{tr}(g^-|\Gamma(\xi^-))}{\det(1-dg_k)} .$$

Now, it is possible to decompose $T_{y_k}(M) = T_k M$ into a direct sum of orthogonal 2-planes

$$T_k M = \bigoplus_{j=1}^{m} E_{j,k}$$

so that $(dg_k)(E_{j,k})$. Let us select a basis $\{e_{j,k}, e'_{j,k}\}$ for each $E_{j,k}$, $1 \leq j \leq m$, so that

$$e_{1,k} \wedge e'_{1,k} \wedge \cdots \wedge e_{m,k} \wedge e'_{m,k}$$

in the 2m-differential form defined by the orientation of M in y_k; as a consequence, the restriction of dg_k to each $E_{j,k}$ is just a rotation by a certain angle $\Theta_{j,k}$, that is to say,

$$(dg_k)(e_{j,k}, e'_{j,k}) = \begin{pmatrix} \cos \Theta_{j,k} & \sin \Theta_{j,k} \\ -\sin \Theta_{j,k} & \cos \Theta_{j,k} \end{pmatrix} \begin{pmatrix} e_{j,k} \\ e'_{j,k} \end{pmatrix} .$$

The set of angles $\{\Theta_{j,k} | 1 \leq j \leq m\}$ is called a *coherent system* for dg_k. Using the properties of the operator τ defined by (3.4) one can show that

$$\operatorname{tr}(g^+|\Gamma(\xi_k^+)) - \operatorname{tr}(g^-|\Gamma(\xi_k^-)) = \prod_{j=1}^{m} \left(e^{-i\Theta_{j,k}} - e^{i\Theta_{j,k}}\right)$$

$$\det(1-dg_k) = \prod_{j=1}^{m} \left(1-e^{-i\Theta_{j,k}}\right)\left(1-e^{i\Theta_{j,k}}\right)$$

(with $i^2 = -1$) and so,

$$(5.8) \qquad \operatorname{Sign}(g;M) = \sum_{k=1}^{r} \operatorname{Sign}(g;M)_k$$

$$= \sum_{k=1}^{r} (-i)^m \prod_{j=1}^{m} \cot\left(\frac{1}{2}\Theta_{j,k}\right) .$$

There are many applications of the G-Signature Theorem expressed by relation (5.8) (for a more general form of the Theorem, please consult [35; III]). For example, Atiyah and Bott proved the following result in [24; II]. For a prime number p $(p \neq 2)$, consider a smooth action of the cyclic group $\mathbb{Z}_p$ on a homology sphere and suppose that such action has only two fixed points; then, the induced representations of $\mathbb{Z}_p$ on the tangent spaces to these two points are isomorphic (Theorem 7.15). In particular, this result allows one to show that two h - cobordant lens spaces are isometric (cf. Milnor [178]). Other interesting applications of the G - Signature theorems can be found in [117]; these will consent the reader to have a better grasp of the relations between Number Theory and the G - Signature theorems.

We conclude the Chapter with an example about *exotic involutions*, also taken from [24; Part II]. This time we shall utilize the Fixed Point Formula concerning the Lefschetz number of an isometry of a Spin-manifold (see 3.11 and 3.12). Let M be a Spin - manifold of dimension $n = 2m$, ζ the Spinor bundle over M and $\vartheta^+ : \Gamma(\zeta^+) \to \Gamma(\zeta^-)$, the Dirac Operator. In addition, let $g : M \to M$ be a Spin - structure preserving isometry whose fixed points are all isolated. Then, it is possible to lift g to a bundle isomorphism $\hat{g} : g^*\zeta \to \zeta$ which induces an automorphism g^+ (resp. g^-) of $\Gamma(\zeta^+)$ (resp. $\Gamma(\zeta^-)$), such that the following diagram commutes:

$$\begin{array}{ccc} \Gamma(\zeta^+) & \xrightarrow{\vartheta^+} & \Gamma(\zeta^-) \\ \Big\downarrow g^+ & & \Big\downarrow g^- \\ \Gamma(\zeta^+) & \xrightarrow[\vartheta^+]{} & \Gamma(\zeta^-) \end{array} .$$

According to the Atiyah-Bott Formula, the Lefschetz number $\mathrm{Spin}(\hat{g};M)$ corresponding to the action of $\hat{g}$ over the Dirac complex is given by

$$(5.9) \qquad \mathrm{Spin}(\hat{g};M) = \sum_{k=1}^{r} \epsilon_k(\hat{g}) \left(\frac{i}{2}\right)^m \prod_{j=1}^{m} \mathrm{cosec}\left(\frac{1}{2}\,\Theta_{j,k}\right) ,$$

where $i^2 = -1$, $\{\Theta_{j,k}\ 1 \leq j \leq m\}$ is a coherent system of angles for dg_k and $\epsilon_k(\hat{g}) = \pm 1$, depending on the fixed point y_k and the lifting $\hat{g}$. Now, for a given complex number t and an odd number $a \geq 3$, let $X_t(a) \subseteq \mathbb{C}^{n+1}$ be the Brieskorn manifold defined by

$$X_t(a) = \left\{ z = (z_o, \ldots, z_n) \in \mathbb{C}^{n+1} \mid z_o^2 + \ldots + z_{n-1}^2 + z_n^a = t \right\} .$$

Consider its intersection with the unit sphere S^{2n+1} (respectively, the unit disc D^{2n+2}) of $\mathbb{C}^{n+1}$, that is to say,

$$S_t(a) = X_t(a) \cap S^{2n+1}$$

$$D_t(a) = X_t(a) \cap D^{2n+2} .$$

We agree to write $X(a)$, $S(a)$ and $D(a)$ for $X_o(a)$, $S_o(a)$ and $D_o(a)$, respectively. Then, if t is sufficiently small, the following statements are true (see [54]).

(1) $S_t(a)$ is the boundary of the $2n$-manifold $D_t(a)$;

(2) if $n \geq 3$, $D_t(a)$ is $(n-1)$-connected (indeed, $D_t(a)$ is homotopically equivalent to a bouquet of n-spheres);

(3) $S_t(a)$ and $S(a)$ are diffeomorphic;

(4) if $n \geq 3$, $S(a)$ is $(n-2)$-connected;

(5) *if* n *is odd, say* $n = 2m+1$, *then* $S(a)$ *is a topological sphere* S^{4m+1} (actually $S(a)$ is diffeomorphic to the standard sphere S^{4m+1}, except for $m \geq 2$ and $a \equiv \pm 3$ (mod.8)).

Let γ be the involution of $\mathbb{C}^{n+1} = \mathbb{C}^{2m+2}$ given by

$$\begin{cases} \gamma \cdot z_k = -z_k & \text{if } 0 \leq k \leq n-1 \\ \gamma \cdot z_n = z_n . \end{cases}$$

Clearly γ preserves $D_t(a)$ and its boundary $S_t(a)$; furthermore, for t small, the action of γ on $S_t(a)$ is isomorphic to its action on $S(a)$. The action of γ on a topological sphere $S(a)$ will be called an *exotic involution*. We are interested in knowing when two exotic involutions, say on $S(a)$

and $S(b)$, are isomorphic. Before studying this problem, we observe that the fixed point set of $D_t(a)$ under the action of γ is given by

$$(D_t(a))^\gamma = \left\{ z \in D_t(a) \mid (\forall\ 0 \leq k \leq n-1)\, z_k = 0 \text{ and } z_n^a = t \right\} ;$$

furthermore, the rotation

$$\rho_a : \mathbb{C}^{n+1} \longrightarrow \mathbb{C}^{n+1}$$

defined by

$$\begin{cases} \rho_a(z_k) = z_k , & \text{if } 0 \leq k \leq n-1 \\ \rho_a(z_n) = \zeta_a \cdot z_n , & \text{where } \zeta_a = \exp\left(\dfrac{2\pi i}{a}\right) \end{cases}$$

commutes with the action of γ on $D_t(a)$ and interchanges the points of $(D_t(a))^\gamma$. Suppose now that the actions induced by γ on $S(a)$ and $S(b)$ are isomorphic. Then, for t sufficiently small, there is a diffeomorphism $f : S_t(a) \to S_t(b)$ which commutes with the actions induced by γ on $S_t(a)$ and $S_t(b)$. Let M be the space obtained by the adjunction of $D_t(a)$ and $D_t(b)$ via the map f ; observe that if $m \geq 1$, M is at least 2 - connected. Thus (cf. 3.11 and the following lines) M is a Spin - manifold of dimension $4m + 2$. Moreover, γ acts on M and respects its Spin - structure; hence, let $\hat{\gamma}$ be a lifting of $\gamma|M$ and $\mathrm{Spin}(\hat{\gamma};M)$, its Lefschetz number. Consider the Spin - structures of $D_t(a)$ and $D_t(b)$; let $\mathrm{Spin}(\hat{\gamma};D_t(a))$ (resp. $\mathrm{Spin}(\hat{\gamma};D_t(B))$) be the Lefschetz number relative to the restriction of γ to $D_t(a)$ (resp. $D_t(b)$). Because

$$\mathrm{Spin}(\hat{\gamma};M) = \mathrm{Spin}(\hat{\gamma};D_t(a)) + \mathrm{Spin}(\hat{\gamma};D_t(b)) ,$$

relation (5.9) implies that

$$\mathrm{Spin}(\hat{\gamma};M) = \sum_{j=1}^{a} \epsilon_j(\hat{\gamma})\left(\frac{i}{2}\right)^{2m+1} + \sum_{k=1}^{b} \epsilon'_k(\hat{\gamma})\left(\frac{i}{2}\right)^{2m+1} .$$

We want to compute $\epsilon_j(\hat{\gamma})$ (or $\epsilon'_k(\hat{\gamma})$). To this end, let $y_j \in D_t(a)$ be a fixed point; since the order of ρ_a is odd and further, ρ_a commutes with the action of γ on $D_t(a)$, it follows that

$$\epsilon_j(\hat{\gamma}) = \epsilon_{\rho_a(j)}(\hat{\gamma}) \quad \text{with} \quad y_{\rho_a(j)} = \rho_a(y_j) \ .$$

But ρ_a is a circular permutation on the fixed point set of $D_t(a)$. Hence, the $\epsilon_j(\hat{\gamma})$'s, $1 \leq j \leq a$, are equal among themselves; the same happens to the $\epsilon'_k(\hat{\gamma})$'s, $1 \leq k \leq b$, and consequently,

$$\text{(5.10)} \qquad \operatorname{Spin}(\hat{\gamma};M) = (\pm a \pm b)\left(\frac{i}{2}\right)^{2m+1} \ .$$

Furthermore, γ is an involution on $\mathbb{C}^{n+1}$ and in view of the construction of M, it follows that $\hat{\gamma}$ has a period equal to at most 4. From this we conclude that the complex number $\operatorname{Spin}(\hat{\gamma};M)$ is a Gaussian integer. Thus, relation (5.10) shows that *if the exotic involutions on* $S(a)$ *and* $S(b)$ *are isomorphic*

$$a \equiv \pm b \pmod{2^{m+1}} \ .$$

In [52], Bredon proved that if $m = 1$ and $a = 3$, the action of γ on the topological 5-sphere $S(3)$ is not isomorphic to the standard antipodal map; the previous remark shows that to be the case for $m \geq 1$ and every topological $(4m+1)$-sphere $S(3)$.

BIBLIOGRAPHY

[1] ADAMS, J.F. - On the non-existence of elements of Hopf invariant one. Ann. of Math. 72 (1960), 20-104.

[2] ADAMS, J.F. - Vector fields on spheres. Ann. of Math. 75 (1962), 603-632.

[3] ADAMS, J.F. - H - spaces with few cells. Topology 1 (1962), 67-72.

[4] ADAMS, J.F. - On the groups J(X). Part I: Topology 2 (1963), 181-195. Part II: ibid 3 (1965), 137-171. Part III: ibid 3 (1965), 193-222. Part IV: ibid 5 (1966), 31-38.

[5] ADAMS, J.F. - Lectures on Lie groups. Benjamin, New York: 1969.

[6] ADAMS, J.F.; ATIYAH, M.F. - K - theory and the Hopf invariant. Quart. J.Math. Oxford (2) 17 (1966), 31-38.

[7] ADAMS, J.F.; WALKER, G. - An example in homothopy theory. Proc. Cambridge Philos. Soc. 60 (1964), 699-700.

[8] ADAMS, J.F.; WALKER, G. - On complex Stiefel manifolds. Proc. Cambridge Philos. Soc. 61 (1965), 81-103.

[9] ADEM, J. - The iteration of the Steenrod squares in algebraic topology. Proc. Nat. Acad. Sci. U.S.A. 38 (1952), 720-726.

[10] ALEXANDROFF, P.; HOPF, H. - Topologie. Springer, Berlin: 1935.

[11] ALLARD, J. - Adams operations in $KO(X) \oplus KSp(X)$. Bol. Soc. Bras. Math (1) 5 (1974), 85-96.

[12] ATIYAH, M.F. - Thom complexes. Proc. London Math. Soc. (3) 11 (1961), 291-310.

[13] ATIYAH, M.F. - Immersions and embeddings of manifolds. Topology 1 (1961), 125-132.

[14] ATIYAH, M.F. - Power operations in K-theory. Quart. J. Math. Oxford (2) 17 (1966), 165-193.

[15] ATIYAH, M.F. - K-theory and reality. Quart. J. Math. Oxford (2) 17 (1966), 367-386.

[16] ATIYAH, M.F. - Algebraic topology and elliptic operators. Comm. Pure Appl. Math. 20 (1967), 237-249.

[17] ATIYAH, M.F. - K-theory. Benjamin, New York: 1967.

[18] ATIYAH, M.F. - Vector fields on manifolds. Westdeutscher Verlag, Köln: 1970.

[19] ATIYAH, M.F. - Riemann surfaces and spin structures. Ann. Sci. Ecole Norm. Sup. 4 (1971), 47-62.

[20] ATIYAH, M.F. - Classical groups and classical differential operators on manifolds. Differential operators on manifolds; C.I.M.E., Varenna 1975. Ed. Cremonese, Roma: 1975.

[21] ATIYAH, M.F.; BOTT, R. - The index problem for manifolds with boundary. Bombay Colloquium on differential analysis; Bombay 1964. Oxford Univ. Press, Oxford: 1964.

[22] ATIYAH, M.F.; BOTT, R. - On the periodicity theorem for complex vector bundles. Acta Math. 112 (1964), 229-247.

[23] ATIYAH, M.F.; BOTT, R. - A Lefschetz fixed point formula for elliptic differential operators. Bull.Amer.Math.Soc. 72 (1966), 245-250.

[24] ATIYAH, M.F.; BOTT, R. - A Lefschetz fixed point formula for elliptic complexes. Part I: Ann. of Math 86 (1967), 374-407. Part II: ibid 88 (1968), 451-491.

[25] ATIYAH, M.F.; BOTT, R.; PATODI, V.K. - On the heat equation and the index theorem. Invent.Math. 19 (1973), 279-330. Errata: ibid 28 (1975), 277-280.

[26] ATIYAH, M.F.; BOTT, R.; SHAPIRO, A. - Clifford modules. Topology 3 (Suppl. 1) (1964), 3-38.

[27] ATIYAH, M.F.; HIRZEBRUCH, F. - Quelques théorèmes de non-plongement pour les variétés différentiables. Bull.Soc. Math. France 87 (1959), 383-396.

[28] ATIYAH, M.F.; HIRZEBRUCH, F. - Vector bundles and homogeneous spaces. Proc. Symp. Pure Math. III. Amer. Math. Soc., Providence: 1961.

[29] ATIYAH, M.F.; HIRZEBRUCH, F. - Analytic cycles on complex manifolds. Topology 1 (1962), 25-45.

[30] ATIYAH, M.F.; HIRZEBRUCH, F. - The Riemann-Roch theorem for analytic embeddings. Topology 1 (1962), 151-166.

[31] ATIYAH, M.F.; HITCHIN, N.J.; SINGER, I.M. - Deformations of instantons. Proc.Nat.Acad.Sci. U.S.A. 74 (1977), 2662-2663.

[32] ATIYAH, M.F.; PATODI, V.K.; SINGER, I.M. - Spectral asymmetry and riemannian geometry. Bull. London Math.Soc. 5 (1973), 229-234.

[33] ATIYAH, M.F.; SEGAL, G.B. - The index of elliptic operators. Part II: Ann. of Math. 87 (1968), 531-545.

[34] ATIYAH, M.F.; SINGER, I.M. - The index of elliptic operators on compact manifolds. Bull.Amer.Math.Soc. 69 (1963), 422-433.

[35] ATIYAH, M.F.; SINGER, I.M. - The index of elliptic operators. Part I: Ann. of Math. 87 (1968), 484-530. Part. III: ibid 87 (1968), 546-604. Part IV: ibid 93 (1971), 119-138. Part V: ibid 93 (1971), 139-149.

[36] ATIYAH, M.F.; TALL, D.O. - Group representations, λ-rings and the J - homomorphism. Topology 8 (1969), 253-297.

[37] ATIYAH, M.F.; TODD, J.A. - On complex Stiefel manifolds. Proc. Cambridge Philos.Soc. 56 (1960), 342-353.

[38] ATIYAH, M.F.; WARD, R.S. - Instantons and algebraic geometry. Comm.Math.Phys. 55 (1977), 117-124.

[39] BECKER, J.C. - Cohomology and the classification of liftings. Trans.Amer.Math.Soc. 133 (1968), 447-475.

[40] BECKER, J.C. - The span of spherical space forms. Amer. J.Math. 94 (1972), 991-1025.

[41] BECKER, J.C.; GOTTLIEB, D.H. - The transfer map and fiber bundles. Topology 14 (1975), 1-12.

[42] BERGER, M.; GAUDUCHON, P.; MAZET, E. - Le spectre d'une variété riemannienne. Lecture Notes in Math. 194. Springer, Berlin-Heidelberg-New York: 1971.

[43] BOREL, A. - Sur la cohomologie des espaces fibrés principaux et des espaces homogènes des groupes de Lie compacts. Ann. of Math. 57 (1953), 115-207.

[44] BOREL, A.; HIRZEBRUCH, F. - Characteristic classes and homogneous spaces. Part I: Amer.J.Math. 80 (1958), 458-538. Part II: ibid 81 (1959), 315-382. Part III: ibid 82 (1960), 491-504.

[45] BOREL, A.; SERRE, J.P. - Groups de Lie et puissances réduites de Steenrod. Amer.J.Math. 75 (1953), 409-448.

[46] BOTT, R. - The stable homology of the classical groups. Ann. of Math. 70 (1959), 313-337.

[47] BOTT, R. - Quelques remarques sur les théorèmes de périodicité. Bull.Soc.Math. France 87 (1959), 293-310.

[48] BOTT, R. - A note on the KO - theory of sphere-bundles. Bull.Amer.Math.Soc. 68 (1962), 395-400.

[49] BOTT, R. - Lectures on K(X). Benjamin, New York: 1969.

[50] BOTT, R. - Some aspects of invariant theory in differential geometry. Differential operators on manifolds; C.I.M.E., Varenna 1975. Ed. Cremonese, Roma: 1975.

[51] BOURBAKI, N. - Topologie générale. Hermann, Paris: 1951.

[52] BRENDON, G.E. - Examples of differentiable group actions. Topology 3 (1965), 115-122.

[53] BRENDON, G.E.; KOSINSKI, A. - Vector fields on π - manifolds. Ann. of Math. 84 (1960), 85-90.

[54] BRIESKORN, E. - Beispiele zur Differentialtopologie von Singularitäten. Invent.Math. 2 (1966), 1-14.

[55] BROUWER, L.E.J. - Über Abbildung von Mannigfaltigkeiten. Math.Ann. 71 (1912) 97-115.

[56] BROWDER, W. - Torsion in H-spaces. Ann. of Math. 74 (1961), 24-51.

[57] BROWDER, W. - On differential Hopf algebras. Trans.Amer. Math.Soc. 107 (1963), 153-176.

[58] BROWDER, W. - Embedding smooth manifolds. Proc. International Congress of Math., Moscow (1966), 712-719.

[59] BROWDER, W.; THOMAS, E. - On the projective plane of an H-space. Illinois J.Math. 7 (1963), 492-502.

[60] BUHŠTABER, V.; MIŠČENKO, A. - K-theory on the category of infinite cell-complexes. Math. USSR - Izv. (3) 2 (1968), 515-556.

[61] CALABI, E.; ECKMANN, B.- A class of compact complex manifolds. Ann. of Math. 58 (1953), 495-500.

[62] CARTAN, H.; EILENBERG, S. - Homological algebra. Princeton Univ. Press, Princeton: 1956.

[63] CARTAN, H.; SCHWARTZ, L. - Théorème d'Atiyah-Singer sur l'indice d'un opérateur differentiel elliptique. Séminaire H.Cartan 1963/64. Ecole Normale Supérieure, Paris: 1965.

[64] CHAN, W.M. - Ph.D. thesis. University of Waterloo 1975.

[65] CHAN, W.M. - HOFFMAN, P. - Vector bundles over suspensions. Canad.Math.Bull. 17 (1974), 483-484.

[66] CHOW, W.L. - On compact complex analytic varieties. Amer. J.Math. 71 (1949), 893-914.

[67] CURTIS, C.W.; REINER, I. - Representation theory of finite groups and associative algebras. Interscience, New York: 1962.

[68] CURTIS, M.; MISLIN, G. - Two new H-spaces. Bull.Amer. Math.Soc. 76 (1970), 851-852.

[69] CURTIS, M.; MISLIN, G. - H-spaces which are bundles over S^7. J. Pure Appl. Algebra 1 (1971), 27-40.

[70] DE RHAM, G. - Variétés différentiables. Hermann, Paris: 1960.

[71] DIECK, TOM,T. - Die Hopfsche Invariante. Gebrüder Bornträgen, Berlin: 1963.

[72] DIEUDONNE, J. - Foundations of modern analysis. Academic Press, New York: 1960.

[73] DOLBEAULT, J.P. - Formes différentielles et cohomologie sur une variété analytique complexe. Part I: Ann. of Math. 64 (1956), 83-130. Part II: ibid 65 (1957), 282-330.

[74] DOLD, A. - Über fasernweise Homotopieäquivalenz von Faserräumen. Math. Z. 62 (1955), 111-126.

[75] DOLD, A. - Erzeugende der Thomschen Algebra $\mathfrak{N}$. Math.Z. 65 (1956), 25-35.

[76] DOLD, A. - Relation between ordinary and extraordinary cohomology. Colloquium on algebraic topology. Aarhus Univ., Aarhus: 1962.

[77] DOLD, A. - Partitions of unity in the theory of fibrations. Ann. of Math. 78 (1963), 223-255.

[78] DOLD, A. - Halbexakte Homotopiefunktoren. Lecture Notes in Math. 12. Springer, Berlin-Heidelberg-New York: 1966.

[79] DOLD, A. - Lectures on algebraic topology. Springer, Berlin-Heidelberg-New York: 1972.

[80] DOLD, A.; LASHOF, R. - Principal quasifibrations and fibre homotopy equivalence of bundles. Illinois J.Math.3 (1959), 285-305.

[81] DOUGLAS, R.R.; SIGRIST, F. - Sphere bundles over spheres and H-spaces. Topology 8 (1969), 115-118.

[82] ECKMANN, B. - Zur Homotopietheorie gefaserter Räume. Comment.Math.Helv. 14 (1941/42), 141-192.

[83] ECKMANN, B. - Systeme von Richtungsfeldern in Sphären und stetige Lösungen komplexer linearer Gleichungen. Comment.Math.Helv. 15 (1942/43), 1-26.

[84] ECKMANN, B. - Stetige Lösungen linearer Gleichungssysteme. Comment.Math.Helv. 15 (1942/43), 318-339.

[85] ECKMANN, B. - Gruppentheoretischer Beweis des Satzes von Hurwitz-Radon über der Komposition quadratischer Formen. Comment.Math.Helv. 15 (1942/43), 358-366.

[86] ECKMANN, B. - Cohomologie et classes caractéristiques. Classi caratteristiche e questioni connesse; C.I.M.E., L'Aquila 1966. Ed. Cremonese, Roma: 1967.

[87] ECKMANN, B. - Continuous solutions of linear equations - some exceptional dimensions in topology. Battelle Rencontres 1967. Benjamin, New York: 1968

[88] EHRESMANN, C. - Sur la théorie des espaces fibrés. Colloque de Topologie algébrique. C.N.R.S., Paris: 1947.

[89] EHRESMANN, C. - Sur les variétés plongées dans une variété différentiable. C.R.Acad.Sci. Paris 226 (1948), 1879-1880.

[90] FUJII, K. - On the K-ring of S^{4n+3}/H_m. Hiroshima Math.J. 3 (1973), 251-265.

[91] FUJII, K.; KOBAYASHI, T.; SUGAVARA, M. - KO-groups of lens spaces modulo powers of two. Hiroshima Math. J. 8 (1978), 469-489.

[92] FUJII, M. - KU-groups of Dold manifolds. Osaka J.Math.3 (1966), 49-64.

[93] FUJII, M. - Ring structures of KU-cohomologies of Dold manifolds. Osaka J.Math. 6 (1969), 107-115.

[94] FUJII, M.; YASUI, T. - KO-cohomologies of the Dold manifolds. Math.J. Okayama Univ. 16 (1973), 55-84.

[95] GELFAND, I.M. - On elliptic equations. Uspehi Math.Nauk SSSR (3) 15 (1960), 121-132/ Russian Math.Surveys (3) 15 (1960), 113-123.

[96] GILKEY, P.B. - Curvature and the eigenvalues of the Laplacian for elliptic complexes. Advances in Math. 10 (1973), 344-382.

[97] GILKEY, P.B. - The index theorem and the heat equation. Math.Lectures Serie 4. Publish or Perish, Boston: 1974.

[98] GOLDBERG, S.I. - Curvature and homology. Academic Press, New York: 1962.

[99] HAEFLIGER, A. - Differentiable embeddings. Bull.Amer. Math.Soc. 67 (1961), 103-112.

[100] HAEFLIGER, A. - Plongements différentiables de variétés dans variétés. Comm.Math.Helv. 36 (1961), 47-82.

[101] HAEFLIGER, A. - Plongements différentiables dans le domaine stable. Comm.Math.Helv. 37 (1962), 155-176.

[102] HAEFLIGER, A.; HIRSCH, M.W. - Immersions in the stable range. Ann. of Math. 75 (1962), 231-241.

[103] HELD, R.; SJERVE, D. - On the homotopy properties of Thom Complexes. Math.Z. 135 (1973/4), 315-323.

[104] HILTON, P. - An introduction to homotopy theory. Cambridge Univ. Press, Cambridge: 1953.

[105] HILTON, P. - Homotopy theory and duality. Gordon and Breach, New York: 1965.

[106] HILTON, P. - General cohomology theory and K-theory. London Math.Soc.Lecture Notes 1. Cambridge Univ.Press, Cambridge: 1971.

[107] HILTON, P.; MISLIN, G.; ROITBERG, J. - Localization of nilpotent groups and spaces. Math.Studies 15. North-Holland, Amsterdam-Oxford: 1975.

[108] HILTON, P.; ROITBERG, J. - On principal S^3-bundles over spheres. Ann. of Math. 90 (1969), 91-107.

[109] HILTON, P.; ROITBERG, J. - On the classification problem for H-spaces of rank two. Comment.Math.Helv. 46 (1971), 506-516.

[110] HILTON, P.; WYLIE, S. - Homology theory. Cambridge Univ. Press, Cambridge: 1960.

[111] HIRSCH, M. - Immersions of manifolds. Trans.Amer.Math. Soc. 93 (1959), 242-276.

[112] HIRZEBRUCH, F. - Doctoral thesis. University of Münster 1950.

[113] HIRZEBRUCH, F. - On Steenrod's reduced powers, the index of inertia, and the Todd genus. Proc.Nat.Acad.Sci. U.S.A. 39 (1953), 951-956.

[114] HIRZEBRUCH, F. - Arithmetic genera and the theorem of Riemann-Roch for algebraic varieties. Proc.Nat.Acad.Sci. U.S.A. 40 (1954), 110-114.

[115] HIRZEBRUCH, F. - Automorphe Formen und der Satz von Riemann-Roch. Symp. Internacional de topologia algebraica, Mexico 1956. Univ.Nac.Aut. de Mexico y UNESCO, Mexico: 1958.

[116] HIRZEBRUCH, F. - Topological methods in algebraic topology. 3 rd. ed., Springer, Berlin-Heidelberg-New York: 1966.

[117] HIRZEBRUCH, F.; ZAGIER, D. - The Atiyah-Singer theorem and elementary number theory. Math.Lectures Serie 3. Publish or Perish, Boston: 1974.

[118] HODGE, W.V.D. - The theory and applications of harmonic integrals. 2 nd ed., Cambridge Univ.Press, Cambridge: 1952.

[119] HOPF, H. - Zum Clifford-Kleinschen Raumproblem. Math. Ann. 95 (1952), 313-339.

[120] HOPF, H. - Vectorfelder in n - dimensionalen Mannigfaltigkeiten. Math.Ann. 97 (1927), 225-260.

[121] HOPF, H. - Über die Abbildungen der dreidimensionalen Sphäre auf die Kugelfläche. Math.Ann. 104 (1931), 637-665.

[122] HOPF, H. - Über die Abbildungen von Sphären auf Sphären niedrigerer Dimension. Fund.Math. 25 (1935), 427-440.

[123] HOPF, H. - Über den Rang geschlossener Liescher Gruppen. Comment. Math.Helv. 13 (1940), 119-143.

[124] HOPF, H. - Über die Topologie der Gruppen-Mannigfaltigkeiten und ihre Verallgemeinerungen. Ann. of Math. 42 (1941), 22-52.

[125] HOPF, H. - Zur Topologie der komplexen Mannigfaltigkeiten. Studies and essays presented to R.Courant. Interscience, New York: 1948.

[126] HÖRMANDER, L. - Pseudo-differential operators and hypoelliptic equations. Proc.Symp.Pure Math. X. Amer.Math. Soc., Providence: 1967.

[127] HOSCHSCHILD, G.; SERRE, J.P. - Cohomology of group extensions. Trans.Amer.Math.Soc. 74 (1953), 110-134.

[128] HÖSLI, H. - Sur les H-espaces à deux générateurs cohomologiques. C.R.Acad.Sci. Paris 270 (1970), 746-749.

[129] HUBBUCK, J.R. - Some results in the theory of H-spaces. Bull.Amer.Math.Soc. 74 (1968), 965-967.

[130] HUBBUCK, J.R. - Generalized cohomology operations and H-spaces of low rank. Trans.Amer.Math.Soc. 141 (1969), 335-360.

[131] HUBBUCK, J.R. - On homotopy-commutative H-spaces. Topology 8 (1969), 119 - 126.

[132] HUBBUCK, J.R. - Primitivity in torsion free cohomology Hopf algebras. Comment.Math.Helv. 46 (1971), 13-43.

[133] HURWITZ, A. - Über die Komposition der quadratischen Formen. Math.Ann. 88 (1923), 1-25.

[134] HUSEMOLLER, D. - Fibre bundles. Mc Graw-Hill, New York: 1966. 2nd ed., Springer, Berlin-Heidelberg-New York: 1975.

[135] IWATA, K. - Span of lens spaces. Proc.Amer.Math.Soc. 33 (1972), 211-212.

[136] JAMES, I. - On spaces with a multiplication. Pacific J. Math. 7 (1957), 1083-1100.

[137] JAMES, I. - Whitehead products and vector fields on spheres. Proc.Cambridge Philos.Soc. 53 (1957), 817-820.

[138] JAMES, I. - The intrinsic join: a study of the homotopy groups of Stiefel manifolds. Proc.London Math.Soc. (3) 8 (1958), 507-535.

[139] JAMES, I. - Cross-sections of Stiefel manifolds. Proc. London Math.Soc. (3) 8 (1958), 536-547.

[140] JAMES, I. - Spaces associated with Stiefel manifolds. Proc. London Math.Soc. (3) 9 (1959), 115-140.

[141] JAMES, I. - The space of the bundle maps. Topology 2 (1963), 45-59.

[142] JAMES, I. - Bundles with special structures. Part I: Ann. of Math. 89 (1969), 359-390.

[143] JAMES, I. - The topology of Stiefel manifolds. Cambridge Univ.Press, Cambridge: 1976.

[144] KÄHLER, E. - Über eine bemerkenswerte Hermitesche Metrik. Abh. Math.Sem.Univ. Hamburg 9 (1939), 173-186.

[145] KAHN, P.J. - Obstructions to extending almost X - structures. Illinois J.Math. 13 (1969), 336-357.

[146] KAMATA, M. - On complex cobordism groups of classifying spaces for dihedral groups. Osaka J.Math. 11 (1974), 367-378.

[147] KAMBE, T. - The structure of K_Λ-rings of the lens space and their applications. J.Soc.Math. Japan 18 (1966), 135-146.

[148] KAROUBI, M. - K - théorie équivariante des fibrés en sphères. Topology 12 (1973), 275-281.

[149] KERVAIRE, M. - Non-parallelizability of the n - sphere for $n > 7$. Proc.Not.Acad.Sci. U.S.A. 44 (1958), 280-283.

[150] KILLING, W. - Über die Clifford-Kleinschen Raumformen. Math.Ann. 39 (1891), 257-278.

[151] KIRCHHOFF, A. - Sur l'existence de certains champs tensoriels sur les sphères. C.R.Acad.Sci. Paris 225 (1947), 1258-1260.

[152] KOBAYASHI, T. - Non-immersion theorems for lens spaces. J.Math. Kyoto Univ. 6 (1966), 91-108.

[153] KOBAYASHI, T. - Immersions and embeddings of lens spaces. J.Sci. Hiroshima Univ. 32 (1968), 285-292.

[154] Kobayashi, T. - Immersions and embeddings of lens spaces. Hiroshima Math.J. 2 (1972), 345-352.

[155] KOBAYASHI, T. - Note on γ-operations in KO-theories. Hiroshima Math.J. 4 (1974), 425-434.

[156] KOBAYASHI, T.; MURAKAMI, S.; SUGAVARA, M. - Note on J-groups of lens spaces. Hiroshima Math.J. 7 (1977), 387-409.

[157] KOBAYASHI, T.; SUGAVARA, M. - Note on KO-rings of lens spaces mod 2^r. Hiroshima Math.J. 8 (1978), 85-90.

[158] KODAIRA, K. - On cohomology groups of compact analytic varieties with coefficients in some analytic sheaves. Proc.Nat.Acad.Sci. U.S.A. 39 (1953), 865-868.

[159] KODAIRA, K. - On Kähler varieties of restricted type. Ann. of Math. 60 (1954), 28-48.

[160] LAM, K. - Fiber homotopic trivial bundles over $\mathbb{C}P^{n-1}$. Proc.Amer.Math.Soc. 33 (1972), 211-212.

[161] LANG, S. - Introduction to differentiable manifolds. Interscience, New York: 1962.

[162] LEHMANN, D. - (Co)-homologies généralisées des espaces $K(\pi,i)$ et des formes sphériques. Bull.Soc.Math. France 98 (1970), 305-318.

[163] LEHMANN, D.; PICCININI, R. - Remarques sur l'application Hom $(G,H) \to [BG,BH]$. Math.Z. 128 (1972), 231-233.

[164] MAC KEAN, H.P.; SINGER, I.M. - Curvature and the eigenvalues of the Laplacian. J.Diff.Geom. 1 (1967), 43-69.

[165] MAC LANE, S. - Homology. Springer, Berlin-Göttingen-Heidelberg: 1963.

[166] MAHAMMED, N. - A propos de la K-théorie des espaces lenticulaires. C.R.Acad.Sci. Paris 271 (1970), 639-642.

[167] MAHAMMED, N. - K-théorie des espaces lenticulaires. C.R.Acad.Sci. Paris 272 (1971), 1363-1365.

[168] MAHAMMED, N. - K-théorie des formes sphériques tétraédriques. C.R.Acad.Sci. Paris 281 (1975), 141-144.

[169] MAHAMMED, N. - K-théorie des formes sphériques. Thèse, Université de Lille 1975.

[170] MAHOWALD, M.E. - On obstruction theory in orientable fibre bundles. Trans.Amer.Math.Soc. 110 (1964), 315-349.

[171] MASSEY, W.S. - Exact couples in algebraic topology. Ann. of Math. 56 (1952), 363-396.

[172] MASSEY, W.S. - Obstructions to the existence of almost complex structures. Bull.Amer.Math.Soc. 67 (1961), 559-564.

[173] MAYER, K.H. - Elliptische Differentialoperatoren und Ganzzahligkeitssätze für charakteristische Zahlen. Topology 4 (1965), 295-313.

[174] MILNOR, J. - On the immersion of n-manifolds in (n+1)-space. Comment.Math.Helv. 30 (1956), 275-284.

[175] MILNOR, J. - A procedure for killing homotopy groups of differentiable manifolds. Proc.Symp.Pure Math. III. Providence: Amer.Math.Soc. 1961.

[176] MILNOR, J. - A survey of cobordiom theory. Enseignement Math. 9 (1962), 16-23.

[177] MILNOR, J. - Topology from the differentiable view point. Univ. Press of Virginia, Charlottesville: 1965.

[178] MILNOR, J. - Whitehead torsion. Bull.Amer.Math.Soc. 72 (1966), 358-426.

[179] MILNOR, J.; MOORE, J. - On the structure of Hopf algebras. Ann. of Math. 81 (1965), 211-264.

[180] MILNOR, J.; STASHEFF, J.D. - Characteristic classes. Annals of Math. Studies 76. Princeton Univ.Press, Princeton: 1974.

[181] MIMURA, M.; NISHIDA, G.; TODA, H. - Localization of CW-complexes and its applications. J.Math.Soc. Japan 23 (1971), 593-624.

[182] MIMURA, M.; NISHIDA, G.; TODA, H. - On the classification of H-spaces of rank 2. J.Math. Kyoto Univ. 13 (1973), 611-627.

[183] MINAKSHISUNDARAM, S.; PLEIJEL, Å. - Some properties of the eigenvalues of the laplace operator on riemannian manifolds. Canad.J.Math. 1 (1949), 242-256.

[184] MISLIN, G.; ROITBERG, J. - Remarks on the homotopy classification of finite-dimensional H-spaces. J.London Math.Soc. 22 (1971), 593-612.

[185] MOSTOW, G.D. - Equivariant embeddings in euclidean spaces. Ann. of Math. 65 (1957), 432-446.

[186] MUNKRES, J. - Elementary differential topology. Annals of Math. Studies 54. Princeton Univ. Press, Princeton: 1963.

[187] NAKAGAWA, R. - Embedding of projective spaces and lens spaces. Sci. Report Tokyo Kyoiku Daigaku 9 (1967), 170-175.

[188] NAKAGAWA, R.; - KOBAYASHI, T. - Non-embeddability of lens spaces mod. 3. J.Math. Kyoto Univ. 5 (1966), 313-324.

[189] NIVEN, I.; ZUCKERMANN, H. - An introduction to the theory of numbers. 3 rd ed., John Whyley, New York: 1972.

[190] NOETHER, F. - Über eine Klasse singulärer Integralgleichungen. Math.Ann. 82 (1921), 42-63.

[191] NOVIKOV, S.P. - Homotopically equivalent smooth manifolds. Jzv.Akad. Nauk SSSR Ser.Mat. 28 (1964), 365-474.

[192] PALAIS, R.S. - Imbedding of compact, differentiable ransformation groups in orthogonal representations. J.Math.Mech. 6 (1957), 673-678.

[193] PALAIS, R.S. - Seminar of the Atiyah-Singer index theorem. Annals of Math. Studies 57. Princeton Univ.Press, Princeton: 1965.

[194] PATODI, V.K. - Curvature and the eigenvalues of the laplace operator. J.Diff.Geom. 5 (1971), 233-249.

[195] PICCININI, R. - A note on a conjecture of Atiyah and Hirzebruch. An.Acad. Brasil Cien. 42 (1970), 159-162.

[196] PITT, D. - Free actions on generalized quaternion groups on spheres. Proc. London Math.Soc. (3) 26 (1973), 1-18.

[197] POINCARE, H. - Sur les combes définies par les équations différentielles. Troisième partie: J.Math. Pures Appl. (4) 1 (1885), 167-244.

[198] PORTEOUS, I.R. - Topological geometry. Von Nostrand, London: 1969.

[199] QUILLEN, D. - The Adams conjecture. Topology 10 (1971), 67-80.

[200] RADON, J. - Lineare Scharen orthogonaler Matrizen. Abh. Math.Sem. Univ. Hamburg 1 (1923), 1-14.

[201] REES, E. - Embedding odd torsion manifolds. Bull. London Math.Soc. 3 (1971), 356-362.

[202] SANDERSON, B. - Immersions and embeddings of projective spaces. Proc. London Math.Soc. 14 (1964), 137-153.

[203] SANDERSON, B.; SCHWARZENBERGER, R. - Non-immersion theorems for differentiable manifolds. Proc. Cambridge Philos. Soc. 59 (1963), 312-322.

[204] SCHEERER, H. - On principal bundles over spheres. Indag. Math. 32 (1970), 353-355.

[205] SCHÖN, R. - Fibrations over a CWh - base. Proc.Amer.Math. Soc. 62 (1977), 165-166.

[206] SCHWARTZ, L. - Opérateurs elliptiques et indices. Séminaire H.Cartan 1963/64. Paris: Ecole Normale Supérieure 1965.

[207] SEELEY, R.T. - Integro-differential operators on vector bundles. Trans.Amer.Math.Soc. 117 (1965), 167-205.

[208] SEELEY, R.T. Complex powers of an elliptic operator. Proc.Symp. Pure Math. X . Amer.Math.Soc.Providence: 1967.

[209] SEELEY, R.T. - Elliptic singular integral equations. Proc.Symp. Pure Math. X. Amer.Math.Soc. Providence: 1967.

[210] SEGAL, G. - Equivarant K - theory. Publ.Math.Inst. Hautes Etudes Sci. Paris 34 (1968), 129-151.

[211] SEIFERT, H.; THRELFALL, W. - Topologische Untersuchung der Discontinuitätsbereiche endlicher Bewegungsgruppen des dreidimensionalen sphärischen Raumes. Math.Ann. 104 (1930), 1-70.

[212] SEIFERT, H.; THRELFALL, W. - Lehrbuch der Topologie Teubner, Leipzig: 1934.

[213] SERRE, J.P. - Homologie singulière des espaces fibrés. Ann. of Math. 54 (1951), 425-505.

[214] SERRE, J.P. - Quelques calculs de groupes d'homotopie. C.R.Acad.Sci. Paris 236 (1953), 2475-2477.

[215] SERRE, J.P. Un théorème de dualité. Comment.Math.Helv. 29 (1955), 9-26.

[216] SHANAHAN, P. - The Atiyah-Singer index theorem. Lecture Notes in Math. 638. Springer, Berlin-Heidelberg-New York: 1978.

[217] SHAPIRO, A. - Obstructions to the imbedding of a complex in an euclidean space. Part I: The first obstruction. Ann. of Math. 66 (1957), 256-269.

[218] SIGRIST, F.; SUTER, U. - Eine Anwendung der K - Theorie in der Theorie der H - Räume. Comment.Math.Helv. 47 (1972), 36-52.

[219] SIGRIST,F.; SUTER, U. - Cross-sections of symplectic Stiefel manifolds. Trans.Amer.Math.Soc. 184 (1973), 247-259.

[220] SIGRIST, F.; SUTER, U. - On immersions $CP^n \hookrightarrow R^{4n-2\alpha(n)}$. Algebraic topology, Proceedings, Vancouver 1977. Lecture Notes in Math. 673. Springer, Berlin-Heidelberg-New York: 1978.

[221] SJERVE, D. - Geometric dimension of vector bundles over lens spaces. Trans.Amer.Math.Soc. 134 (1968), 545-558.

[222] SJERVE, D. - Vector bundles over orbit manifolds. Trans. Amer.Math.Soc. 138 (1969), 97-106.

[223] SMALE, S. - A classification of immersions of the two-sphere. Trans.Amer.Math.Soc. 90 (1959), 281-290.

[224] SMALE, S. - The classification of immersions of spheres in euclidean spaces. Ann. of Math. 69 (1959), 327-344.

[225] SPANIER, E. - Algebraic topology. Mac Graw-Hill, New York: 1966

[226] STASHEFF, J. - On extensions of H - spaces. Trans.Amer. Math.Soc. 105 (1962), 126-135.

[227] STASHEFF, J. - A classification theorem for fibre spaces. Topology 2 (1963), 239-246.

[228] STASHEFF, J. - Manifolds of the homotopy type of (non-Lie) groups. Bull.Amer.Math.Soc. 75 (1969), 998-1000.

[229] STASHEFF, J. - H - spaces from a homotopy point of view. Lecture Notes in Math. 161. Springer, Berlin-Heidelberg-New York: 1970.

[230] STEENROD, N. - Cohomology invariants of mappings. Ann. of Math. 50 (1949), 954-988.

[231] STEENROD, N. - The topology of fibre bundles. Princeton Univ.Press, Princeton: 1951.

[232] STEENROD, N.; WHITEHEAD, J.H.C. - Vector fields on the n - sphere. Proc.Nat.Acad.Sci. USA 37 (1951), 58-63.

[233] STIEFEL, E. - Richtungsfelder und Fernparallelismus in n - dimensionalen Mannigfaltigkeiten. Comment.Math.Helv. 8 (1935/36), 305-355.

[234] SUTHERLAND, W.A. - An example concerning stably complex manifolds. J.London Math.Soc. (2) 6 (1973), 348-350.

[235] SUZUKI, M. - On finite groups with cyclic Sylow subgroups for all odd primes. Amer.J.Math. 77 (1955), 657-691.

[236] SWITZER, R. - Algebraic topology: Homotopy and Homology. Springer, Berlin-Heidelberg-New York: 1970.

[237] SZCZARBA, R.H. - On tangent bundles of fibre spaces and quotient spaces. Amer.J.Math. 86 (1964), 685-697.

[238] THOM, R. - Espaces fibrés en sphères et carrés de Steenrod. Ann.Sci. Ecole Norm.Sup. 69 (1952, 109-182.

[239] THOMAS, E. - Vector fields on low dimensional manifolds. Math.Z. 103 (1968), 85-93.

[240] THOMAS, E. - Vector fields on manifolds. Bull.Amer.Math. Soc. 75 (1969), 643-683.

[241] TODA, H. - Le produit de Whitehead et l'invariant de Hopf. C.R.Acad.Sci. Paris 241 (1955), 849-850.

[242] TODA, H. - Composition methods in homotopy groups of spheres. Annals of Math. Studies 49. Princeton Univ.Press, Princeton: 1962

[243] TODA, H. - On homotopy groups of S^3 - bundles over spheres. J.Math. Kyoto Univ. 2 (1963), 193-207.

[244] UCCI, J. - Immersions and embeddings of Dold manifolds. Topology 4 (1965), 283-293.

[245] UCHIDA, F. - Immersions of lens spaces. Tohoku Math.J. 4 (1966), 383-397.

[246] UCHIDA, F. - K - theory of lens-like spaces and S^1-actions on $S^{2n+1} \times S^{2m+1}$. Tohoku Math.J. 25 (1973), 201-211.

[247] VINCENT, G. - Les groupes linéaires finis sans points fixes. Comment.Math.Helv. 20 (1947), 117-171.

[248] WALL, C.T.C. - All 3 - manifolds embed in 5 - space. Bull.Amer.Math.Soc. 71 (1965), 564-567.

[249] WEIL, A. - Variétés kählériennes. Hermann, Paris: 1958.

[250] WHITEHEAD, G. - On the homotopy groups of spheres and rotation groups. Ann. of Math. 43 (1942), 132-146.

[251] WHITEHEAD, G. - On families of continuous vector fields on spheres. Ann. of Math. 47 (1946), 779-785.

[252] WHITEHEAD, G. - Generalized homology theories. Trans. Amer.Math.Soc. 102 (1962), 227-283.

[253] WHITEHEAD, J.H.C. - Combinatorial homotopy I.Bull.Amer. Math.Soc. 55 (1949), 213-245.

[254] WHITNEY, H. - Differentiable manifolds. Ann. of Math. 37 (1936), 645-680.

[255] WHITNEY, H. - The self-intersections of a smooth n-manifold in 2n-space. Ann. of Math. 45 (1944), 220-246.

[256] WHITNEY, H. - The singularities of a smooth n-manifold in (2n-1) space. Ann. of Math. 45 (1944), 247-293.

[257] WILKERSON, C. - Spheres which are loop spaces mod p. Proc.Amer.Math.Soc. 39 (1973), 616-618.

[258] WILKERSON, C. - K-theory operations in mod p loop spaces. Math.Z. 132 (1973), 29-44.

[259] WOLF, J. - Sur la classification des variétés riemanniennes homogènes à courbure constante. C.R.Acad.Sci. Paris 250 (1960), 3443-3445.

[260] WOLF, J. - Spaces of constant curvature. Mac Graw-Hill, New York: 1967.

[261] WOODWARD, L.M. - Vector fields on spheres and a generalization. Quart.J.Math. Oxford (2) 24 (1973), 357-366.

[262] WU, WEN-TSÜN - Sur la structure presque complexe d'une variété differentiable réelle de dimension 4. C.R.Acad. Sci. Paris 227 (1948), 1076-1078.

[263] WU, WEN-TSÜN - On the isotopy of C^r-manifolds of dimension n in euclidean (2n+1)-space. Sci. Record 2 (1958), 271-275.

[264] YASUO, M. - γ-dimension and products of lens spaces. Mem.Fac.Sci. Kyushu Univ.Ser. A.31 (1977), 113-126.

[265] YOSHIDA, T. - A remark on vector fields on lens spaces. J.Sci. Hiroshima Univ. 31 (1967), 13-15.

[266] YOSHIDA, T. - Note on the span of certain manifolds. J.Sci. Hiroshima Univ. 34 (1970), 13-15.

[267] ZABRODSKY, A. - Homotopy associativity and finite CW-complexes. Topology 9 (1970), 121-128.

[268] ZABRODSKY, A. - The classification of simply connected H-spaces with three cells. Part I: Math.Scand. 30

(1972), 193-210. Part II: ibid 30 (1972), 211-222.

[269] ZABRODSKY, A. - On the construction of new finite CW H - spaces. Invent.Math. 16 (1972), 260-266.

[270] ZASSENHAUS, H. - Über endliche Fastkörper. Abh.Math. Sem. Univ. Hamburg 11 (1936), 187-220.

INDEX